Remote Sensing and Spatial Science: Principles and Applications

Remote Sensing and Spatial Science: Principles and Applications

Edited by Matt Weilberg

SYRAWOOD
PUBLISHING HOUSE

New York

Published by Syrawood Publishing House,
750 Third Avenue, 9th Floor,
New York, NY 10017, USA
www.syrawoodpublishinghouse.com

Remote Sensing and Spatial Science: Principles and Applications
Edited by Matt Weilberg

International Standard Book Number: 978-1-68286-527-9 (Hardback)

Cataloging-in-Publication Data

Remote sensing and spatial science : principles and applications / edited by Matt Weilberg.
    p. cm.
Includes bibliographical references and index.
ISBN 978-1-68286-527-9
1. Remote sensing. 2. Space optics. 3. Space sciences. I. Weilberg, Matt.
G70.4 .R46 2018
621.367 8--dc23

TABLE OF CONTENTS

PREFACE

Remote sensing refers to the use of the satellite, which is used to detect objects on Earth. Remote sensing is mainly classified into active or passive remote sensing. It is mainly used in fields such as oceanography, hydrology, geography, etc. This book aims to equip students and experts with the advanced topics and upcoming concepts in the area of remote sensing and spatial science. A number of latest researches have been included in this book to keep the readers up-to-date with the global concepts in this area of study. For someone with an interest and eye for detail, this book covers the most significant topics in the field of remote sensing and spatial science.

This book unites the global concepts and researches in an organized manner for a comprehensive understanding of the subject. It is a ripe text for all researchers, students, scientists or anyone else who is interested in acquiring a better knowledge of this dynamic field.

I extend my sincere thanks to the contributors for such eloquent research chapters. Finally, I thank my family for being a source of support and help.

Editor

1

UNDERWATER CALIBRATION OF DOME PORT PRESSURE HOUSINGS

E. Nocerino [a], F. Menna [a], F. Fassi [b], F. Remondino [a]

[a] 3D Optical Metrology unit, Bruno Kessler Foundation (FBK), via Sommarive 18, Trento 38123, Italy
Email: (nocerino, fmenna, remondino)@fbk.eu
[b] Politecnico di Milano, ABC Dep. 3DSurvey Group, via Ponzio 31, Milano 20133, Italy – Email: francesco.fassi@polimi.it

Commission V, WG 1

KEY WORDS: Underwater, Photogrammetry, Camera calibration, Decentring distortion

ABSTRACT

Underwater photogrammetry using consumer grade photographic equipment can be feasible for different applications, e.g. archaeology, biology, industrial inspections, etc. The use of a camera underwater can be very different from its terrestrial use due to the optical phenomena involved. The presence of the water and camera pressure housing in front of the camera act as additional optical elements. Spherical dome ports are difficult to manufacture and consequently expensive but at the same time they are the most useful for underwater photogrammetry as they keep the main geometric characteristics of the lens unchanged. Nevertheless, the manufacturing and alignment of dome port pressure housing components can be the source of unexpected changes of radial and decentring distortion, source of systematic errors that can influence the final 3D measurements. The paper provides a brief introduction of underwater optical phenomena involved in underwater photography, then presents the main differences between flat and dome ports to finally discuss the effect of manufacturing on 3D measurements in two case studies.

1. INTRODUCTION

As in case of more traditional aerial and terrestrial photogrammetry, also in the field of underwater photogrammetry advances and progresses have gone hand in hand with improvements in photography. Nowadays the number of demanding applications is constantly growing mainly thanks to technical achievements in diving apparatus, photographic equipment and underwater manned and unmanned vehicles. Stating the peculiarities of a hostile environment such as the underwater world, probably even more then in air, an insight of the basic principles of underwater photography and equipment is fundamental for a successful approach to underwater photogrammetry. Testing and investigating the geometrical characteristics of underwater consumer grade photographic equipment when used for photogrammetric applications would be advisable if accuracy and reliability matter. Professional results always rely on the control of all the technical parameters involved. The knowledge about photographic equipment and its behaviour in different conditions is the first step to be investigated.

Moving from this consideration, in the first part of this contribution the main physical properties of water are explained in relation to how they affect underwater photography. Then, a brief history on the evolution on underwater camera equipment is presented, highlighting the difference between the two types of lens ports (flat and spherical). The study is then focused on the geometric and optic characterization of a consumer grade pressure housing with a dome port (NiMAR NI303D and NI320), in which a DSLR Nikon D300 with a Nikkor 24 mm is mounted. Both simulations and tests underwater are carried out and described. The theoretical part of the research study is conducted using freely available WinLens 3D Basic and Predesigner and software application by Qioptiq Photonics and Matlab scripts developed had-hoc. Underwater tests are performed both on a small archaeological find and an elongated object to analyse different underwater camera calibration procedures. In the conclusion, specific aspects which deal with photogrammetric acquisitions are considered and practical suggestions provided.

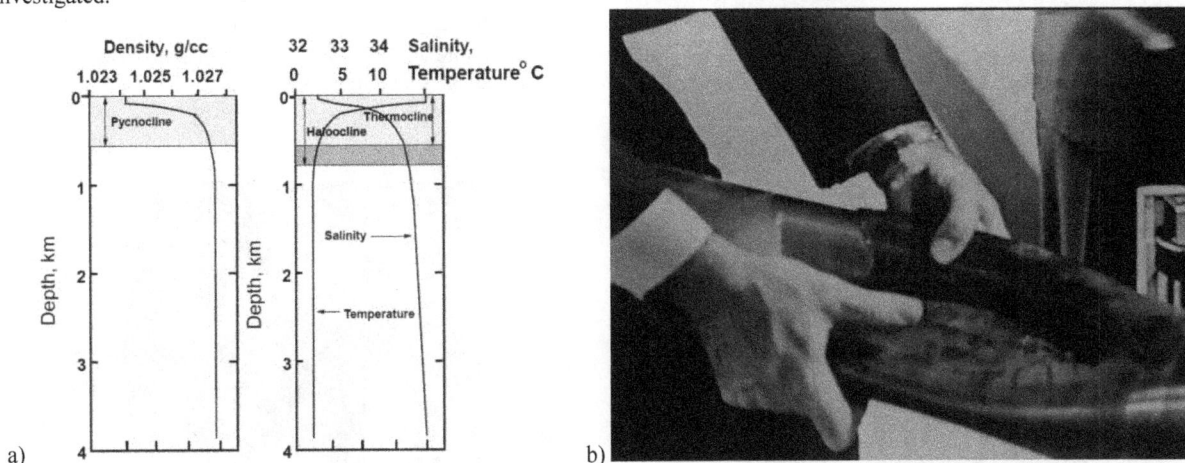

Figure 1. (a) Water density, salinity and temperature variation with depth (Ocean Stanford Edu). (b) Effect of pressure at high depth: Steel waterproof camera housing collapsed at a depth of about 6000 m (frame extracted from EDC).

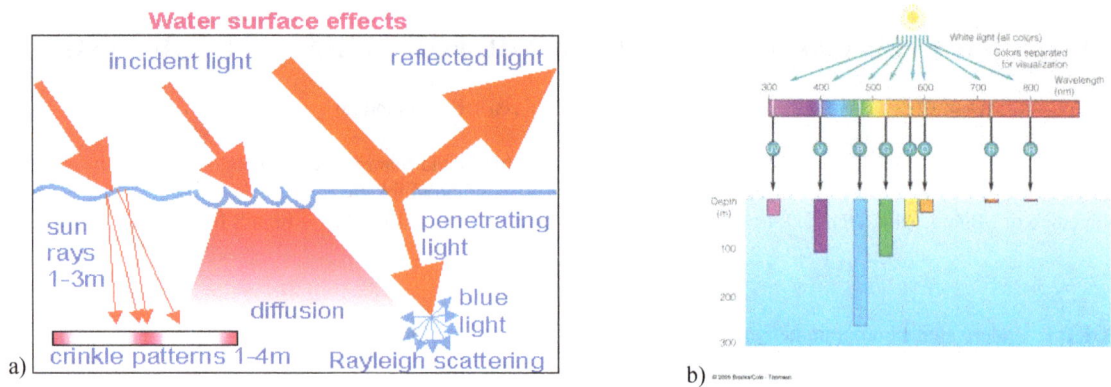

Figure 2. (a) Water surface effects on incident sunlight (Floor, 2005); (b) selective light absorption of colour (wavelength) in the open ocean (Kennesaw State University).

2. PROPERTIES OF WATER AND THEIR INFLUENCE ON UNDERWATER PHOTOGRAPHY AND PHOTOGRAMMETRY

Water is a medium inherently different from air and the first essential difference resides in the medium density. Seawater is nearly 800 times denser than air, and this influences the image formation underwater altering the path of optical rays. Density of seawater is not constant through the depth, being a function of temperature, salinity and pressure (Fig. 1a). These quantities are all correlated: density increases as temperature decreases, salinity increases as pressure increases, pressure increases linearly with depth (every 10 m the pressure increases of 1 atmosphere, equal to 1.033 N/cm2). It is fundamental to properly consider the effect of pressure. Although this factor is extremely critical for very deep underwater inspections (see Fig. 1b), pressure variation with depth affects any underwater optical system at whatever depth. Internal arrangement may be altered and subjected to changes as the working depth varies.

Other optical phenomena are to be accounted for underwater, all related to the so called inherent optical properties that govern propagation of light in water.

2.1 Light absorption

When sunlight reaches the sea surface, the great amount of the radiation penetrates and it is absorbed almost the 94% in open ocean, source National Snow and Ice Data Centre – NSIDC), little is reflected. The amount of light that is reflected upward depends strongly on the height of the sun (place on Earth, time of day and season) and the condition of the sea. A rough sea absorbs more light whereas a mirror-like sea reflects more (Fig. 2a). Sunlight reflections casted by the sea surface should be

firmly avoided for photogrammetric applications because they could affect the extraction of automatic interesting features, as well as produce poor quality object texture. Figure 3 shows the same scene with (a) and without (b) sunlight reflections. Water acts as a selective filter: the great amount of light entering the sea is absorbed (it is converted in heat) within the first meters; only 1% of light entering the sea reaches 100 m. The different components of light, characterised by different wavelengths, are absorbed differently. Longer wavelengths in the visible spectrum (red, orange) together with UV are absorbed first, short wavelengths are absorbed last. The maximum penetration depth depends on water composition: in turbid coastal waters light rarely penetrates deeper than 20 m; while in the open ocean blue light penetrates even more than 200 m (Fig. 2b) and after that depth there is almost no light.

To restore the full range of colours in marine environment, the use of artificial light source (strobes or flashes) is required. The use of artificial light source is also crucial to compensate for light attenuation due to the absorption that limits the visibility distance.

2.2 Turbidity, scattering and backscattering

Turbidity in water is due to suspended particles (phytoplankton, organic matter, pollution, etc.) that cause the light to be scattered. The more the particles, the higher the turbidity. Turbidity of water is generally quantified using the Secchi distance, an old and simple method introduced in 1865. A circular disk divided in four alternating sections, two white and two black, is immersed in water from a boat and the distance at which the disk is not more visible is defined as one Secchi distance.

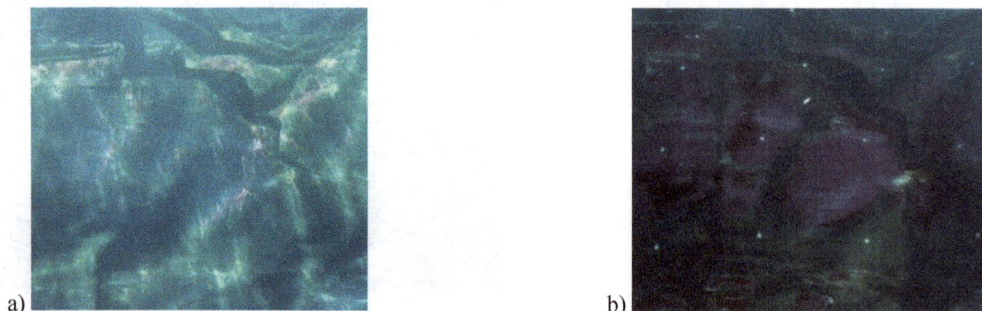

Figure 3. Effect of sunlight reflections: the two images show the same submerged area with (a) and without (b) reflections.

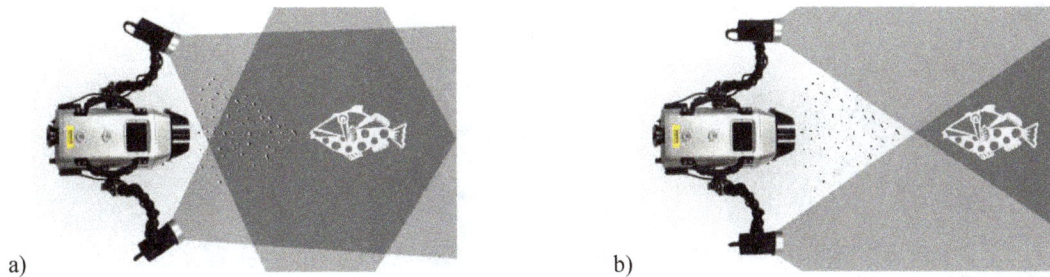

Figure 4. Backscattering: improper (a) and proper (b) position of strobe lights (Backscatter).

Scattering or diffuse reflection is an optical phenomenon that arises when the light rays are randomly deviated from their straight paths. Scattering limits image quality, reducing the contrast and producing blurred images.

When strobes are used, also backscattering can be introduced: it is similar to scattering with the difference that the light from the artificial source is reflected from the particles mainly back to the camera. To reduce backscattering, strobes should be carefully positioned, avoiding to point them directly to the subject (Fig. 4). The closer the flash to the camera, the more backscatter is produced. The closer to the subject the picture is taken, the less water and particles are present between the camera and subject, the less backscatter is produced.

As a consequence of the reported considerations, it is evident that taking photographs underwater is strictly conditioned by the medium characteristics, among which turbidity is the main limiting constraint. Moreover, scatter and backscatter reduce contrast of the scene and the final quality of the image. Considering the special lighting setup needed underwater, even in very clear water, yet the image acquisition can be difficult. Indeed when a Secchi distance corresponds to several meters a strong illumination would be required to light the object and a wider baseline would be necessary between the two lateral light sources. In these cases the system of cameras and strobe lights would require more than one diver.

3. UNDERWATER PHOTOGRAPHY EQUIPMENT

Since the first developments of photography it has been clear that to take pictures underwater proper equipment shall be used. The first underwater camera systems were developed by the French marine biologist Louis Boutan at the end of the nineteenth century. In the more compact and portable version (Figure 5a) the system incorporated two dual, electrical lamps for enhancing illumination thus reducing the exposure time. The breakthrough in underwater photography was represented in 1943 by the invention of the first open-circuit, self-contained underwater breathing apparatus (or "SCUBA"), which revolutionized the world of diving, and the development of pearliest portable underwater camera systems in the '50, like the Rolleimarin underwater housing for the medium format professional twin-lens Rollei 6x6cm camera (Fig. 5b) and Nikon's Nikonos series cameras (Fig. 5c). In the 1970s, mainly pushed by Al Giddings, a worldwide renowned underwater cinematographer, the use of hemispherical dome ports, optimized for underwater and wide-angle optic, started to become more and more popular (Encyclopedia). At the end of the 1990s, the revolution brought by compact digital still cameras has marked irreversibly the world of photography, in general, and that of underwater photography, in particular. A variety of functional and fancy, professional and consumer-grade waterproof housings has been released on the market for any type of digital cameras, featuring two types of lens port, flat and spherical (Fig. 5d-e-f).

3.1 Flat and dome lens port

Flat and hemispherical dome ports main characteristics are summarised in table 1. As shown, they significantly differ from each other and influence the image acquisition and formation. The dissimilar effect and behaviour is due to their shapes: flat ports, being flat surfaces between two distinct media characterised by different refractive indices (i.e., water outside the waterproof housing and air inside) obeys to the Snell's law. As a consequence, the optical rays deviate from the original path, when pass through the flat port from the water to the camera-lens system inside the pressure housing, and are bent towards the port surface normal (refraction). On the contrary, hemispherical dome ports are composed by two spherical surfaces (external and internal), which theoretical should have the same centre of curvature. The dome port thickness is provided by the difference of radii of curvature of the two surfaces and should be manufactured as much uniform as possible. If the centre of lens (lens entrance pupil, EP, which also represents the perspective centre) is correctly placed in the ideally unique centre of the spherical surfaces, the light rays enter the dome port almost perpendicularly and go to the EP without refraction. In order to verify such conditions, the manufacture of dome port lenses should be highly accurate, thus being more demanding and expensive than the production of flat ports.

3.2 Geometric and optical characterization of a hemispherical dome port

In Menna et al. (2016), the geometric and optic characterization of the 7" NI320 dome port (Fig. 5f) produced by the Italian company NiMAR is presented. The work aims at understanding how deviations of the actual manufacturing from the ideal spherical and concentric shape of the dome surfaces would influence the optics of the system. A reverse engineering process of the dome is carried out, showing that the curvature centres of the outer and inner surfaces have a misalignment less than 1.5 mm, with the maximum component along the optical axis equal to 7.7mm. The dome can be mounted on the NiMAR NI303D waterproof polycarbonate case (Fig. 5f) for Nikon D300 DSLR camera and is designed to work with different lenses whose focal length ranges from 20mm to 35mm. With the possibility of using several focal lengths, it should be expected that the position of the EP will vary accordingly.

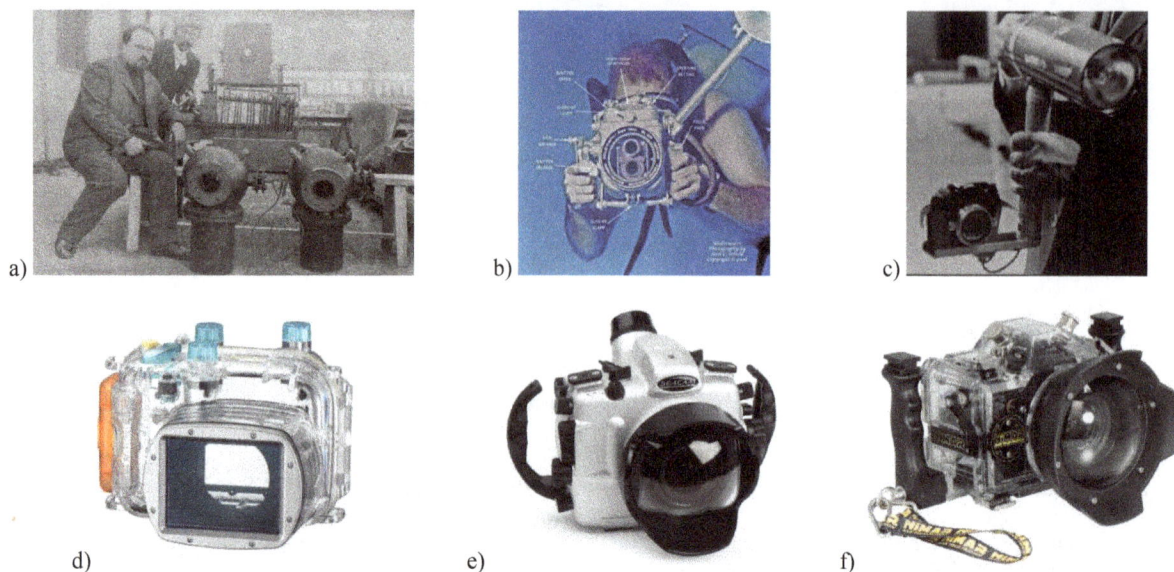

a)   b)   c)

d)   e)   f)

Figure 5. (a) Boutan's underwater camera (Fadedanblurred). (b) Rolleimarin underwater housing with flash (Pbase). (c) Nikonos Calypso camera with underwater electronic flash (frames extracted from EDC). (d-e-f) Waterproof housings for digital cameras; from left to right: Canon with flat port for compact cameras, Seacam and NiMAR with dome port for DSLR cameras.

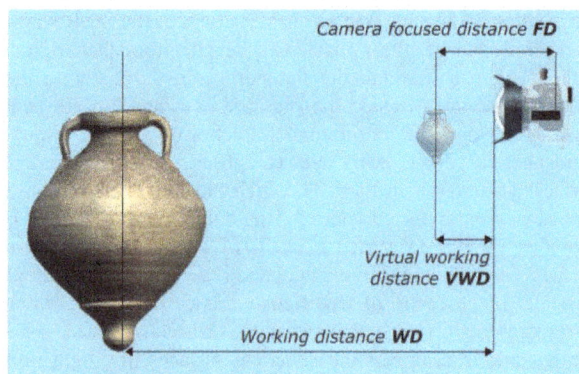

Figure 6. Effect of a dome port: a subject located at a working distance (WD) appears smaller and much closer to the camera - virtual working distance (VWD).

An experimental investigation is realised to locate the EP inside the NI303D pressure case and with the respect to curvature centre of the NI320 dome port.

Two Nikkor lenses are tested, namely the 24mm f/2. 8AF-D and 35mm f/2.0 AF-D. The EP position results maximum 5.4mm ahead of the curvature centre of the dome surfaces (i.e., closer to the dome), whit a misalignment in the plane perpendicular to the optical axis of about 1mm. Using the geometric data reported above, the optics of the investigated dome port is studied using the optical ray tracing software Winlens.

Table 2 shows the values of working distances (WD) from a real object point underwater versus its virtual image distance (or subject to entrance pupil distance DS2EP) from the entrance pupil (supposed to be placed in the dome centre). Virtual working distances (VWD), as defined in Figure 6, are also reported. Note that for simplicity the distances are reported positive. It is worth to note that the virtual image of a real object underwater is compressed in a very narrow virtual space just 20 cm deep in front of the dome glass. Moreover if the minimum focusing distances for the two Nikkor AF 24 and 35 mm lenses

are considered, it results that the minimum working distances WD for the two lenses is respectively −750 mm and −300 mm. For closer objects, additional close up lenses should be mounted to the front of the camera in order to produce sharp images.

4. UNDERWATER AND IN-AIR SYSTEM CALIBRATION

To investigate the influence of the dome port in real conditions, the camera system, which underwater comprises the camera + lens together with the pressure housing + lens port, is calibrated in a swimming pool. As in classical photogrammetry, calibration is fundamental to assure accurate and reliable measurements of 3D objects. Several algorithms and procedures have been proposed in literature for underwater camera calibration, both with flat and dome lens ports. All the different approaches fall into two main classes: (a) the ray tracing method which aims at rigorously and explicitly modelling the effect of refraction and any deviation from the ideal straight path described by light rays, and (b) the implicit absorption of the optical effects due to water and lens port adopting the standard pinhole camera model and a terrestrial-like self-calibration approach. A detailed and critical review on underwater camera calibration is out of the scope of this research work, but the interested readers can refer to Shortis (2015). The calibration method adopted in this investigation follows approach (b). Calibration of underwater camera system at the predominant working conditions would provide more accurate and reliable results. Moreover, dome ports should introduce little refraction effects that can be handled using a classical photogrammetric self-calibration approach. Two photogrammetric acquisitions for self-calibration are realized, one underwater and one in air. The camera with the 24 mm at f/8.0 is focused at 1m, setting fixed during the acquisition (autofocus disabled) to not change the interior orientation parameters.

	FLAT PORT	**HEMISPHERICAL DOME PORT**
Description	Flat plane of optically transparent glass or plastic	Concentric lens acting as additional optical element (negative or diverging lens)
Field of view (FOV) WRT the camera-lens system	Reduced	Equal
Focal length WRT the camera-lens system	Increased (by a factor equal to approximately the ratio between the refraction indices of water and air)	Equal
Magnification WRT the camera-lens system	Increased (by a factor equal to approximately the ratio between the refraction indices of water and air)	Equal
Effect on the observed object	The object appears closer to the camera by a factor equal to approximately the ratio between the refraction indices of water and air.	An upright, smaller virtual image of the object is formed at a distance from the dome surface equal to 3 times the curvature radius of the dome. The camera-lens system focuses on this virtual image.
Maximum FOV	Limited to 96°	Not limited
Lens distortion	Pincushion distortion	No significant distortion
Other effects	Chromatic aberration	• Increase of Depth of field (DOF) by a factor equal to approximately the ratio between the refraction indices of water and air. • Spherical aberration • Field curvature
Costs	Cheaper	More expensive
Typical use	For compact digital cameras	For DSLR cameras

Table 1. Characteristics of flat and dome ports.

WD (mm)	200	300	400	500	750	1000	3000	5000	10000	Infinity
DS2EP (mm)	164.1	184.9	199.7	210.8	229.4	240.8	269.7	276.8	282.5	288.5
VWD (mm)	80.8	101.6	116.4	127.5	146.1	157.5	186.4	193.5	199.2	205.2

Table 2. Real object distance versus its virtual image underwater for the NiMAR NI320 dome port.

Between the two calibrations, the camera is not removed from the pressure housing to keep the system stable as much as possible. An ad-hoc underwater test-field made of a planar aluminium board with resolution and photogrammetric coded targets is used (Fig. 7) both underwater in the pool and for calibration in air. Table 3 reports the camera calibration parameters obtained from the two calibrations. As shown in Figure 8, the lens displays quite a pronounced barrel radial distortion both in air (red) and in water (blue). As previously anticipated by the reverse engineering of the dome, the advanced position of the entrance pupil of the lens respect to the dome centre introduces a small pincushion compensation effect resulting in a less negative overall distortion (less barrel). A significant variation in the principal distance between in air and underwater calibrations is also observed. This change is expected as the closer is the lens to the dome surface, the less spherical is the portion of the surface of the dome the camera

looks trough. The extreme limit is when the lens front is very close to the dome inner surface and the entrance pupil is much more ahead than in the case study of this paper: in this case the dome portion in the field of view of the camera approaches the one of a flat port with a consequent increase of the principal distance by a factor of about 1.33 as reported in Table 1.

Decentring distortion is introduced, due to the offset in the plane perpendicular to the optical axis between lens entrance pupil and dome surface centre. In air the decentring distortion parameters are not statistically significant thus are not adjusted for. As it can be observed in the graph, its magnitude in water is anyway very small compared to the radial component, as expected due to the smaller in-plane than along the axis misalignment. The in-plane offset can also explain the difference in the coordinates of the principal points. Figure 9 shows the difference between distortion maps in air and in water.

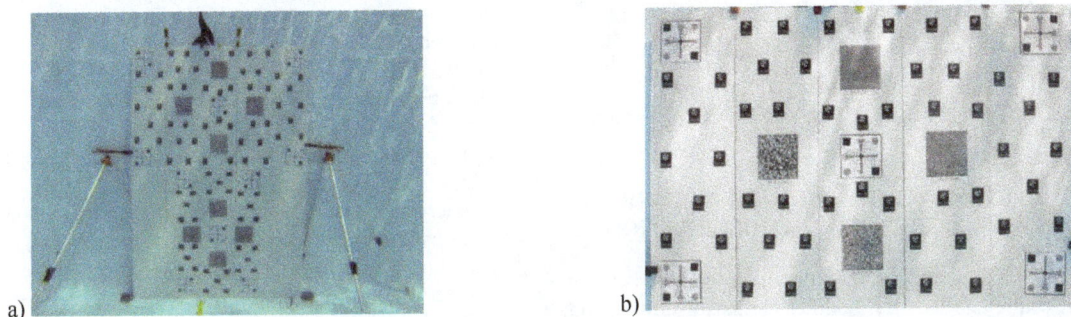

a)   b)

Figure 7. Test-field used in the swimming pool with stand with resolution and photogrammetric targets.

Camera Calibration Parameters	AIR		UW	
	value	std. dev.	value	std. dev.
Principal distance (mm)	25.801	0.006	26.208	0.002
Principal Point x_0 (mm)	−0.026	0.002	−0.058	0.003
Principal Point y_0 (mm)	−0.144	0.003	−0.207	0.002
K_1	1.842e-004	1.2e-006	1.663e-004	6.1e-007
K_2	−3.030e-007	7.4e-009	−2.582e-007	3.4e-009
K_3	-	-	-	-
P_1	-	-	6.582e-006	1.2e-006
P_2	-	-	1.620e-005	8.7e-007

Table 3. Comparison between camera calibration in water (UW) and in air. Only statistically significant additional parameters are computed.

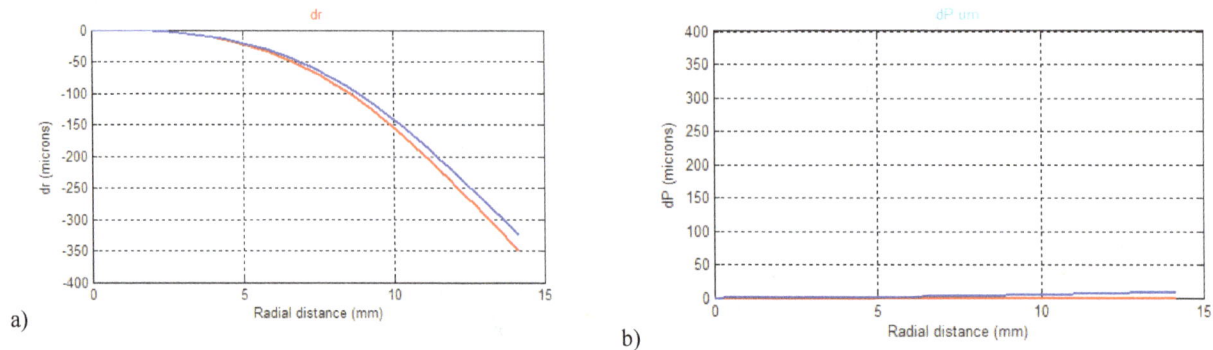

Figure 8. (a) Radial and (b) decentering distortion curves: the curves in red are related to the camera calibration in air, the curves in blue to the camera calibration underwater.

A distortion map displays according to a colour scale map the difference between the ideal pixel position (no distortion) and the actual pixel position due to the influence of radial and decentring distortions determined through camera calibration. As depicted in the figure and expected, the maximum difference is reached at the borders, whose magnitude is comparable with the differences highlighted in the distortion curves. An asymmetric behaviour can be also observed, likely due to the small in-plane misalignment between the lens entrance pupil and dome surface centre of curvature, slightly bigger along the Y axis.

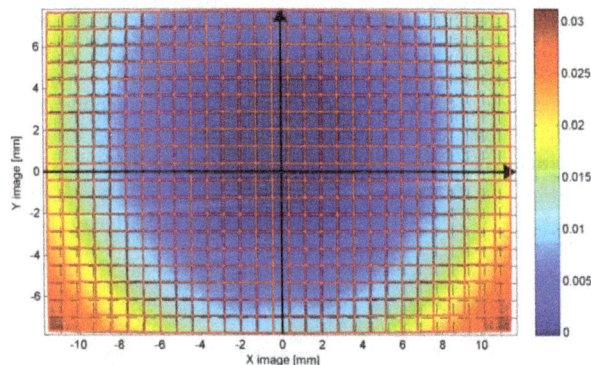

Figure 9. Difference between in air and in water distortion maps.

5. INFLUENCE OF CALIBRATION PARAMETERS ON UNDERWATER 3D OBJECT RECOSTRUCTION

The previous sections have shown that misalignment between dome port centre and lens EP modifies the lens distortion and, in particular, can introduce a decentring component. While up to now the investigation has been concentrated in image space, the aim of the following analysis is to expand the results in object space. In particular, the study aims at understanding if neglecting the decentring distortion produces relevant deformation in the reconstructed object. In the followings, two peculiar case studies are presented: the first experiment is designed to resemble a typical cultural heritage acquisition, with a small object surveyed at a big image scale with a circular and closed camera network; the second case involves the acquisition of an elongated body, surveyed with an aerial-like camera network with the inclusion of convergent images.

5.1 Ancient amphora – circular camera network

The ancient amphora (height $\approx$ 50cm) shown in Figure 10a is employed as test object and is measured and reconstructed both in air and underwater. The acquisition in air, considered the reference or ground truth to verify the underwater results, is carried out using a DSLR Nikon D750 camera with a 50 mm lens and a ground sample distance (GSD) of about 0.1 mm. The system used underwater is the NiMAR pressure housing with dome lens port and the Nikon D300 with 24mm lens presented before. The acquisition of the amphora is realised in the pool immediately after (Fig. 10b), in the same environmental conditions and camera settings (i.e., fixed focusing distance) of the underwater calibration (Section 4), obtaining a GSD of about 0.3 mm.

Different calibration procedures for the underwater acquisition are tested and verified against the reference object:

a. Full pre-calibration: the images are oriented using as camera calibration parameters the full set (Table 3) derived from the underwater system calibration presented in Section 4.

b. Pre-calibration with only radial distortion: the images are oriented using another set of parameters derived from the underwater calibration, solving only for radial distortion.

c. Structure from motion (SfM) self-calibration: a self-calibrating bundle adjustment is performed to estimate in one-step solution both camera calibration parameters with only radial distortion and image orientation.

The dense image matching is performed both for the reference and three underwater calibration approaches at ¼ of the original image resolution (i.e., double of the original GSD). From the dense point clouds, polygonal mesh models are generated with a mean spatial resolution of 0.5 mm. Figure 11a and Figure 11b show a detail of the in-air reference model and underwater calibration procedure (a). The comparisons with the reference model do not highlight either significant deformation in the geometry or substantial differences between the different underwater calibration approaches. As an example, the differences between the in-air model and the model from method (a) and (c) are shown in Figure 11c and Figure 11d, respectively. In all the cases 95% of differences are within ±0.5 mm.

5.2 Elongated ship gash – aerial-like camera network

In 2012, the Italian cruise ships "Costa Concordia" partially sunk off the coast of a small island in the Mediterranean Sea after the collision against a rock. The produced 60m long gash was situated on the above-the-water side of the stranded ship and extended at the current waterline 4m above and 4m below the sea surface. The technique developed for surveying and modelling the ship part interested by the collision (Fig. 12a) is detailed presented in previous works (Menna et al., 2013; Nocerino, 2015). Here, the analysis is focused on the underwater camera calibration and its influence in object space when an elongated object is measured. For the Costa Concordia, the same underwater camera system under investigation (NiMAR pressure housing with dome lens port and the Nikon D300 with 24mm lens) is used and also in this case, decentring distortion parameters are statistically significant. About 800 underwater images are taken according to a photogrammetric aerial-like strip scheme, with 4 overlapping strips at different depths, assuring a forward overlap of ca. 80% along strip and a sidelap of ca. 40% between two adjacent strips. Convergent images are also included in the camera network. The mean object distance is 3m, providing a GSD of about 0.7 mm. The underwater mesh in Figure 12a is obtained including the decentring distortion. To show the influence of neglecting decentring component in the bundle adjustment, in Figure 12b the Euclidean distances in meters for the processing with and without decentring distortion parameters are presented as colour map. The differences reach a maximum value of about 6 cm.

6. CONCLUSIONS

Despite being more complex to build and assemble to camera pressure housings when compared to flat ports, dome ports show numerous advantages from a photogrammetric point of view as they preserve the main geometrical characteristics that rule the image formation. As far as the photogrammetric planning is concerned when using such ports it is fundamental to consider the variation in the focusing distance, modified due to the projection of a virtual image right in front of the port surface. A pre-defined focusing distance cannot be easily established if not using optical simulation software such as Winlens and particular care must be taken to avoid blurred images due to out of focus issues. The paper presented both from a theoretical point of view and from field experiments that the focal length (hence the field of view), the principal point position and the radial and decentring distortions are all preserved if the dome port and camera housing are properly

manufactured and aligned. Misalignments in the order of few millimetres between the dome port and lens entrance pupil as well as difference between the inner and outer radii of the spherical surfaces of the dome do not produce departures from the geometrical model used in standard photogrammetry. The two case studies presented showed that the choice of the proper set of calibration parameters must be driven by the application of interest. For example, few pixels of maximum decentring distortion can be negligible for simple modelling tasks where the accuracy is not of primary importance while this is not the case when large objects are to be reconstructed. The archaeological case study represented by the amphora showed that the small maximum magnitude of decentring distortion do not produce statistically significant differences in the generated 3D models if they are not accounted for during camera calibration. On the contrary, for more complex case studies such as the one of the 60 m long Costa Concordia gash, the inclusion or exclusion of decentring distortion parameters produces relative differences as high as 6 cm.

ACKNOWLEDGEMENTS

The authors thanks NiMAR which supported this research by providing photographic underwater equipment and useful insights about pressure housings manufacturing techniques.

REFERENCES

Backscatter, http://www.backscatter.com/learn/article/article.php?ID=15

EDC, The Edgerton Digital Collections project, http://edgerton-digital-collections.org/videos/hee-fv-019

Encyclopedia, http://encyclopedia.jrank.org/articles/pages/1193/Underwater-Photography.html

Fadedanblurred, http://fadedandblurred.com/articles/the-worlds-first-underwater-photographer-louis-boutan

Floor, J. A., 2005. Water and Light in Underwater Photography. http://www.seafriends.org.nz/phgraph/water.htm

Kennesaw State University, http://science.kennesaw.edu/~jdirnber/BioOceanography/Lectures/LecPhysicalOcean/LecPhysicalOcean.html

Menna, F., Nocerino, E., Troisi, S. and Remondino, F., 2013. A photogrammetric approach to survey floating and semi-submerged objects. *SPIE Optical Metrology*, Vol. 8791, pp. 87910H-87910H

Menna, F., Nocerino, E., Fassi, F. and Remondino, F., 2016. Geometric and optic characterization of a hemispherical dome port for underwater photogrammetry. *Sensors*, Vol. 16(1), 48

NiMAR S.r.l., http://www.nimar.it/

Nocerino, E., 2015. *A full photogrammetric approach for surveying semi-submerged and floating objects*. PhD thesis.

Ocean Stanford Edu, ocean.stanford.edu/bomc/chem/lecture_03.pdf

Qioptiq Photonics, http://www.qioptiq.com/

Pbase, http://www.pbase.com/image/64660924

Shortis, M., 2015. Calibration techniques for accurate measurements by underwater camera systems. *Sensors*, 15(12), pp.30810-30826

Shortis, M., Harvey, E. and Seager, J., 2007. A review of the status and trends in underwater videometric measurement. SPIE Vol. 6491, pp. 1-26

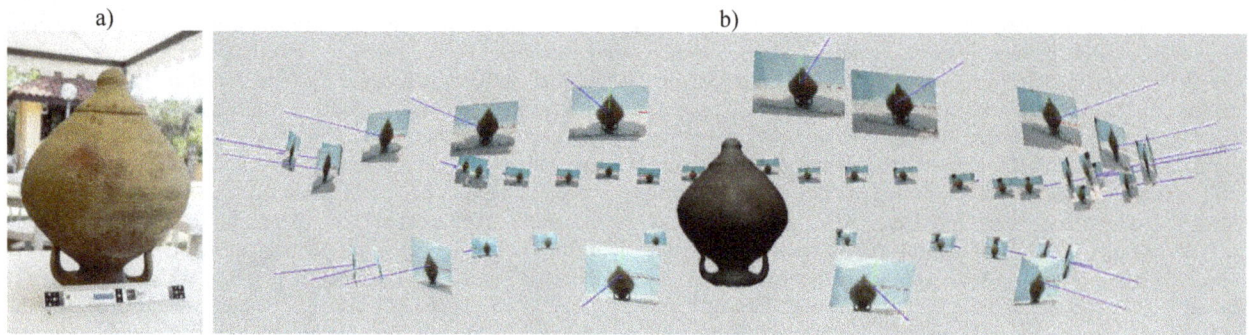

Figure 10. (a) Test object (amphora) pictured in air. (b) Underwater camera network.

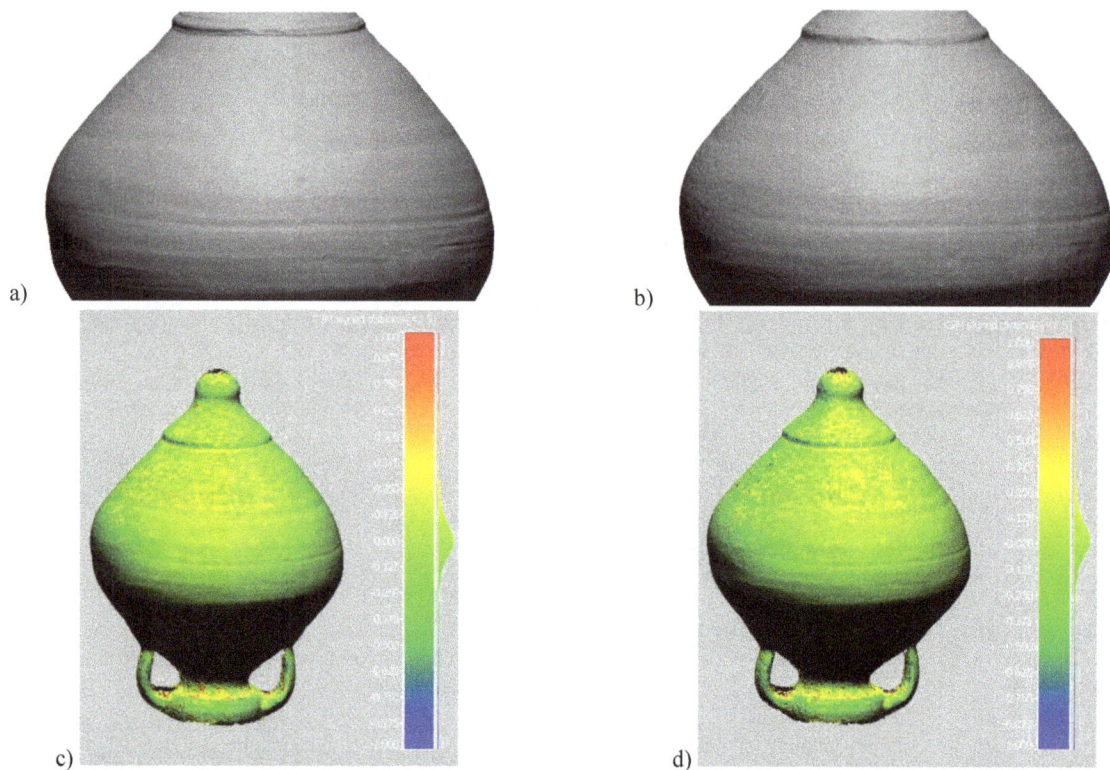

Figure 11. Particular of reference (a) and underwater (b) mesh model. Colour maps of Euclidean distances [mm] between the reference model and the one obtained with the underwater calibration approach a (c) and the underwater calibration approach c (d).

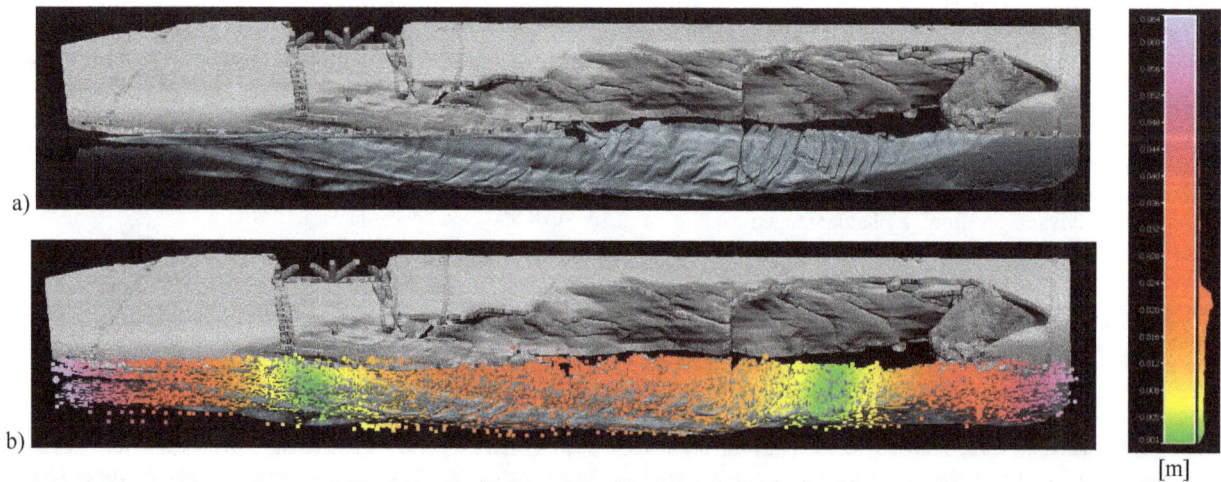

Figure 12. (a) Mesh model of the Costa Concordia gash. (b) Colour maps of Euclidean distances [m] between tie points of the underwater part with and without decentring distortion parameters.

2

A DIFFERENT WEB-BASED GEOCODING SERVICE USING FUZZY TECHNIQUES

P. Pahlavani [a], R.A. Abbaspour [a], A. Zare Zadiny [a, *]

[a] Dept. of Surveying and Geomatics Eng., College of Engineering, University of Tehran, Tehran, Iran
(Pahlavani, Abaspour, Zare_zardiny)@ut.ac.ir

Commission VI, WG VI/4

KEY WORDS: Geocoding, Address, Fuzzy, Nearness, Overlay Function

ABSTRACT

Geocoding - the process of finding position based on descriptive data such as address or postal code - is considered as one of the most commonly used spatial analyses. Many online map providers such as Google Maps, Bing Maps and Yahoo Maps present geocoding as one of their basic capabilities. Despite the diversity of geocoding services, users usually face some limitations when they use available online geocoding services. In existing geocoding services, proximity and nearness concept is not modelled appropriately as well as these services search address only by address matching based on descriptive data. In addition there are also some limitations in display searching results. Resolving these limitations can enhance efficiency of the existing geocoding services. This paper proposes the idea of integrating fuzzy technique with geocoding process to resolve these limitations. In order to implement the proposed method, a web-based system is designed. In proposed method, nearness to places is defined by fuzzy membership functions and multiple fuzzy distance maps are created. Then these fuzzy distance maps are integrated using fuzzy overlay technique for obtain the results. Proposed methods provides different capabilities for users such as ability to search multi-part addresses, searching places based on their location, non-point representation of results as well as displaying search results based on their priority.

1. INTRODUCTION

Geocoding - the process of finding position based on descriptive data such as address or postal code - is considered as one of the most commonly used spatial analyses. Growth of cities on one hand and increase access to internet with the ability to display interactive maps on mobile phones on the other hand, has increased general users interest in geocoding. This subject has motivated online map providers to present geocoding as one of their basic capabilities. Online geocoding services such as Google Maps, Bing Maps, Yahoo Maps, Map Quest, Geocoder US and Open Route Service and so on are among the most popular services of this kind. Despite the diversity of geocoding services, users usually face limitations using them. In linguistic geocoding among humans, proximity to a certain location has critical importance, for example a user wants to find the banks near to a specific location; however in existing geocoding services this concept is not modelled appropriately. In geocoding services, the address entered by the user is only matched with the corresponding address in database in order to search a place. In fact, this search method is based on descriptive data (and not on spatial data) which leads to inability of services to consider spatial proximity in address finding. For example, when a user searches the phrase "Fatemi-Bank" in Google Maps, only the banks that their descriptive address contains "Fatemi" are shown. However, a bank which is in 100 meters from Fatemi Street but doesn't probably have "Fatemi" in its address is not shown for the user (Figure1).

In addition, existing geocoding services enable users to only search one- or two-part addresses while they don't show any results if multipart addresses (over two parts) are entered by users. As well as, there are also some limitations in display results. Existing services, display searching results as a point,

regardless what a user searched (a street, an area or a point of interest). Moreover, geocoding services do not usually consider any priorities when they display the results.

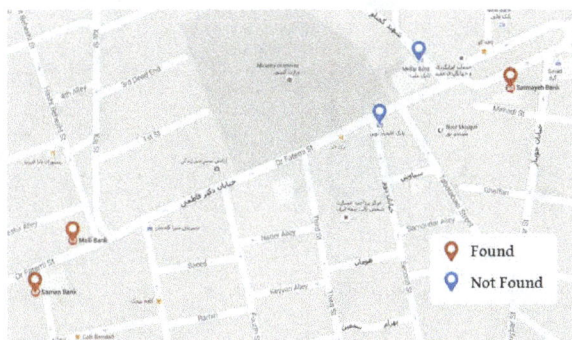

Figure 1. Result of Search "Fatemi-Bank" in Google Maps

Resolving these limitations can enhance efficiency of the above mentioned services. This paper proposes the fuzzy technique in order to deal with the existing limitations. Accordingly, studies have been performed in this field are reviewed first and then a new algorithm is investigated for fuzzy geocoding. Next, a web-based system is designed and implemented according to this proposed method. Finally the most significant results are presented.

2. LITERATURE REVIEW

Various methods have been presented for geocoding so far. In existing geocoding process, the address is found only based on address matching. Therefor many researches have worked on

---

* Corresponding author

efficiency of this matching. For example Peter Christen et.al in [1] described a geocoding system that used a learning address parser based on Hidden Markov Models to separate free-form address into components, and a rule-based matching engine to determine the best set of candidate matches to a reference file. In another case, Daras K et.al in [2] presented a fuzzy matching algorithms for geocoding historical addresses.

These algorithms and methods usually use fuzzy concept only for process the descriptive data (address matching). But there are not any researches that use fuzzy concept on spatial data processing in geocoding. In this paper, fuzzy technique is used in related to spatial data in addition to descriptive data.

3. PROPOSED METHOD

In this paper fuzzy technique is used in order to deal with the limitations mentioned in introduction. Fuzzy technique can model proximity concept in address finding and remove the limitations in display the searching results. In this paper, nearness to places is defined by fuzzy membership functions and multiple fuzzy distance maps are created. Then these fuzzy maps are integrated using fuzzy overlay technique for obtain the results. Fuzzy overlay is used to determine the locations that best meet the nearness criteria, that is have high likelihood of membership in all fuzzy distance maps. In order to implement the proposed method, a web-based system is designed. The overall structure of this system is displayed in figure 2.

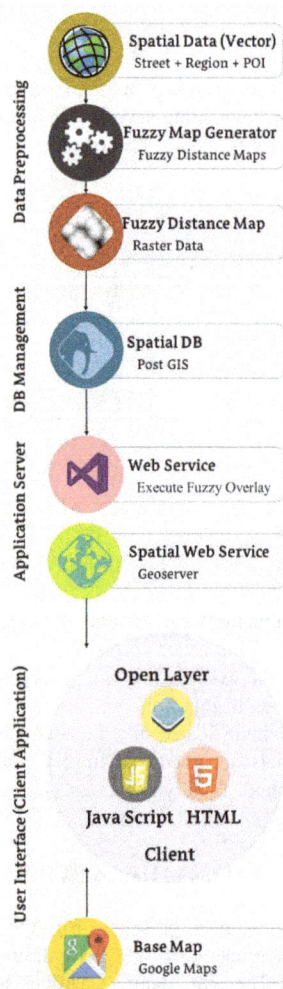

Figure 2. Proposed method

This proposed method contains four main parts. These part are data pre-processing, database management, application server and user interface. In continue each parts of proposed method is explained.

3.1 Data Pre-Processing

Data that used for implementation are related to Tehran Sixth district including network of roads and streets (linear layer), urban areas (polygon layer) and point of interest such as schools, hospitals (point layer). These data are in shape file format. For using data in this method, spatial data must be imported in spatial database but before it, these data need to pre-process. Data pre-processing include of two step. First step is create distance map. In this step, the Euclidean distance map is created based on existing spatial data. These maps show degrees of proximity to the related data discretely. The maximum distance parameter in create of distance map is variable depending on geometric of data. This distance is 50, 70, and 100 meters for point, polyline, and polygon data layers respectively. Figure 3 display an example of distance map for a point layer.

Figure 3. Euclidean Distance Map

Nearness concept is defined with definition of a fuzzy system in step two. This process is performed through introduction of a fuzzy membership function. Introduced membership function is applied on Euclidean distance map. In this implementation for create fuzzy distance map, the Gaussian membership function is selected. A Gaussian membership function is specified by two parameters $\{c, \sigma\}$

$$gaussian(x; c, \sigma) = \exp[-\frac{(x-c)^2}{2\sigma^2}] \qquad (1)$$

A Gaussian membership function is determined by c and σ; c represents the membership function centre and σ determines the membership function width. In this implementation width and midpoint equal to 0.1 and 0. Figure 4 displays an example of fuzzy distance map for a point layer.

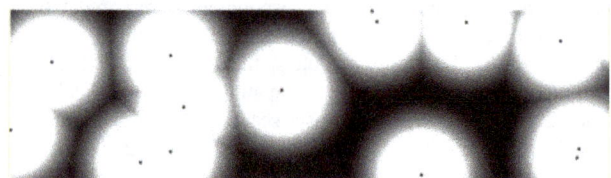

Figure 4. Fuzzy Distance Map

In this system, a windows form application is implemented in Microsoft Visual Studio 2010 to create the distance map and fuzzy distance maps from each vector data and finally save the results as tiff files.

3.2   Database Management

After Data pre-processing phase, for each spatial data including streets, urban areas and point of interest a fuzzy distance map is generated. The address finding process is done based on these raster data. For implementation the web based system, it is necessary to load data in spatial database. The database that used for this implementation is PostGIS. For import raster data (fuzzy distance maps) to PostGIS database, raster2pgsql loading tools is used. Raster2pgsql converts a raster file into a series of SQL commands that can be loaded into a database. The output of this command may be captured into a SQL file, or piped to the psql command, which will execute the commands against a target database. After loading raster data in database, visual check of the raster data is done with Quantum GIS.

3.3   Application Server

When user want to find an address in this system, client application sends a request for application server. This request contains the address. Target address including one or more parts that these parts are separated by '-'. Application server selects data related to each part of address by an address matching. "Like" command is used for selection the spatial data corresponding to each part of address entered by user. While using this command, there is no need to enter the exact name equal to what exists in database.

In this paper, an ASP Web Service is implemented as application server. This web service connects to Post GIS and selects the data (fuzzy distance map) related to each part of address and finally integrated these fuzzy distance maps using fuzzy overlay technique. Fuzzy overlay, combine fuzzy membership raster's data together, based on selected overlay function. The following lists described the most commonly used fuzzy overlay function:

- The Fuzzy AND overlay function will return the minimum value of the sets the cell location belongs to. This technique is useful when we want to identify the least common denominator for the membership of all the input criteria. Fuzzy AND uses the following function (Equation 2) in the evaluation:

$$\mu_{AND} = \min(\mu_1, \mu_2, ..., \mu_n) \qquad (2)$$

- The Fuzzy OR overlay function will return the maximum value of the sets the cell location belongs to. This technique is useful when we want to identify the highest membership values for any of the input criteria. Fuzzy OR uses the following function (Equation 3) in the evaluation:

$$\mu_{OR} = \max(\mu_1, \mu_2, ..., \mu_n) \qquad (3)$$

- The Fuzzy Product overlay function will, for each cell, multiply each of the fuzzy values for all the input criteria. The resulting product will be less than any of the input, and when a member of many sets is input, the value can be very small. It is difficult to correlate the product of all the input criteria to the relative relationship of the values. Fuzzy Product uses the following function (Equation 4) in the evaluation:

$$\mu_{Multiplication} = \prod_{i=1}^{n} \mu_i \qquad (4)$$

- The Fuzzy Sum overlay function will add the fuzzy values of each set the cell location belongs to. The resulting sum is an increasing linear combination function that is based on the number of criteria entered into the analysis. Fuzzy Sum uses the following function (Equation 5) in the evaluation:

$$\mu_{Summation} = 1 - \prod_{i=1}^{n} (1 - \mu_i) \qquad (5)$$

- The Fuzzy Gamma function is an algebraic product of Fuzzy Product and Fuzzy Sum, which are both raised to the power of gamma. The generalize function is as Equation 6. When Gamma is 1 the result is the same as Fuzzy Sum. When Gamma is 0 the result is the same as Fuzzy Product.

$$\mu_{Combination} = (\mu_{Summation})^{\gamma} \times (\mu_{Multiplication})^{1-\gamma} \qquad (6)$$

In this implementation, for integration of all Fuzzy distance maps, Gamma overlay function is used with different values for γ coefficient. For each γ value, a new result is obtained. In this implementation γ coefficient can be changed by the user optionally. Application server perform Gamma function on selected raster data (fuzzy distance maps) by ST_MapAlgebra command. ST_MapAlgebra, returns a one-band raster given one or two input raster's, band indexes and one or more user-specified SQL expressions. An algebraic expression involving the two raster's and Post GIS defined functions/operators that will define the result pixel value. After executing the ST_MapAlgebra command, a raster is generated and store in database. Finally, Geoserver as a spatial web service received the target raster from Post GIS and sent it for client.

3.4   User Interface

For implementation the client application, the Open Layers 3 is customized. Open Layers is an open source java script library to load, display and render maps from multiple sources on web pages based on HTML5 and CSS3. Figure 5 display a snapshot of client application.

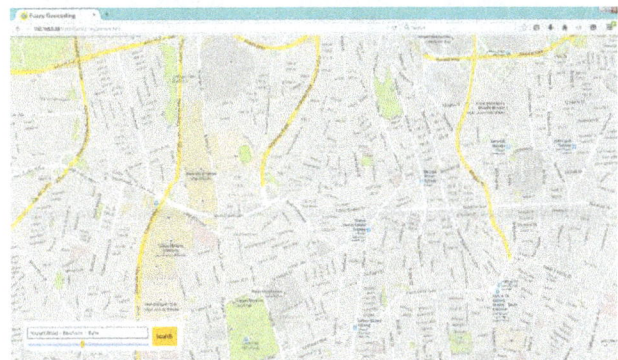

Figure 5. User Interface of Web-based Geocoding System

User entered the address in textbox and select γ coefficient. Client application sent address for ASP web service and receive result from Geoserver as a WMS Layer. Finally the result is displayed on base map (Google Maps).

4. EXPERIMENTAL RESULT

The following scenarios show some instances of system outputs:

4.1 Scenario 1

Suppose that a user wants to search "Karim khan – Bank" in the system. Figure 6 displays the output of this search for $\gamma = 0.7$.

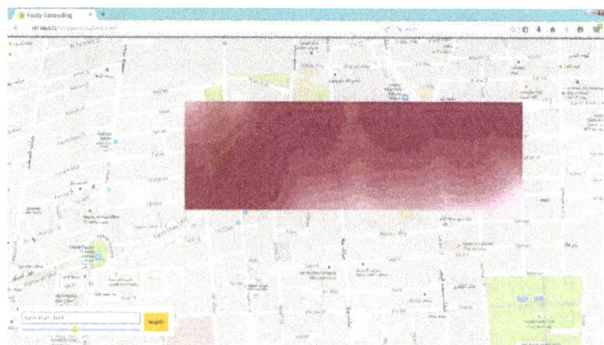

Figure 6. Output of Scenario 1

4.2 Scenario 2

For evaluation, the outputs of proposed method can be compared with output of most commonly used online geocoding services such as Google Maps, Yahoo Maps, Tehran.ir maps and Open Route Service. Result of find "Karim khan - Bank" in these geocoding services is considered in figures 7-10.

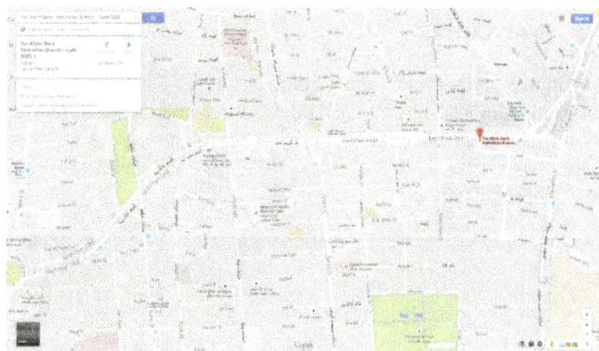

Figure 7. Output of Search Address in Google Maps

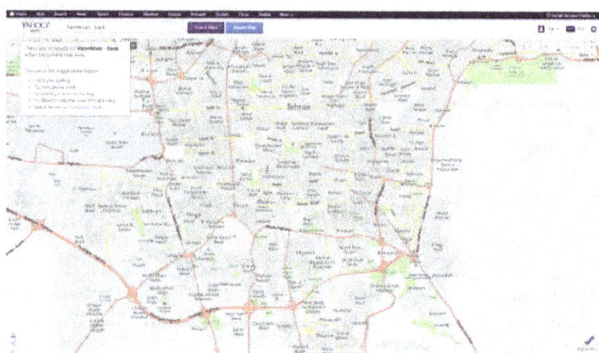

Figure 8. Output of Search Address in Yahoo Maps

Figure 9. Output of Search Address in Open Route Service

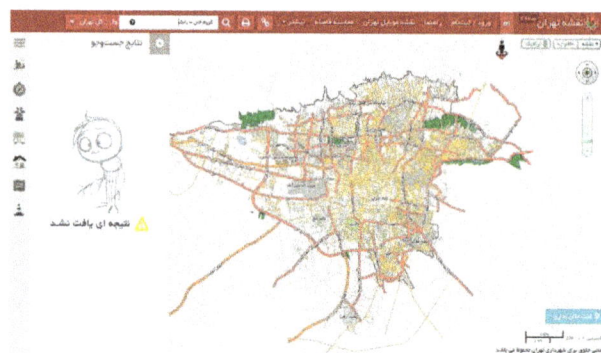

Figure 10. Output of search address in Tehran.ir maps

4.3 Scenario 3

User can change γ coefficient and receive different outputs from the application service. The figure 11 indicates system outputs for different γ coefficient.

5. CONCLUSION

As it was mentioned in introduction, users face some limitations when they use available online geocoding services. This paper proposes the idea of integrating fuzzy technique with geocoding process to resolve these limitations. Accordingly, a new algorithm is proposed for geocoding, while a web-based system is also implemented. The system provides the following capabilities for users:

- The ability to search multi-part addresses: Users can search a multi-part address including district, street, and a specific location's name in the system.
- Searching places based on their location: With defining nearness concept in this service, it is possible to search address based on nearness to a specific location.
- Non-point representation of results: As it was observed in implementation outputs, search results are shown as an area for the user.
- Displaying search results based on their priority: A specific priority has been defined based on different degrees of transparency and according to implementation outputs so that highlighted areas represent high priority search results.

Integration of this service with existing geocoding services available can enhance them. Moreover, defining a weight for each part of an address can provide more accurate and closer results to what users want. Furthermore, system speed improvement can be suggested as a research focus for future investigations.

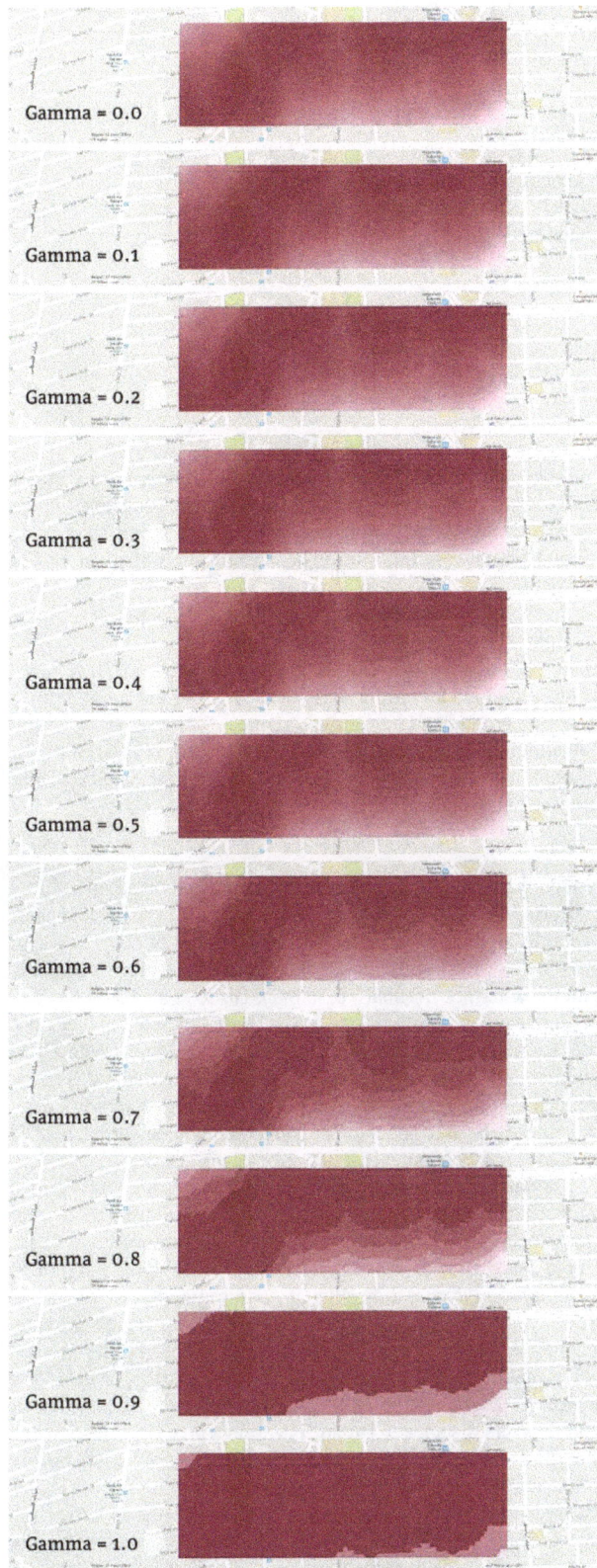

Figure 11. Different Outputs for Different γ (from 0 to 1)

REFERENCES

Peter Christen, Tim churches and Alan Willmore. A Probabilistic Fuzzy Geocoding system based on a National Address File.

Daras K, Feng Z and Dibben C. 2014. HAG-GIS: A spatial framework for geocoding historical addresses.

PostGIS 2.1.9dev Manual, the PostGIS Development Group Chapter 5. Raster Data Management, Queries, and Applications

Web Sites:

- https://maps.yahoo.com
- http://openrouteservice.org
- https://maps.google.com
- http://map.tehran.ir
- http:// openlayers.org

Revised July 2015

3

AUOTOMATIC CLASSIFICATION OF POINT CLOUDS EXTRACTED FROM ULTRACAM STEREO IMAGES

M. Modiri [a]*, M. Masumi [b], A. Eftekhari [c]

[a] Professor at Malek Ashtar University of Technology, Esfahan, Iran- mmodiri@ut.ac.ir
[b] M.s degree of Information Technology Management,- mm.slm.masumi@gmail.com
[c] M.s of degree of Remote Sensing, Dept. of surveying and Geomatics engineering, University of Tehran, Tehran, Iran -
akrameftekary@gmail.com

Commission VI, WG VI/4

KEY WORDS: Point Clouds, Ultracam Images, point's classification, clustering, non-ground points, DEM

ABSTRACT

Automatic extraction of building roofs, street and vegetation are a prerequisite for many GIS (Geographic Information System) applications, such as urban planning and 3D building reconstruction. Nowadays with advances in image processing and image matching technique by using feature base and template base image matching technique together dense point clouds are available. Point clouds classification is an important step in automatic features extraction. Therefore, in this study, the classification of point clouds based on features color and shape are implemented.

We use two images by proper overlap getting by Ultracam-x camera in this study. The images are from Yasouj in IRAN. It is semi-urban area by building with different height. Our goal is classification buildings and vegetation in these points.

In this article, an algorithm is developed based on the color characteristics of the point's cloud, using an appropriate DEM (Digital Elevation Model) and points clustering method. So that, firstly, trees and high vegetation are classified by using the point's color characteristics and vegetation index. Then, bare earth DEM is used to separate ground and non-ground points.

Non-ground points are then divided into clusters based on height and local neighborhood. One or more clusters are initialized based on the maximum height of the points and then each cluster is extended by applying height and neighborhood constraints. Finally, planar roof segments are extracted from each cluster of points following a region-growing technique.

1. INTRODUCTION

Three-dimensional (3D) information of the earth one of the most important data on land management and its related events. Areal stereo imaging is one of the first methods that used in collection 3D information. Today, both improvements in camera technology and the rise of new matching approaches triggered the development of suitable software tools for image based 3D reconstruction by research groups and vendors of photogrammetric software. Based on dense pixel-wise matching, the photogrammetric generation of dense 3D point clouds and Digital Surface Models from highly overlapping aerial images has become feasible(Shorter and Kasparis 2009). These point clouds include of many artificial features as buildings, vegetation, pipe line, etc. that automatic features extraction from point clouds are interested by researchers.

In between, 3D building models are used for variety of applications such as urban planning, city modeling, disaster management, etc.

Accurate building models can be created manually by using aerial images, LIDAR data, building blueprints, and other data sources but often rely on costly and time consuming processing(Bandyopadhyay, van Aardt et al. 2013).

There are much research in classification and segmentation of point clouds that used to extract and detect vegetation and roof plan especially by LIDAR data but generating dense and high accurate point clouds by photogrammetric methods and aerial images is low price in comparison with LiDAR data(Omidalizarandi and Saadatseresht 2013). In most studies, methods of surface extraction can be categorized in two main

groups. Firstly, surface parameters can be estimated directly by clustering or finding maximum parameter in the parameter space. Secondly, point clouds can be segmented on the basis of proximity of the point clouds or similarity measures like locally estimated surface normal(Vosselman, Gorte et al. 2004).

In this article, an algorithm is developed based on the color characteristics of the point's cloud, using an appropriate DEM (Digital Elevation Model) and points clustering method. Firstly, trees and high vegetation are classified by using the point's color characteristics and vegetation index. Then, bare earth DEM is used to separate ground and non-ground points.

The remaining not classified non-ground points are then divided into clusters based on height and local neighborhood. One or more clusters are initialized based on the maximum height of the points and then each cluster is extended by applying height and neighborhood constraints. Planar roof segments are extracted from each cluster of points following a region-growing technique. Planes are initialized using coplanar points as seed points and then grown using plane compatibility tests. If the estimated height of a point is similar to height different limit, or if it's normal distance to a plane is within a predefined limit, then the point is added to the plane. Once all the planar segments are extracted, the common points between the neighboring planes are assigned to the appropriate planes based on the plane intersection line, locality and the angle between the normal at a common point and the corresponding plane. Tree planes which are small in size and randomly oriented is removed by applying a rule-based procedure.

---

* Corresponding author

2. METHOLOGY

The steps followed in the methodology used are shown in Figure 1.

The proposed method first separates vegetation from the input Point clouds data by applying vegetation index. Then ground and non-ground points are divided. The non-ground points, including objects above the ground such as buildings and trees but trees are separated, are further processed for building extraction. We cluster points first and extract planar roof segments from each cluster of points using a region-growing technique in object space.

The extracted segments are refined based on the relationship between each pair of neighboring planes. Points on the neighboring planar segments are accumulated to form individual building regions.

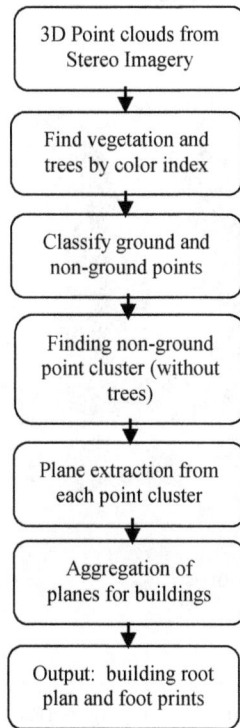

Figure1: the proposed building extraction method

2.1 Vegetation and trees detection by using RGB texture in point clouds

Since we used RGB images to extract point clouds, the three bands from the color image are utilized to calculate two different indices(Gitelson 2004).

Ratio Index (RI): This is the ratio of green radiance (ρ_{green}) and the blue radiance (ρ_{blue}).

$$RI = \frac{\rho_{green}}{\rho_{blue}} \qquad 1$$

Visible Atmospherically Resistant Index (VARI): This index is expressed as,

$$VARI = \frac{\rho_{green} - \rho_{red}}{\rho_{green} + \rho_{red} - \rho_{blue}} \qquad 2$$

Both the index images have gray scale values between 0-255, which were then converted to the binary images based on a threshold value; this value was determined as a mid-range value between highest and lowest gray level for each individual index image. The intersection of these two index images produced a vegetation mask image(Bandyopadhyay, van Aardt et al. 2013).

2.2 Finding non-ground points

DEM (Digital Elevation Model) can be easily used to separate ground and non-ground points. If a bare-earth DEM is not available, one can be generated from the LIDAR point cloud data. We assume that the bare-earth DEM is given as an input to the proposed method. For each 3D point, the corresponding DEM height is used as the ground height H_g. A height threshold $T_h = H_g + h_c$, where h_c is a height constant that separates low objects from higher objects, is then applied to the data. In this study, $h_c = 1$ m has been set(Awrangjeb and Fraser 2014).

2.3 Points clustering

All the non-ground points are now processed to generate clusters of points based on height.

Firstly, all these points are not belonged to any clusters. Considering the maximum data height as the current height, points at this height (within a specified tolerance) are found and one or more clusters are initialized depending on their locality. Then each of the cluster is extended until no points can be added to the cluster. Points in each cluster are marked so that they are not assigned to another cluster. Once all the clusters initialized from the current height are finalized, the points that are not yet assigned to any clusters are then processed to generate more clusters. The next current height will be the maximum height of these unassigned points.

Let the current maximum height be h_m and the set of points that has similar height (i.e., within $h_m \pm T_f$, where $T_f = 0.2$ m(Awrangjeb, Zhang et al. 2013) to allow the error in data generated heights) is S_c. One or more clusters ξ_i, where $i \geq 1$, are initialized using S_c based on the locality. For a point $P \epsilon \xi_i$, there is at least one neighboring point Q in the same cluster such that the 2D Euclidean distance $|P.Q| \leq H_d$. If P and Q are in two different clusters then $|P.Q| > H_d$.

In order to extend a cluster ξ_i, let R be a neighbor of P, where $P \epsilon \xi_i$ but R has been neither assigned to any cluster. R is considered to be a neighbor of P, if $|P.R| \leq H_d$ and their height difference is within V_d. R is added to ξ_i, which is iteratively extended until no R is found as a neighbor of P(Awrangjeb, Lu et al. 2014).

Eventually, an individual building, a roof plane or a roof section are represented by each of the extended clusters. Thereafter, an extended cluster ξ_i is considered a valid cluster if it is larger than 1 m2 in area. The area of ξ_i is thus roughly estimated by counting the number of black pixels multiplied by the pixel size.

2.4 Roof Plan Extraction

Segmentation is an efficient method for roof plan extraction.
The segmentation method has two stages:
2.4.1 Normal and Flatness estimation
The normal for each point was estimated by fitting a plane to its neighboring points. The neighboring points can be selected based on K nearest neighbor (KNN) or fixed distance neighbor (FDN) methods(Rabbani 2006).

The estimation of normal vectors and flatness measure can be performed by eigen-analysis of the covariance matrix of the point positions. Let $p \epsilon P$ be the sample point in the point clouds, where

p consists of x, y and z coordinates, and $\bar{p}$ be the centroid of the neighborhood of p, i.e,

$$\bar{p} = \frac{1}{|N_p|} \sum_{i \in N_p} p_i \qquad 3$$

The 3x3 covariance matrix C for the sample point p is given by

$$C = \frac{1}{|N_p|} \sum_{i \in N_p} (p_i - \bar{p})(p_i - \bar{p})^T \qquad 4$$

If $\lambda_0 \leq \lambda_1 \leq \lambda_2$ are the eigenvalues sorted in the ascending order, then the eigenvector corresponding to the smallest eigenvalue, i.e., λ_0, defines the normal vector at any point p (Bandyopadhyay, van Aardt et al. 2013). The flatness or surface variation₁ at any point p can be estimated as:

$$F = \frac{\lambda_0}{\lambda_0 + \lambda_1 + \lambda_2} \qquad 5$$

2.4.2   Region Growing

This step uses the point normal and their flatness values to group points into smooth surfaces. Two constraints are followed in the region-growing algorithm(Rabbani 2006):
I: Points belonging to a segment should be locally connected. The constraint is enforced by including the nearest neighbor in the region-growing process.
ii: Points belonging to a segment should form a smooth surface, i.e., surface normal of the points should vary within a predefined threshold (θ_{th}). Region-growing starts with calculating the flatness (F) of each point in the data set. Among all points, the one with minimum flatness is considered as a seed point and other points, based on their surface normal orientation, are iteratively added to the region(Bandyopadhyay, van Aardt et al. 2013).
Here we wanted to prevent over-segmentation of surfaces, but we also did not want all points to form a single segment. Thus, depending of the rooftop structure, we set a threshold angle between surface normal of points (θ_{th}) at 10.
Once all the planar segments are extracted, the common points between the neighboring planes are assigned to the appropriate planes based on the plane intersection line, locality and the angle between the normal at a common point and the corresponding plane. Tree planes which are small in size and randomly oriented is removed by applying a rule-based procedure(Awrangjeb, Lu et al. 2014).

3.   DATA SETS AND RESULTS

3.1   Data

We use two images by proper overlap getting by Ultracam-x camera in this study. The images are from Yasouj in IRAN. It is semi-urban area by building with different height. Also Bare earth DEM by 1 meter resolution is used. Dense point clouds data are extracted from aerial stereo images. Orthophoto of studied area and point clouds of this are shown in Figure 2, 3.

Figure 2: Orthophoto of studied area

Figure 3: 3D view of point clouds

3.2   Results

By applying proposed method, at first step, vegetation and trees are extracted from point clouds by applying RI (Rational Index) indicator. We consider 1.2 for threshold value. The results of this step is shown in figure 4.

Figure4: white points are vegetation and trees in point clouds that extracted by RI and black points are unclassified points.

At second step, bare earth DEM is used for separation ground and non-ground from data that process at previous step. The classified points in three class vegetation, ground and non-ground is presented as follow:

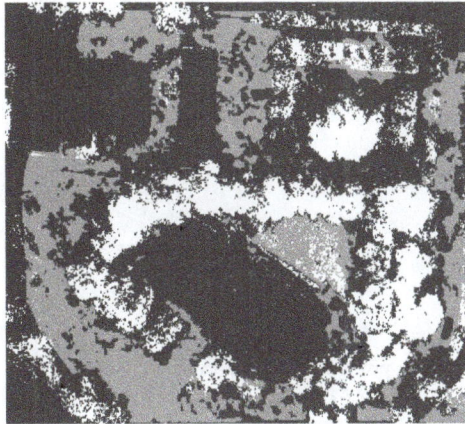

Figure5: points by grey color are ground points, black points are non-ground and white points are in ground vegetation and trees.

The next step in proposed methodology is clustering non-ground points by explained tactic in 2.3 section. Cluster points are shown in Figure 6.

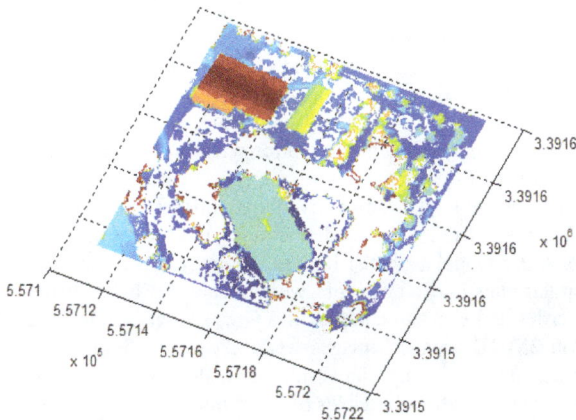

Figure6: points in different clusters are shown.

After applying segmentation method that described, roof plans are detected. The results shown in Figure 7.

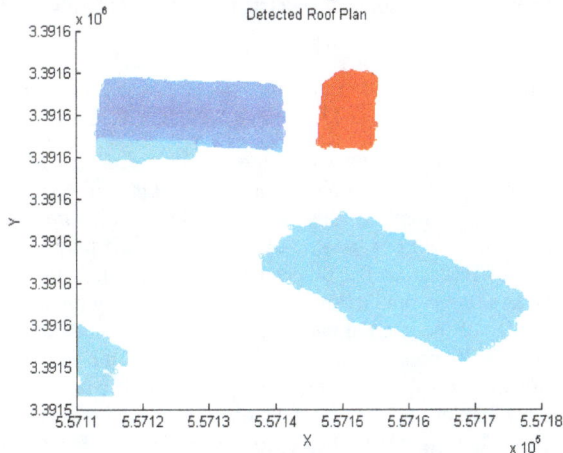

Figure7: Roof plan detected from point clouds.

4. CONCLUSIONS

Our classification algorithm, based on the using dense point clouds extracted from areal stereo images by RGB texture, resulted in a good classification result for both buildings and vegetation in a semi-urban environment.

Visual comparison between ortho-image and results that shown in Figure 4 and 7, is demonstrated that the roof plans and vegetation have been identified with acceptable accuracy.

But we need ground truth for better evaluation. Also the results of this study can be compared by LIDAR data if it's available in the future studies. Finally, we want to implement buildings extraction model for extract boundary box of building for preparing data to input in GIS in the next studies.

REFERENCES

Awrangjeb, M. and C. S. Fraser (2014). "Automatic segmentation of raw LiDAR data for extraction of building roofs." Remote Sensing 6(5): 3716-3751.

Awrangjeb, M., et al. (2014). "Automatic building extraction from LIDAR data covering complex urban scenes." ISPRS-International Archives of the Photogrammetry, Remote Sensing and Spatial Information Sciences 1: 25-32.

Awrangjeb, M., et al. (2013). "Automatic extraction of building roofs using LIDAR data and multispectral imagery." ISPRS Journal of Photogrammetry and Remote Sensing 83: 1-18.

Bandyopadhyay, M., et al. (2013). Classification and extraction of trees and buildings from urban scenes using discrete return LiDAR and aerial color imagery. SPIE Defense, Security, and Sensing, International Society for Optics and Photonics.

Gitelson, A. A. (2004). "Wide dynamic range vegetation index for remote quantification of biophysical characteristics of vegetation." Journal of plant physiology 161(2): 165-173.

Omidalizarandi, M. and M. Saadatseresht (2013). "SEGMENTATION AND CLASSIFICATION OF POINT CLOUDS FROM DENSE AERIAL IMAGE MATCHING." The International Journal of Multimedia & Its Applications 5(4): 33.

Rabbani, T. (2006). Automatic reconstruction of industrial installations using point clouds and images, NCG Nederlandse Commissie voor Geodesie Netherlands Geodetic Commission.

Shorter, N. and T. Kasparis (2009). "Automatic vegetation identification and building detection from a single nadir aerial image." Remote Sensing 1(4): 731-757.

Vosselman, G., et al. (2004). "Recognising structure in laser scanner point clouds." International archives of photogrammetry, remote sensing and spatial information sciences 46(8): 33-38.

4

THE PERFORMANCE EVALUATION OF MULTI-IMAGE 3D RECONSTRUCTION SOFTWARE WITH DIFFERENT SENSORS

V. Mousavi [a], M. Khosravi [a], M. ahmadi [a], N. Noori [a], A. Hosseini naveh [a], M. Varshosaz [a]

a K.N.Toosi University of Tecknology, Faculty of Surveying Engeeniring ,Tehran,Iran
vmoosavy@mail.kntu.ac.ir, khosravi.msd@gmail.com, ahmadii.eng@gmail.com
negar_noori123@yahoo.com, haghshenas.shiva@yahoo.com, ali_hosseini_naveh@yahoo.com
varshosazm@kntu.ac.ir

KEY WORDS: sensor, software, Image-based 3D reconstruction, Data comparison, point cloud, three-dimensional model

ABSTRACT

Today, multi-image 3D reconstruction is an active research field and generating three dimensional model of the objects is one the most discussed issues in Photogrammetry and Computer Vision that can be accomplished using range-based or image-based methods. Very accurate and dense point clouds generated by range-based methods such as structured light systems and laser scanners has introduced them as reliable tools in the industry. Image-based 3D digitization methodologies offer the option of reconstructing an object by a set of unordered images that depict it from different viewpoints. As their hardware requirements are narrowed down to a digital camera and a computer system, they compose an attractive 3D digitization approach, consequently, although range-based methods are generally very accurate, image-based methods are low-cost and can be easily used by non-professional users. One of the factors affecting the accuracy of the obtained model in image-based methods is the software and algorithm used to generate three dimensional model. These algorithms are provided in the form of commercial software, open source and web-based services. Another important factor in the accuracy of the obtained model is the type of sensor used. Due to availability of mobile sensors to the public, popularity of professional sensors and the advent of stereo sensors, a comparison of these three sensors plays an effective role in evaluating and finding the optimized method to generate three-dimensional models. Lots of research has been accomplished to identify a suitable software and algorithm to achieve an accurate and complete model, however little attention is paid to the type of sensors used and its effects on the quality of the final model. The purpose of this paper is deliberation and the introduction of an appropriate combination of a sensor and software to provide a complete model with the highest accuracy. To do this, different software, used in previous studies, were compared and the most popular ones in each category were selected (Arc 3D, Visual SfM, Sure, Agisoft). Also four small objects with distinct geometric properties and especial complexities were chosen and their accurate models as reliable true data was created using ATOS Compact Scan 2M 3D scanner. Images were taken using Fujifilm Real 3D stereo camera, Apple iPhone 5 and Nikon D3200 professional camera and three dimensional models of the objects were obtained using each of the software. Finally, a comprehensive comparison between the detailed reviews of the results on the data set showed that the best combination of software and sensors for generating three-dimensional models is directly related to the object shape as well as the expected accuracy of the final model. Generally better quantitative and qualitative results were obtained by using the Nikon D3200 professional camera, while Fujifilm Real 3D stereo camera and Apple iPhone 5 were the second and third respectively in this comparison. On the other hand, three software of Visual SfM, Sure and Agisoft had a hard competition to achieve the most accurate and complete model of the objects and the best software was different according to the geometric properties of the object.

1. INTRODUCTION

Nowadays, due to advances in technology, generating of 2D and 3D products with significant geometric accuracy and detail, has become possible. A detailed three-dimensional model for representation and documentation of different objects are needed in different branches of engineering.(Hess, Robson, & Hosseininaveh Ahmadabadian, 2014)
Three-dimensional modelling of scenes and objects in different scales is performed using both image-based methods (passive) and range-based methods (active). Range-based or active methods (Vosselman & Maas, 2010) like structured light systems and laser scanner are common systems for point cloud generation. Image-based or passive methods (Remondino & El-Hakim, 2006) which have made significant progress with the development of computer (Furukawa, Curless, Seitz, & Szeliski, 2010; Pollefeys et al., 2008) are currently regarded as automatic methods for image orientation (Remondino, Del Pizzo, Kersten,

& Troisi, 2012) and three-dimensional reconstruction in different scales (Haala, 2013; Lafarge & Mallet, 2012).

A variety of algorithms can be used for three-dimensional modelling of different objects or scenes. One of the most common algorithms in this field is SFM (Structure from Motion) algorithm (Agarwal et al., 2011). For this purpose, an archive of images of the object or scene is needed. In SFM method, a large number of images is oriented without any knowledge of the internal parameters of the camera (Barazzetti, Scaioni, & Remondino, 2010) . Orientation of the images is performed automatically through a series of specific points with identical features. One of the most commonly used operators in this operation is SIFT (Scale Invariant Feature Transform) (Fahmi, 2011) , which extracts identical features between images with different positions, scales and lighting.

Different software, based on this algorithm or similar algorithms, are available as open source (VisualSFM, Bundler,

Apero, Insight3D, etc.), commercial (Agi Soft, Photo Modeller) and Web service (Microsoft Photosynth Autodesk123D Catch Beta, My 3D Scanner, Hyper3D, Arc3D, etc.) for users (Remondino et al., 2012). Web-based three-dimensional software services create three-dimensional models quickly and without any special knowledge. (Cavas-Martínez et al., 2014)

In previous studies, effective factors such as network design (Ahmadabadian, Robson, Boehm, & Shortis, 2014) are examined for reducing image distortion of the object. But the type of sensor used has not been evaluated so far. Due to the availability of mobile sensors for everyone and propagation of professional sensors and the advent of stereo sensors, comparing these three types of sensors is an effective step in the evaluation and finding optimal methods for creation detailed three-dimensional models. The question here is whether the type of sensors used in the model can affect the accuracy of the model?

The purpose of this paper is to identify the best software and sensors for generating a complete and accurate three-dimensional model from small objects. For this purpose, a detailed overview of recent research in this field is presented in section 2. According to the mentioned three types of sensors and appropriate identified software, precise geometric evaluation is performed based on evaluation standards of point clouds in section 4.

2. REVIEW OF PREVIOUS WORK

Active sensors have been used in various fields since 2000. Although these sensors have made significant progress in recent years, a lot of problems occur when using them; for instance, they are heavy and non-portable and they cost a lot. (Remondino, Spera, Nocerino, Menna, & Nex, 2014)

Today, the integration of computer vision and photogrammetry methods has led to creation of image-based modelling processes.(Remondino & El-Hakim, 2006) In this method, multiple images of a scene turn into point cloud by correspondence algorithms and then a uniform three-dimensional model of this point cloud is generated. Perhaps it can be said that correspondence is one of the key steps in image-based modelling. Finally, using this method we can extract semantic information from images. In recent years, due to the development of computer and related technology and the emergence of some algorithms like SFM, image-based methods is more economical. SFM algorithm includes simultaneous determination of internal and external parameters of the camera and reconstructing the three-dimensional structure of the scenes. In image-based methods first the correspondence between two images is examined by extracting key features. to do this, the area-based or pixel-based algorithms are used. SIFT (Scale Invariant Feature Transform) and SURF (Speeded Up Robust Feature) (Bay, Tuytelaars, & Van Gool, 2006) are the most common algorithms. Then, using SFM algorithm or its integration with other algorithms like DMVR (Dense Multi View Reconstruction), orientation of cameras and internal and external parameters of the camera are calculated and a sparse point cloud is created. Many systems use Bundle Adjustment (Engels, Stewénius, & Nistér, 2006) in order to improve the accuracy of camera direction and reduce image distortion. In some software, additionally, an algorithm is added to create a dense point cloud.
Examining the image-based methods and also comparing these methods with range-based methods has been in the spotlight of researchers in recent years.

These days, with the advances in technology and the advent of a variety of algorithms, different methods and software for 3-D modeling is available: free Open source software, commercial software which is for professional modeling, and web-based software that make 3-D modeling very simple. Each of these applications use various algorithms in different stages of modeling process and as a result, network design, the number of images, system and even the time needed for processing a model is different. In this article we tried to use all three kinds of software and also different algorithms in the modeling process. Table 1 shows the summary of the software used in recent researches.

Software	Image Matching	Spars Reconstruction	Dense Reconstruction	Author
Visual SFM	SIFT	SFM	CMVS/ PMVS	(Wu, 2011)
Agisoft	SIFT-like	SFM	SGM	2014
SURE	-	-	SGM	(Rothermel, Wenzel, Fritsch, & Haala, 2012)
Arc3D	SURF	SFM	MVS	K.U.Leuven (2005)
Photosynth	SIFT	SFM	-	Microsoft (2008)

Table 1 - Used Software in recent studies

Preparing a three-dimensional model of an object depends on photogrammetric methods, geometric accuracy of the obtained model, the quality of the object texture, resolution of the images and network design (Skarlatos & Kiparissi, 2012). So the quality and resolution of the image used, is an important factor in three-dimensional model generation. A good camera certainly brings better results in comparison with a cheaper one.
The greater the number of pixels in an image, the more the number of created points in final model, and thus, the time needed for model generation increases. However, due to the heavy cost of professional cameras, the question is that which resolution is sufficient to provide an appropriate model. If the camera lens is better in its image quality, the amount of light received by the CCD becomes more and the lens distortions becomes less. Both these factors improve accuracy of the modeling. The amount of noise positioned to image by sensor or its efficiency in terms of its stability in internal parameters of the camera is an important issue which has not been studied in recent years yet. Due to the availability of mobile sensors for everyone and popularity of professional sensors and emergence of stereo sensors. Comparing these three is an effective step in evaluating and finding an optimal procedure to produce detailed three-dimensional models. Therefore, in this study, three different sensors are used in the process of point cloud generation and three-dimensional modelling.

3. MATERIALS AND METHODS OF THE STUDY

In order to evaluate the performance of different software as well as various sensors in terms of optimal model generation, four different objects have been used in the evaluation process.

Three of the objects have regular shape and the one of them is a statue with a more complexity in geometric shape.

Figure 1 –pictures of studied objects

3.1 Data collection

Data collection consists of two main steps. The first step involves the preparation of three-dimensional model of objects by active scanner sensor ATOS Compact Scan 2M which is related to Gom Inspect company. It conducts three-dimensional model by ToF[1] method.

Atos compact scan 2M
The number of camera pixels: 2 x 2 000 000
The dimensions of the sensor: 340 mm x 130 mm x 230 mm

The second step is image acquisition. In this stage, the network design is performed according to different algorithms, so that it is compatible with all software and algorithms. In this stage, 69 images are taken in a ring with the convergence angle of about 5 degrees.

3.2 Image acquisition stage

If we divide image sensors to three categories, these categories include professional sensors, emerging new stereo sensors and current mobile sensors.

In this study, one sensor is chosen from these three categories and each of them are evaluated. Among professional cameras, Nikon D3200 and among stereo cameras, Fujifilm FinePix Real3D W3 and among mobile cameras, IPhone 5 was chosen.

Figure 2 - Sensors that used in this article

Photogrammetry algorithms in processing step may face two problems: one of them is original image quality (noise, low radiometric quality of the images, shadows, etc.), second is the surface of the object (the shiny surface of an object or lack of texture). (Remondino et al., 2014)

In relation to the quality of the original image or the amount of noise, three kinds of sensors are investigated in this study and the amount of noise and the efficiency of each sensor for 3-D model generation are examined. But 3-D model generation of a bright object [2] is still one of the problems in photogrammetry that needs more research. To solve this problem, in this study, a solution of titanium dioxide powder and alcohol is used. The reason for using this solution is the fine particles of the powder, therefore it doesn't harm the accuracy of the model.

In the process of point cloud generation of an object, at first matching process between original images is needed. Matching refers to determining the key and corresponding points between images. So according to this definition, key points must be found in images. But in objects with regular geometric shape or without texture on its surface, this process fails or if done, the result is a totally sparse point cloud. To create an artificial texture on these kinds of objects, optical methods are usually used. This means that a texture is fitted on the surface by a projector. The minimum number of the projectors is two, but in this study, three projectors were used to prevent the shadow.

Also, due to this fact that objects used in this study are metal and each metal has a thermal expansion coefficient, it is necessary to keep the ambient temperature fixed for image acquisition and also similar temperature while generating 3-D model from active sensor. In this study this temperature was 22 °C.

3.3 The process of three-dimensional modelling in software

In the process of creating three-dimensional model of mentioned objects, at first the surface of the object is covered with titanium dioxide and then by an appropriate network design, images are taken. For projecting texture to the surface of each objects, three video projectors are used which are located at 120 degrees of each other and a specific pattern is projected on the object. According to the dimensions of the object and the distance of projectors, the distance between the camera and the object is calculated as well as the number of image stations according to this distance. Distance from the camera to the desired object varies around 70 to 100 cm and 68 image stations was considered. This process was conducted for each sensor and each object and at the end, 48 data sets were obtained. Each of software was run 12 times for 4 objects and 3 sensors and the results were compared with true data. The results of the comparison are examined in the next section.

4. RESULTS

After the mentioned stages for each object and each sensor, raw data is obtained. These raw data are run in various software and the point cloud for each object is achieved in three different sensors. Then, these different sensors should be compared in terms of accuracy and quality of the obtained model.

object	Evaluation criteria between two point clouds
Stone statue	Fitting two plates to each cloud point and comparing the distance between these two in two different cloud points
Sphere	fitting a sphere to each cloud point and comparing obtained spheres
cylinder	Fitting a cylinder to a section of cloud point and comparing two cylinders
hexagon	Fitting two plates to opposite sides of each other for each cloud point and comparing the distance between these two in two cloud points

---

[1] Time of flight

[2] shiny

4.1 The comparison of sensors

According to the selected criteria for each object, a special shape is fitted to objects and the comparison has been made between them. For each point cloud, standard deviation of all points of the fitted shape is considered as quality criteria for the obtained point cloud. The result of the comparison is illustrated in following tables.

Accuracy	Agi Soft	VSFM	SURE	Arc3D
Stone statue	0.41	0.48	0.39	No model
Sphere	0.17	0.09	0.10	No model
Cylinder	0.03	0.15	0.16	No model
Hexagon	0.08	0.08	0.17	No model

Table 2- Accuracy for Iphone5 sensor

Accuracy	Agi Soft	VSFM	SURE	Arc3D
Stone statue	0.40	0.48	0.1	0.37
Sphere	0.17	0.09	0.07	0.25
Cylinder	0.03	0.14	0.09	0.38
Hexagon	0.03	0.07	0.06	0.07

Table 3 - Accuracy for Nikon D3200 sensor

Accuracy	Agi Soft	VSFM	SURE	Arc3D
Stone statue	0.39	0.59	0.40	0.90
Sphere	0.17	0.13	0.01	No model
Cylinder	0.06	0.14	0.33	0.43
Hexagon	0.13	0.49	0.18	No model

Table 4 - Accuracy for Fuji film FinePix Real 3D W3 sensor

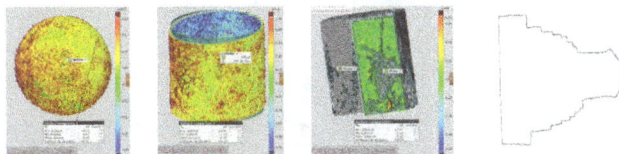

Figure 3 – Models that obtained from Stereo sensor and SURE algorithm

5. DISCUSSION AND CONCLUSION

Today, due to advances in computer science, three-dimensional modeling of objects and scenes at different scales has become very popular. Research has been done in this field. Three-dimensional modeling algorithms vary and they are examined and compared by researchers in various fields in recent years.

As noted above, the geometric accuracy of an obtained model depends on several general parameters. Examining these parameters alongside algorithms can help the accuracy of the obtained model. One of the parameters involved in the model is resolution or image quality which has not been investigated in the recent years. Therefore, in addition to comparing several algorithms for three-dimensional modeling, we also examine the effect of different sensors in the obtained model. In general, three types of sensors are available in the market, which are all used in this study.

According to the conducted survey we can clearly see that professional sensors lead to better accuracy and density in the final model and then stereo sensors and mobile sensors have lower accuracy and produce models with more noise. But the obtained accuracies in different sensors have slight differences relative to each other. In general, according to the table of accuracy in different sensors and algorithms in similar condition, we can see that how much the sensors used in three-dimensional modeling depend on the used sensor or algorithm.

According to the mentioned conditions, there is a limitation for modeling objects with the image-based method. Some of these limitations are the brightness of the object, darkness of the object or lack of texture in the surface. These limitations create some bugs in modeling and sometimes make it fail.

In the future we can eliminate these limitations by providing methods to model dark or bright objects or solve the problem of objects without texture.

REFERENCES

Agarwal, S., Furukawa, Y., Snavely, N., Simon, I., Curless, B., Seitz, S. M., & Szeliski, R. (2011). Building rome in a day. *Communications of the ACM, 54*(10), 105-112.

Ahmadabadian, A. H., Robson, S., Boehm, J., & Shortis, M. (2014). Stereo-imaging network design for precise and dense 3d reconstruction. *The Photogrammetric Record, 29*(147), 317-336.

Barazzetti, L., Scaioni, M., & Remondino, F. (2010). Orientation and 3D modelling from markerless terrestrial images: combining accuracy with automation. *The Photogrammetric Record, 25*(132), 356-381.

Bay, H., Tuytelaars, T., & Van Gool, L. (2006). Surf: Speeded up robust features *Computer vision–ECCV 2006* (pp. 404-417): Springer.

Cavas-Martínez, F., Pérez-Sánchez, C., Adrián-Sáez, J., Cañavate, F., Nieto, J., & Fernández-Pacheco, D. (2014). USE OF DIGITAL IMAGES FOR THREE-DIMENSIONAL RECONSTRUCTION OF MECHANICAL COMPONENTS.

Engels, C., Stewénius, H., & Nistér, D. (2006). Bundle adjustment rules. *Photogrammetric computer vision, 2.*
Fahmi, H. M. (2011). Scale Invariant feature transform. *noppa. lut. fi.*

Furukawa, Y., Curless, B., Seitz, S. M., & Szeliski, R. (2010). *Towards internet-scale multi-view stereo.* Paper presented at the Computer Vision and Pattern Recognition (CVPR), 2010 IEEE Conference on.

Haala, N. (2013). The landscape of dense image matching algorithms.

Hess, M., Robson, S., & Hosseininaveh Ahmadabadian, A. (2014). A contest of sensors in close range 3D imaging: performance evaluation with a new metric test object. *ISPRS-International Archives of the Photogrammetry, Remote Sensing and Spatial Information Sciences, 1*, 277-284.

Lafarge, F., & Mallet, C. (2012). Creating large-scale city models from 3D-point clouds: a robust approach with hybrid representation. *International Journal of Computer Vision, 99*(1), 69-85.

Pollefeys, M., Nistér, D., Frahm, J.-M., Akbarzadeh, A., Mordohai, P., Clipp, B., . . . Merrell, P. (2008). Detailed real-time urban 3d reconstruction from video. *International Journal of Computer Vision, 78*(2-3), 143-167.

Remondino, F., Del Pizzo, S., Kersten, T. P., & Troisi, S. (2012). Low-cost and open-source solutions for automated image orientation–A critical overview *Progress in cultural heritage preservation* (pp. 40-54): Springer Berlin Heidelberg.

Remondino, F., & El-Hakim, S. (2006). Image-based 3D modelling: A review. *The Photogrammetric Record, 21*(115), 269-291.

Remondino, F., Spera, M. G., Nocerino, E., Menna, F., & Nex, F. (2014). State of the art in high density image matching. *The Photogrammetric Record, 29*(146), 144-166.

Rothermel, M., Wenzel, K., Fritsch, D., & Haala, N. (2012). *Sure: Photogrammetric surface reconstruction from imagery.* Paper presented at the Proceedings LC3D Workshop, Berlin.

Skarlatos, D., & Kiparissi, S. (2012). Comparison of laser scanning, photogrammetry and SfM-MVS pipeline applied in structures and artificial surfaces. *ISPRS Annals of the Photogrammetry, Remote Sensing and Spatial Information Sciences, 3*, 299-304.

Vosselman, G. V., & Maas, H.-G. (2010). *Airborne and terrestrial laser scanning*: Whittles.

Wu, C. (2011). VisualSFM: A visual structure from motion system. *URL: http://homes. cs. washington. edu/~ ccwu/vsfm, 9.*

5

A NOVEL IHS-GA FUSION METHOD BASED ON ENHANCEMENT VEGETATED AREA

S. Niazi [a,*], M. Mokhtarzade [b], F. Saeedzadeh[c]

[a] Faculty of Geodesy and Geomatics Engineering, K. N. Toosi University of Technology, Valiasr street, Tehran, Iran- (sniazi@mail.kntu.ac.ir)
[b] Faculty of Geodesy and Geomatics Engineering, K. N. Toosi University of Technology, Valiasr street, Tehran, Iran- (m_mokhtarzade@kntu.ac.ir)
[a] Faculty of Geodesy and Geomatics Engineering, K. N. Toosi University of Technology, Valiasr street, Tehran, Iran- (fsaedzadeh@yahoo.com)

KEY WORDS: Pan sharpening, IHS-GA method, Enhance Vegetated area

ABSTRACT

Pan sharpening methods aim to produce a more informative image containing the positive aspects of both source images. However, the pan sharpening process usually introduces some spectral and spatial distortions in the resulting fused image. The amount of these distortions varies highly depending on the pan sharpening technique as well as the type of data. Among the existing pan sharpening methods, the Intensity-Hue-Saturation (IHS) technique is the most widely used for its efficiency and high spatial resolution. When the IHS method is used for IKONOS or QuickBird imagery, there is a significant color distortion which is mainly due to the wavelengths range of the panchromatic image. Regarding the fact that in the green vegetated regions panchromatic gray values are much larger than the gray values of intensity image. A novel method is proposed which spatially adjusts the intensity image in vegetated areas. To do so the normalized difference vegetation index (NDVI) is used to identify vegetation areas where the green band is enhanced according to the red and NIR bands. In this way an intensity image is obtained in which the gray values are comparable to the panchromatic image. Beside the genetic optimization algorithm is used to find the optimum weight parameters in order to gain the best intensity image. Visual and statistical analysis proved the efficiency of the proposed method as it significantly improved the fusion quality in comparison to conventional IHS technique. The accuracy of the proposed pan sharpening technique was also evaluated in terms of different spatial and spectral metrics. In this study, 7 metrics (Correlation Coefficient, ERGAS, RASE, RMSE, SAM, SID and Spatial Coefficient) have been used in order to determine the quality of the pan-sharpened images. Experiments were conducted on two different data sets obtained by two different imaging sensors, IKONOS and QuickBird. The result of this showed that the evaluation metrics are more promising for our fused image in comparison to other pan sharpening methods.

1. INTRODUCTION

Most of the newest remote sensing systems, such as Landsat8, SPOT, IKONOS, QuickBird, EO-1 and ALOS provide sensors with one high spatial resolution panchromatic (PAN) and several multispectral (MS) bands simultaneously. Meanwhile, an increasing number of applications (e.g. feature detection, change monitoring, land cover classification, etc.) often demand the use of images with both high spatial and high spectral resolution. As a result, the fusion of high resolution PAN image and low resolution multi-spectral images has become a powerful solution and many pan-sharpening methods have been proposed over the last two decades (Pohl et al. 1998, Lau et al. 2000, Wang et al. 2005).

The pan-sharpening of low resolution multispectral (MS) and high resolution panchromatic (PAN) satellite images is a very important concern for numerous remote sensing applications such as classification, segmentation, and object detection. An effective pan-sharpening technique is a useful tool not only for increasing the interpretability of human observers but also for improving the accuracy of image analysis such as feature extraction, modeling and classification (Yang et al. 2012). The general idea behind image sharpening is to preserve the spectral values of the MS image in the pan-sharpened image and to improve the spatial resolution simultaneously (Konus and Ehlers. 2007). Different algorithms may lead to different pan sharpening qualities (Zhang. 2004).

To date, various algorithms for image pan-sharpening have been developed in order to combine the spatial information of high resolution PAN image with the spectral information of a lower resolution MS image to produce high resolution MS image (Alparone et al. 2007). In the remote sensing community, probably the most popular pan sharpening methods are the Intensity-Hue-Saturation (IHS)(Schetsellar. 1998, Choi et al. 2000, Choi. 2006, Myungjin. 2006), the Principal Component Analysis (PCA) (Chavez Sarp and Kwarteng. 1989, Shettigara. 1992, Vrabel et al. 1996, Shah. 2008, Yang and Gong. 2012), the Brovey transform (Earth Resource Mapping Pty Ltd. 1990, Chaves. 1991, Du et al. 2007, Bovolo et al. 2010) and Wavelet-based image fusion(Otazu et al. 2005, Zhang. 2004).

IHS based methods are among the most common fusion methods due to their simple computation, high spatial resolution and efficiency. Many modifications have been proposed to enhance its spectral quality. These pan-sharpening techniques are performed on the pixel level because of the minimum information loss during the sharpening process, so the digital classification accuracy of the pixel level fusion is the highest (Zhang. 2008).

Although IHS is computationally fast and can quickly merge massive volumes of data, there may be color distortions in its resulted fused images. In particular, the high difference between the gray values of PAN and Intensity image cause the large spectral distortion in the fused images of IHS method (Choi et al. 2006). In this paper, we will introduce a novel IHS -GA pan

---

* Corresponding author

sharpening method to address this issue. In order to evaluate the spectral and spatial quality of the proposed method, several metrics were computed, such as ERGAS (L. Wald. 2000), SAM (Goetz et al. 1992), SID (Chang et al. 1999), RASE (Ranchin et al. 2000), Correlation Coefficients (Chavez et al. 1989), RMSE and spatial.

2. IHS Pan-Sharpening Technique

IHS (Intensity-Hue-Saturation) (Carper et al. 1990, Tu et al. 2004), probably is the most common image fusion technique for remote sensing applications and is used in many commercial pan-sharpening softwares. This method converts a color image from RGB space to the IHS color space. In next step the I (intensity) band is replaced by the panchromatic image. Before fusing the images, a histogram matching is performed on the multispectral and the panchromatic image. The MS image is then converted to IHS color space using the following linear transformation (Palsson et al. 2012):

$$I = \alpha_1 R + \alpha_2 G + \alpha_3 B \qquad (1)$$

$$\begin{bmatrix} I \\ H \\ S \end{bmatrix} = \begin{pmatrix} \alpha 1 & \alpha 2 & \alpha 3 \\ \dfrac{-\sqrt{2}}{6} & \dfrac{-\sqrt{2}}{6} & \dfrac{2\sqrt{2}}{6} \\ \dfrac{1}{\sqrt{2}} & \dfrac{-1}{\sqrt{2}} & 0 \end{pmatrix} \begin{bmatrix} R \\ G \\ B \end{bmatrix} \qquad (2)$$

As the common practice equal coefficients of $\alpha_i = 1/3; (i = 1, 2, 3)$ are used. In this way the entire fusion process in the next step is to scale the Pan image so that it has the same mean and variance as the intensity component (Palsson et al, 2012):

$$P = \frac{\sigma_I}{\sigma_P}(P - \mu(P)) + \mu(I) \qquad (3)$$

Where σ and μ are the standard deviation and mean ,respectively.

The intensity component is then replaced with the appropriately scaled Pan Image and finally the inverse IHS transformation is taken to get the fused image. This process is equivalent to (Palsson et al. 2012):

$$\begin{bmatrix} F(R) \\ F(G) \\ F(B) \end{bmatrix} = \begin{bmatrix} R + Pan - I \\ G + Pan - I \\ B + Pan - I \end{bmatrix} \qquad (4)$$

Where F(i); (i=R,G,B) are the fused multi spectral image. Implementing the IHS fusion method in this manner is very efficient and is called the Fast IHS technique (FIHS) (Tu et al. 2001), making IHS ideal for the large volumes imageries produced by satellite sensors.

Ideally the fused image produced by Fast IHS would have a higher spatial resolution and sharper edges than the original color image. At the same time it is desired to have no distributed spectral content in the fused image. However, because the panchromatic image was not created from the same wavelengths of light as the RGB image, this technique produces a fused image with some color distortions (Choi. 2008). This problem becomes worse with higher difference between the intensity band and panchromatic image. Various modifications are proposed to the IHS method to address these problems.

One of the first modifications of the IHS method extends the IHS method from three bands to four by incorporating an infrared

component (Tu et al. 2005). Regarding the fact that the panchromatic sensors pick up infrared light (IR) in addition to visible wavelengths, this modification caused the intensity image to be better matched with the panchromatic image. To address this issue, researchers have extended this method for other multispectral images by using $\alpha_i = 1/N$ where N is the number of bands (Zhang et al. 2004; Choi et al. 2006).

A similar modification of IHS, called the 'Fast Intensity-Hue-Saturation fusion technique with Spectral Adjustment' (FIHS-SA) method, incorporates four bands with weighting coefficients on the green and blue bands to minimize the difference between intensity and the panchromatic images. These weighting coefficients were calculated experimentally a fuse IKONOS images (Tu et al. 2005). In 2008 Choi expanded this work and experimentally determined the coefficients for the red and infrared bands for IKONOS images (Choi. 2008). The green and blue band coefficients were taken from the 2005 paper by Tu. Since these coefficients were calculated using IKONOS data, these parameters are not ideal for fusing QuickBird images (Tu et al. 2005).

3. Methodology

In this section a new method is developed for the sake of spatial and spectral distortion reduction in the IHS fusion technique. In the proposed method, the optimum weight parameters of equation (5), are found via Genetic Algorithms. The optimized weight parameters are then used to produce the intensity image.

$$I = \alpha_1 R + \alpha_2 G + \alpha_3 B + \alpha_4 NIR \qquad (5)$$

Figure 1 illustrates the flowchart of the proposed image fusion method.

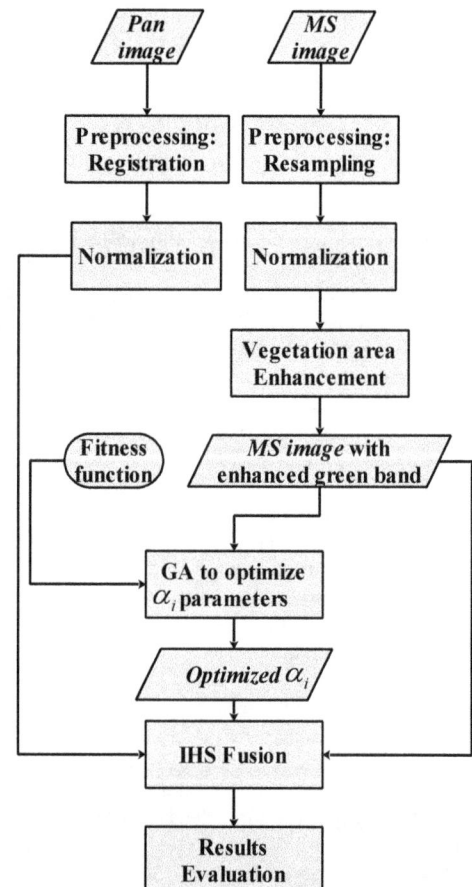

Figure 1. IHS-GA method

As shown in Figure 1, the inputs of the proposed Genetic Algorithm are MS and Pan Images. To this goal some preprocessing are necessary. The MS image was resampled to the PAN image pixel size by nearest neighbor method and then two images were registered by the ENVI4.8 software. After normalization, the process of the weight parameters computation was started by the Genetic Algorithm. After finishing Genetic Algorithm the PAN and Enhanced MS images were fused by the optimum weight parameters in the IHS method.

According to Figure 1, this method has two main steps which are: 1) production of a new Green band (enhanced green band), 2) Genetic algorithm for estimating the weight parameters.

Step 1 .Traditional IHS fusion method suffers from color distortion when dealing with IKONOS or QuickBird images. NDVI (normalized difference vegetation index) is the most common vegetation index that is computed from the NIR and RED bands (Equation 6).

$$NDVI = \frac{NIR - RED}{NIR + RED} \qquad (6)$$

Where NIR and R stand for the spectral reflectance measurements acquired in the near-infrared and red bands, respectively. NDVI varies between -1 and 1. It is related to the fraction of photo-synthetically active radiation. Vegetated areas typically have NDVI values greater than zero. The higher the NDVI, the denser and more greener the vegetation.

In this paper vegetated areas are first found via thresholding on the NDVI and then green band is enhanced in these areas. For this enhancement the idea of (Miloud et al. 2010) is used with some modifications. In this method the difference of NIR and Red bands are applied to correct the Green band. This concepts are presented in the following pseudo code.

Begin / /The process to make the Enhanced vegetated band .

For all image pixels(i= 0,1,...,N; j= 0,1,...,M) in MS image

Compute NDVI(i,j)= $\frac{NIR(i,j) - RED(i,j)}{NIR(i,j) + RED(i,j)}$

if NDVI(i,j) > *Threshold*

Green$_{new}$ (i,j) = Green$_{old}$ (i,j) + [$NIR(i,j) - RED(i,j)$]^2;

else

Green$_{new}$ (i,j) = Green$_{old}$ (i,j);

end // if

end // For

end / /of the Algorithm

In our experiments a variety of thresholds were tested and optimum values of 0.5 and 0.25 were selected for IKONOS and QuickBird images in that order.

Step 2. IHS method converts a color image from RGB space to the IHS color space. In next step I (intensity) band is replaced by the panchromatic image. Intensity image is produced as a linear combination of MS bands (Equation 6) and, thus, its weight parameters have a direct effect on the final fusion result. In previous researches these parameters were set to equal values but in recent studies different experimental values are also used for each data set (Tu et al. 2001, Choi. 2008). In the proposed method weight parameters are estimated by Genetic Algorithm. To this goal, the Green$_{new}$ band with the other bands (Red, NIR, and Blue) are imported to the binary genetic algorithm to achieve the best optimum weight parameters for computing the fused image. Higher amounts of GA cost function reflect more desired results. In our proposed method the cost function was designed as the

linear combination of some metrics shown in the following relation:

$$F = RMSE + ERGAS + (1 - QAVE) + RASE + SAM + SID + (1 - Spatial) \quad (7)$$

The definition of these metrics are explained in below.

- ERGAS (Wald. 2000) is an acronym in French for "Erreur relative globale adimensionnelle de synthese" which translates to "relative dimensionless global error in synthesis".
- A Universal Image Quality Index (Q-average) models any distortion as a combination of three different factors: loss of correlation, luminance distortion, and contrast distortion (Bovik. 2002).
- The relative average spectral error (RASE) characterizes the average performance of the method of image fusion in the spectral bands (Ranchin et al. 2000).
- The Spectral Angle Mapper (SAM) (Goetz et al. 1992) is a metric that calculates the spectral similarity between two spectral vectors as a spectral angle.
- Spectral Information Divergence (SID) (Chang et al. 1999) originates from information theory. Each pixel spectrum is viewed as a random variable and SID measures the difference or discrepancy of the probabilistic behaviors between two spectral vectors, taken from the MS image and final fused image, respectively.
- The Spatial metric used in this paper is based on computing the correlation coefficient between the high-frequency data of each MS band and the high frequency data of the Pan image.

The initial random population size was set to 50 and the maximum number of generations was selected equal to 100. The chromosome length was 28. The optimum α_i parameters were searched in the [0, 1] interval in order to ensure a wide state space. The stopping criterion was chosen as the equally of the secuently migration results.

4. Performance evaluation of the pan-sharpened images

Two datasets were used in this study. The first one was a Quick Bird subset in an urban area and the other was an IKONOS image of a semi urban area containing agricultural field.

Spatial quality can be judged visually, but color changes more difficult to be recognized in this manner (Rahmani et al. 2008). The spectral quality of pan-sharpened images was determined according to the changes in colors of the fused images as compared to the MS reference images. In this study, seven metrics (Correlation Coefficient, ERGAS, RASE, RMSE, SAM, SID and Spatial Coefficient) were used in order to determine the quality of the pan-sharpened images. Before the analysis, each MS image was resampled to the equivalent size of its corresponding PAN image (Rahmani. 2008, Alperton. 2008, and Zhang. 2008).

5. RESULT
5.1. IKONOS data set

The first study area was an IKONOS image taken from a semi urban region located in Shahryar (Longitude: 51˚3', Latitude: 35˚39') Tehran, Iran. The contained results are presented in table 1 where in addition to the proposed IHS-GA four well known extensions of IHS are also evaluated to give us the comparison opportunity. This table content is also depicted in Figure 2 for a more comprehensive view. Figure 3 presents the obtained fused images.

method	C.C	ERGAS	RASE	RMSE	SAM	SID	S.C
FAST IHS (Tu et al. 2001)	0.001	5.810	0.962	8.442	3.533	0.005	0.569
Adaptive IHS (Rahmani et al. 2008)	0.003	3.812	1.384	1.214	0.810	0.003	0.844
Ikonos IHS (Choi. 2008)	0.008	5.563	1.056	0.927	4.983	0.008	0.881
IHS-GA	0.087	0.454	0.106	0.428	1.308	0.001	0.972
refrence	0.000	0.000	0.000	0.000	0.000	0.000	1.000

Table 1: Assessing the fusion methods (IKONOS data set)

method	C.C	ERGAS	RASE	RMSE	SAM	SID	S.C
FAST IHS (Tu et al. 2001)	0.002	9.335	0.994	1.046	1.878	0.008	0.977
Adaptive IHS (Rahmani et al. 2008)	0.012	4.279	1.075	1.131	1.069	0.011	0.815
Ikonos IHS (Choi. 2008)	0.002	4.272	1.075	1.131	1.118	0.010	0.882
IHS-GA	0.546	3.821	0.424	0.446	0.123	0.005	0.909
refrence	0.000	0.000	0.000	0.000	0.000	0.000	1.000

Table 2: Assessing the fusion methods (Quickbird data set)

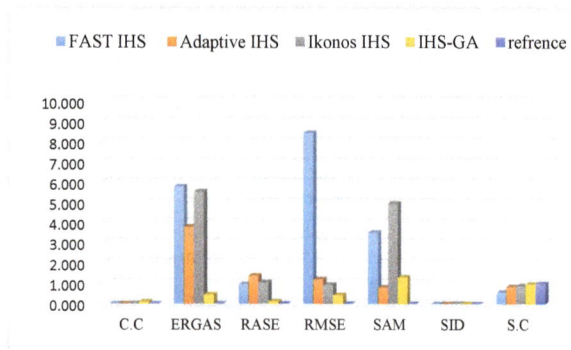

Figure 2. chart of results(IKONOS data sat)

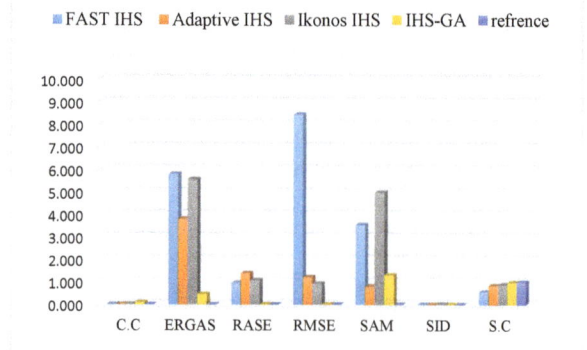

Figure 4. chart of results(QuickBird data sat)

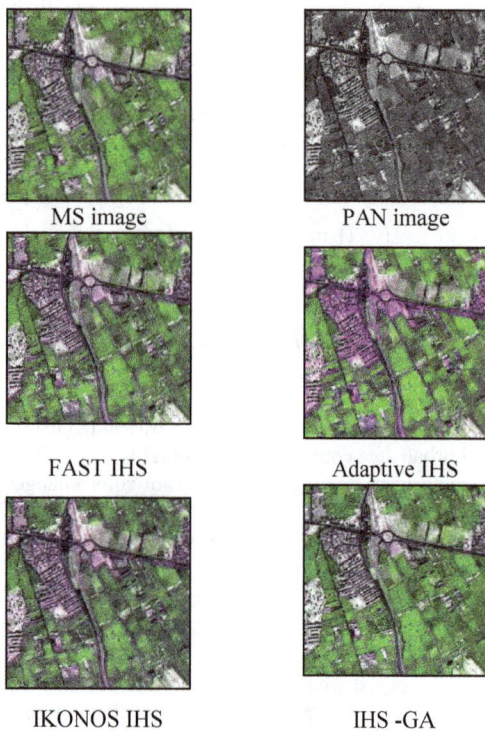

MS image

PAN image

FAST IHS

Adaptive IHS

IKONOS IHS

IHS -GA

Figure 3. outputs of methods(IKONOS data sat)

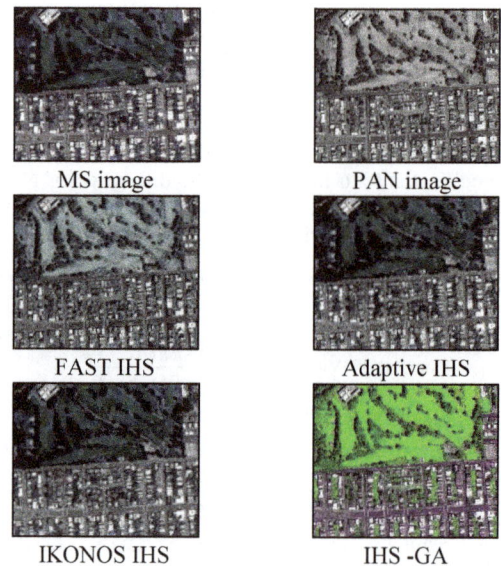

MS image

PAN image

FAST IHS

Adaptive IHS

IKONOS IHS

IHS -GA

Figure 5. outputs of methods(Quickbird data set)

5.2. QuickBird data set

The second study area is an urban San Francisco, California, USA. The dataset consists of a low-resolution (2.4m) multispectral image with four bands R, G, B and NIR and a high-resolution (0.6m) panchromatic image of resolution. The contained results for this data set are presented in table2 where in addition to the proposed IHS-GA four well known extensions of IHS are also evaluated to give us the comparison opportunity. This table content is also depicted in Figure 4 for a more comprehensive view. Figure 5 presents the obtained fused images of the proposed method.

In general, the performance of a pan-sharpening method can be subjectively and objectively evaluated. For subjective evaluation, the true color display (Figures 3&5) of the objects in IHS-GA fused images (Enhanced vegetation areas, Building and roads) are visually evaluated for being clear and sharp as well as having natural colors similar to those of the original MS image.

According to In Figures 3&5, the IHS-GA method has clearly sharped the original image and greatly improved its spatial quality. On the contrary spectral distortion can be seen in the FAST IHS and IKONOS IHS results of both datasets.

The noticeably color improvement of IHS-GA method is mainly due to the enhancement of vegetation areas. A more detailed look shows that vegetation areas are much more clear and sharp in the fused image of IHS-GA method.

Next we look at the metrics to evaluate the spectral and spatial quality. The values in Tables 1-2 correspond to the images in Figures 3&5 respectively. In all Tables, the values of the metrics are closer to the reference value when using the IHS-GA method. The Adaptive IHS fused image perform better than the FAST HIS and IKONOS IHS. The performance of IHS-GA is concededly better than its components. Note that, the spectral distortions of the IKONOS data are generally more than QuickBird data set. This is because the IKONOS data is taken from semi urban region and there are fewer strong edges to be sharpened.

According to the results, the two points can be concluded. First, the Fast IHS method produces the images with very high spatial quality whereas the spectral quality depends on the various types of used data. Second, we cannot definite choose the best fusion method, for the spectral quality of results depended on the inputs data. Meanwhile, the evaluation metrics indicate that in IHS-GA the mentioned problems are partiality solved.

6. CONCLUSION

In contrast to previous studies related to fusion techniques (e. g. Rahmani. 2008, Zhang. 2008), which only used spectral metrics in their researches, it is better to consider the both spatial and spectral ones. In this study, the different spatial and spectral metrics were used. The IHS pan-sharpening method results good spatial quality and is a commonly used algorithm for its speed and simplicity .To improve its spectral quality we proposed a novel IHS-GA fusion method based on enhancement vegetation area. The merging of these techniques (Enhanced Vegetation Area, GA and IHS method) improves the spectral quality of the results while keeping its spatial qualities. The performance evaluation metrics confirmed the capacity of the IHS-GA method.

7. REFERENCES

Alparone, L., Alazzi, B., Baronti, S., Garzelli, A., Nencini, F., Selva, M., 2008. *Multispectral and panchromatic data fusion assessment without reference*. In: Photogrammetric Engineering & Remote Sensing.

Bovolo F. Bruzzone L. Capobianco L. Garzelli A. Marchesi S. (2010). *Analysis of effect of pan-sharpening in change detection on VHR Images*. IEEE Transaction on Geoscience and Remote Sensing Letters, 7 (1): 53-57. doi: http://dx.doi.org/10.1109/LGRS.2009.2029248.

Bovik, Alan C., and Zhou Wang. "*A Universal Image Quality Index*." IEEE Signal Processing Letters (2002).

Carper, Lillesand, and Kiefer, (1990). *The use of intensity-hue-saturation transformations for merging spot panchromatic and multispectral image data*, Photogramm. Eng. Remote Sens. vol. 56, no. 4, pp. 459–467.

Chavez P.S. Kwarteng A.Y. (1989). *Extracting spectral contrast in Landsat Thematic Mapper image data using selective principal component analysis*. Photogrammetric Engineering and Remote Sensing, 55: 339-348.

Chavez P.S. Sides S.C. Anderson J.A. (1991). *Comparison of three different methods to merge multiresolution and multispectral data: TM & Spot Pan*. Photogrammetric Engineering & Remote Sensing, 57: 295-303.

Chang, (1999).*Spectral information divergence for hyperspectral image analysis*, inProc. IEEE Int. Geoscience and Remote Sensing Symp. IGARSS'99, vol. 1, pp. 509–511.

Choi M. (2006). *A new intensity-hue-saturation fusion approach to image fusion with a tradeoff parameter*. IEEE Transaction on Geoscience and Remote Sensing, 44 (6 :(1672-1682. doi:

http://dx.doi.org/10.1109/TGRS.2006.869923.

Du Q. Younan N.H. King R. Shah V.P. (2007). *On the performance evaluation of pansharpening techniques*.IEEE Transaction on Geoscience and Remote Sensing Letters, 4 (4): 518-522. doi: http://dx.doi.org/10.1109/LGRS.2007.896328.

Earth Resource Mapping Pty Ltd. (1990). The Brovey transform explained, EMU Forum, 2 (11). Available at: http: // www. ermapper.com/ forum_new/emuf2-11.htm # aiticle_5 (last accessed 05.05.2013).

.Goetz, Yuhas,and.Boardman, (1992). *Discrimination among semi-arid landscape endmembers using the Spectral Angle Mapper (SAM) algorithm*, Summaries 3rd Annu. JPL Airborne Geoscience Workshop, vol. 1, pp. 147–149.

Klonus S. Ehlers M. (2007). *Image fusion using the Ehlers spectral characteristics preserving algorithm*.GIScience and Remote Sensing, 44 (2): 93-116.

Lau, W., King, B. A., Li, Z., (2000). *The influences of image classification by fusion of spatially oriented images*. International Archives of Photogrammetry and Remote Sensing, 33 (Part B7), pp. 752-759.

Luciano Alparone, Lucien Wald, Jocelyn Chanussot, Claire Thomas, Paolo Gamba, et al. (2007). *Comparison of Pansharpening Algorithms: Outcome of the 2006 GRS-S Data Fusion Contest*. IEEE Transactions on Geoscience and Remote Sensing, Institute of Electrical and Electronics Engineers (IEEE), 2007, 45 (10), pp.3012-3021. <10.1109/TGRS.2007.904923>. <Hal-00177641>

Miloud Chikr El Mezouar, Nasreddine Taleb, Kidiyo Kpalma, and Joseph Ronsin. (2010). *An Improved Intensity-Hue-Saturation for a High-Resolution Image Fusion Technique Minimizing Color Distortion*. IJICT, Vol. 3, No. 1, February 2010 / ISSN: 0973-5836 / Serials Publications, India.

Myungjin C. (2006). *New Intensity-Hue-Saturation Fusion Approach to Image Fusion with a Tradeoff Parameter*. IEEE Transaction on Geoscience and Remote Sensing, 44 (6): 1672-1682. doi: http://dx.doi.org/10.1109/TGRS.2006.869923.

Otazu and Gonzalez-Ausicana. (2005).*Introduction of Sensor Spectral Response into Image Fusion Methods: Application to Wavelet-Based Methods*. IEEE Transactions of Geoscience and Remote Sens. vol. 43, pp. 2376-2385.

Palsson, F., Sveinsson, J. R., Benediktsson, J. A., & Aanaes, H. (2012). *Classification of pansharpened urban satellite images*. Selected Topics in Applied Earth Observations and Remote Sensing, IEEE Journal of, 5(1), 281-297.

Pohl, C., Genderen, J. L., (1998). *Multisensor image fusion in remote sensing*: concepts, methods and applications. International Journal of Remote Sensing, 19(5), pp. 823-854.

Rahmani, Sh., Strait, M., Merkurjev, (2008). *Evaluation of pan-sharpening methods*.

Ranchin and Wald, (2000).*Fusion of high spatial and spectral resolution images*: The Arsis concept and its implementation, Photogramm. Eng. Remote Sens. vol. 66, no. 1, pp. 49–61.

Shettigara V.K. (1992). *A generalized component substitution technique for spatial enhancement of multispectral images using a higher resolution data set*. Photogrammetry Engineering and Remote Sensing, 58: 561-567.

Shah V.P. Younan Nh. King R.L. (2008). *An Efficient Pan-Sharpening Method via a Combined Adaptive PCA Approach and Contourlets*. IEEE Transaction on Geoscience and Remote Sensing, 46 (5): 1323-1335. doi: http://dx.doi.org/10.1109/TGRS.2008.916211.

Schetsellar E.M. (1998). *Fusion by the IHS transform: Should we use cylindrical or spherical coordinates*. International Journal of Remote Sensing, 19 (4): 759-765. doi: http://dx.doi.org/10.1080/014311698215982.

Su, Shyn and P. Huang. (2001).*A New Look at IHS-like Image Fusion Methods*. Information Fusion, Vol. 2. pp. 177-186.

Tu, T., Haung, Su, Sh., Shyu, H., P.S., Hung, (2001). *A new look at HIS-like image fusion methods*. In: Information Fusion.

Tu, Huang, Hung, and C.P. Chang, (2004).*A fast intensityhue-saturation fusion technique with spectral adjustment for IKONOS imagery*, IEEE Geosci. Remote Sens. Lett. vol. 1, no. 4, pp. 309–312.

Tu, T., Lee, Y., Chang, Ch., Huang, P., (2005). *Adjustable intensity-hue-saturation and brovey transform fusion technique for IKONOS/Quickbird imagery.* In: Optical Engineering.

Vrabel J. (1996). *Multispectral Imagery Band Sharpening* Study. Photogrammetric Engineering & Remote Sensing, 62: 1075-1083.

Wald, (2000). *Quality of high resolution synthesized images: Is there a simple criterion*, in Proc. 3rd Conf. Fusion of Earth Data: Merging Point Measurements, Raster Maps and Remotely Sensed Images,T. Ranchin and L. Wald, Eds. SEE/URISCA, Nice, France.

Yang S. Wang M. Jiao L. (2012). *Fusion of multispectral and panchromatic images based on support value transform and adaptive principal component analysis*. Information Fusion, 13 (3): 177-184. Doi http://dx.doi.org/10.1016/j.inffus.2010.09.003.

Yang W. Gong Y. (2012). *Multi-spectral and panchromatic images fusion based on PCA and fractional spline wavelet*. International Journal of Remote Sensing, 33 (22): 7060-7074. doi: http://dx.doi.org/10.1080/01431161.2012.698322.

Zhang Y. (2004). *Understanding image fusion*. Photogrammetric Engineering & Remote Sensing, 70: 657-661.

Zhang, Y., (2008). *Methods for image fusion quality assessment-A Review, comparison and analy*sis. In: The International Archives of the Photogrammetry, Remote Sensing and Spatial Information Sciences.

6

3D SURFACE GENERATION FROM AERIAL THERMAL IMAGERY

Behshid Khodaei [b], Farhad Samadzadegan [a], Farzaneh Dadras Javan [a], Hadiseh Hasani [a, *]

[a] School of Surveying and Geospatial Engineering, College of Engineering, University of Tehran, Tehran, Iran - (samadz, fdadrasjavan, hasani)@ut.ac.ir
[b] Miaad Andishe Saz, Research and Development Company, Tehran, Iran – khodaee@masgie.com

Commission VI, WG VI/4

KEY WORDS: Thermal Imagery, DSM, UAV, Bundle Adjustment, Dense

ABSTRACT

Aerial thermal imagery has been recently applied to quantitative analysis of several scenes. For the mapping purpose based on aerial thermal imagery, high accuracy photogrammetric process is necessary. However, due to low geometric resolution and low contrast of thermal imaging sensors, there are some challenges in precise 3D measurement of objects. In this paper the potential of thermal video in 3D surface generation is evaluated. In the pre-processing step, thermal camera is geometrically calibrated using a calibration grid based on emissivity differences between the background and the targets. Then, Digital Surface Model (DSM) generation from thermal video imagery is performed in four steps. Initially, frames are extracted from video, then tie points are generated by Scale-Invariant Feature Transform (SIFT) algorithm. Bundle adjustment is then applied and the camera position and orientation parameters are determined. Finally, multi-resolution dense image matching algorithm is used to create 3D point cloud of the scene. Potential of the proposed method is evaluated based on thermal imaging cover an industrial area. The thermal camera has 640×480 Uncooled Focal Plane Array (UFPA) sensor, equipped with a 25 mm lens which mounted in the Unmanned Aerial Vehicle (UAV). The obtained results show the comparable accuracy of 3D model generated based on thermal images with respect to DSM generated from visible images, however thermal based DSM is somehow smoother with lower level of texture. Comparing the generated DSM with the 9 measured GCPs in the area shows the Root Mean Square Error (RMSE) value is smaller than 5 decimetres in both X and Y directions and 1.6 meters for the Z direction.

1. INTRODUCTION

Thermal imaging sensors have been recently applied for object detection and identification in wide range of applications, such as surveillance and reconnaissance (Leira et al. 2015). Nevertheless, due to low geometric resolution and low contrast of thermal imaging sensors, it is mostly used for interpretation and monitoring purposes (Berni et al. 2009). However, thermal mapping is getting more important in civil applications, as thermal sensors may be applied in foggy weather and night times which is not possible for visible camera (Kuenzer and Dech, 2013). For high-resolution mapping and specific products generation from aerial thermal imagery, high accuracy photogrammetric process is necessary. Therefore, geometric calibration and 3D mathematical modelling of thermal imagery becomes evident for all precise applications.

Camera calibration is a key aspect to obtain geometrical information through photogrammetry process. Therefore, wide range of literature focus on geometrical calibration of thermal imaging sensor. The main difference of them is the designed calibration field (Luhmann et al. 2013; Vidas et al. 2012; López et al. 2015). A wooden plate with the warm up lamps (Luhmann et al. 2013) or foil targets (López et al. 2015) is common test field. However the quality of the targets is quite poor and the centre of the targets cannot be measured with high precision. Vidas et al. proposed a mask based approach that placed an opaque mask in front of warm background (Vidas et al. 2012). According to simplicity and accuracy of mask based calibration test field, this method is chosen for implementation.

The applications of remotely sensed thermal imagery has been investigated. Berni et al. (2009) use thermal and visible cameras mount on UAV for monitoring of vegetation. Detection of people and vehicle is another application of thermal camera (Gąszczak et al. 2011). Pless et al. (2012) in German aerospace center (DLR) generate orthophoto and 3D surface model based on thermal imaging. Two thermal cameras with cooled detector (mid-wave and long wave) are tested for photogrammetric process. Moreover geometric and radiometric calibration is performed before flight. Another study proposed the method to generate multi-temporal thermal orthophoto from UAV data which can be used for spatial analysis of temperature distribution. It shows SIFT is well suited for feature detection in low resolution thermal imagery (Pech et al. 2013). Oblique thermal imagery are also used for the automatic extraction of building geometry. Both the geometry and the thermal information are used in the generation of building 3D models (Lagüela et al. 2014). The classification of roof surface for the installation of solar panel is performed by visible and thermal data fusion (Lopez et al. 2015).

The purpose of this paper is to generate 3D object surface using thermal images acquired with a small UAV equipped with an infrared thermal camera. In following the 3D surface generation methodology including geometric camera calibration, flight planning, bundle block adjustment and dense matching are described. Then experimental results obtained from an industrial field are presented and discussion and conclusion over results are presented.

---

* Corresponding author

2. METHODOLOGY

The proposed method includes three main steps: geometric thermal camera calibration before flight, flight planning and imaging and 3D surface generation form acquired thermal video.

2.1 Thermal Camera Calibration

In the first step, thermal camera is geometrically calibrated in the laboratory. In this study, the calibration grid is constructed based on emissivity differences between the background and the targets. In this way, target detection can be conducted automatically.

The calibration pattern consists of a grid of regularly sized circles cut out of a thin opaque material which is held in front of a powered computer monitor. By this way, the calibration pattern is easily identifiable in the thermal domain and geometric camera calibration is performed (Vidas et al. 2012).

For camera calibration the corrected image coordinates x_{cor} and y_{cor} can be calculated from the measured coordinates x_{meas} and y_{meas} via equation (1):

$$\begin{cases} x_{cor} = x_{meas} - x_p + x \times \dfrac{dr}{r} + P_1 \times (r^2 + 2x^2) + 2\,P_2 xy \\ y_{cor} = y_{meas} - y_p + y \times \dfrac{dr}{r} + P_1 \times (r^2 + 2y^2) + 2\,P_1 xy \end{cases} \quad (1)$$

where $r = \sqrt{x^2 + y^2}$, $x = x_{meas} - x_p$, $y = y_{meas} - y_p$, x_p and y_p are principal point offsets in x and y image coordinate direction and dr is determined by equation (2) as bellow:

$$dr = K_1 \times r^3 + K_2 \times r^5 + K_3 \times r^7 + K_4 \times r^9 + K_5 \times r^{11} \quad (2)$$

K_i is the term of order $(2i + 1)^{th}$ for radial distortion correction. P_1 and P_2 are the coefficients of decentering distortion (Fraser, 1997).

Constructed test field consists of a number of circular and coded targets which their centres must be found automatically. In order to improve center detection of targets, contrast enhancement is performed. Then targets centers are automatically detected and calibration parameters are determined by self-calibration.

2.2 Flight planning

Flight planning is the pre-request step of the photogrammetric mission. There are some parameters to be identified including flight height and speed, camera focal length, capturing rate, *etc.* These parameters is defined based on required map scale, topography, *etc* (Mouget et al 2014).

In standard photogrammetry at least 60% overlap along the flight axis and about 20% sidelap of flight strips should be covered. In the case of UAV missions, due to instability of the platform, its better to increase overlap and sidelap to 80% and 60% respectively.

2.3 3D Surface Generation

In order to generate 3D surface from thermal video, sequences of frames are extracted from video, then tie points are generated by SIFT algorithm. Bundle block adjustment is then applied and the camera position and orientation parameters are determined. Finally, multi-resolution dense image matching algorithm is used to create 3D points cloud of the scene.

2.3.1 Frame Extraction: Data processing starts with thermal frames extraction from acquired airborne video. According to the small size of the thermal frames, to prepare the required overlap and sidelap, the rate of the frame extraction should be high.

2.3.2 Interior Orientation: Geometric calibration parameters computed in camera calibration step are applied and thermal images are resampled to distortion-free images.

Bundle Block Adjustment: In order to solve bundle adjustment of the selected frames, tie points are extracted by SIFT. It is well suited for the feature detection and description because it is able to handle the special characteristics of the input data such as low image contrast as well as rotation, translation and scale differences due to a more instable platform. For computing exterior orientation parameters, two arbitrary images with corresponding tie points are selected for relative orientation computations. For outlier detection and relative orientation computations, Essential matrix with RANSAC algorithm are applied. Next image then added to oriented subset images till all images are oriented relatively. Bundle adjustment is regularly applied on the already oriented images to avoid divergence. After relative orientation of all images, the absolute orientation is performed based on the GCPs and Helmert transformation (Pierrot and Clery, 2006).

2.3.3 3D Surface Generation: Dense image matching is carried out to generate DSM. However, low spatial resolution, low contrast and low level of textures in thermal imagery, poses some challenges in the procedure. For dense image matching, surface reconstruction problem formalized as a minimization of an energy function. The Energy function $E(Z)$ is represented in the equation (3).

$$E(Z) = \sum_{x,y} (A(x, y, Z(x, y)) + \alpha \\ \times \sum_{u,v \in V} P_{u,v}^{ds} |Z(x, y) - Z(x + u, y + v)|) \quad (3)$$

where Z is the disparity image, The first term is the sum of all pixel matching costs, α is a priori weighting coefficient, V is a neighbourhood of the pixel (usually 4 or 8 neighbourhood) and $P_{u,v}^{ds}$ is a weighting function in the neighbourhood V.

This minimization problem is solved based on multi-resolution implementation of a Cox&Roy optimal flow image matching algorithm (Roy & Cox, 1998). This multi-resolution approach is necessary for achieving reasonable processing times on extended areas and improving robustness by restraining matching ambiguities. Finally DSM is obtained by dense matching on thermal images.

2.3.4 3D Surface Evaluation: For quality assessment of the obtained DSM, two scenarios are proposed: region wise scenario based on available dense DSM which derived from visible imagery and point wise scenario based on the coordinates of GCPs determined on the available DSM. In the region based scenario, histogram of the difference between the obtained DSM and available DSM are computed. In the point based scenario, the position of GCPs are measured in the obtained DSM and compare with known coordinate, then statistical analysis is performed and the RMSE and Mean Absolute Error (MAE) are computed to quantitate the accuracy of thermal derived-DSM. (Hobi & Ginzler, 2012).

3. EXPERIMENTAL RESULTS

Potential of the proposed DSM generation method is evaluated based on thermal imaging cover an industrial area.

3.1 Dataset

Thermal video is acquired by TC688 camera which is mounted on small UAV. The specifications of camera are presented in Table 1.

Table 1. Thermal camera specifications

Parameters	Values
Detector type	Uncooled FPA
Pixel pitch	17 μm
IR resolution	640×480
Focal length	25
Frequency	50 Hz
Sensitivity	<65mk@f/1.0

The altitude of the flight was 100 m above the ground which results in a ground sampling distance of about 7 cm. The flight lines were planned for the images to have sidelap of 60% between the image stripes. According to the frame rate of the camera, flight velocity, flight height and the 80 % overlap, the frames are extracted. Figure 1 demonstrates some extracted frames.

Figure 1. Frame extracted from thermal video

3.2 Geometric Calibration

Geometric calibration of thermal camera with planar pattern is performed in laboratory. The test field consists of a grid of 28 regularly sized circle with 4 coded targets cut out of a thin non-conductor material. The size of mask pattern is 44×24 cm^2, diameter of squares are 20 mm which spaced with 50 mm separation. Figure 2 shows the thermal imagery acquired from mask based calibration pattern. The pattern is held in front of a power on computer monitor.

Figure 2. Thermal imagery for calibration

By starching, contrast of thermal images are increased. Then sub-pixel accurate positions of the centre of each circle is defined (Figure 3).

Figure 3. Automatic detection of targets centres

Self-calibration is performed to compute camera calibration parameters. The results of camera calibration are presented in Table 2.

Table 2. Thermal camera specification

Parameters	Values	SD
C	25.8655	0.525
x_p	0.0085	0.235
y_p	-0.9934	0.235
K_1	-5.54162e-004	1.1728e-003
K_2	1.46497e-004	1.17294e-004
K_3	-8.38681e-004	1.0917e-005
K_4	2.10540e-007	3.0611e-007
K_5	-1.89923e-009	3.1147e-009
P_1	-7.2475e-004	3.984e-004
P_2	-6.5166e-004	6.079e-004

3.3 3D Surface Generation

In order to orient extracted frames, tie points are extracted between unordered frames. For these purpose each frame is considered with respect to all other frames to determine tie points. Figure 4 shows two samples of tie points extracted from thermal imagery based on SIFT algorithm.

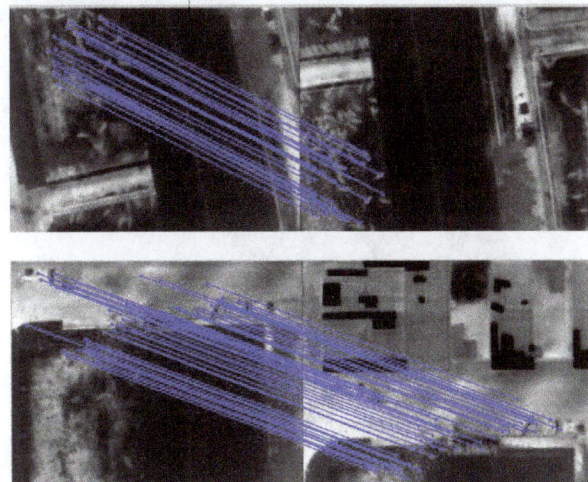

Figure 4. Tie points extraction

Relative orientation is performed based on tie points. The initial value of orientation parameters are determined by essential matrix whereas bundle adjustment is used for relative orientation. Obtained results of relative orientation for four flight strips are illustrated in Figure 5.

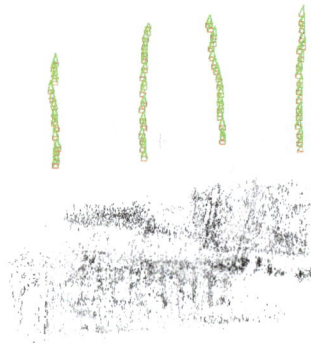

Figure 5. Results of bundle adjustment

The next step is to compute the absolute orientation using some measured GCPs. Here the GCPs are extracted from an available ortho-image. Figure 6 shows the positions of GCPs on the ortho-image.

Figure 6. GCPs on ortho-photo

After absolute orientation step, the dense image matching procedure should be done to generate final products. These products are thermal DSM, thermal ortho-photo and thermal points cloud. Figure 7 shows the generated thermal ortho-photo.

Figure 7. Thermal orhto-photo

As it's obvious from Figure 7, buildings are formed correctly with the sharp edges. However, some parts of generated ortho-photo couldn't form correctly especially in the middle of the area. The existence of noise in the thermal video leads to miss connectivity in the middle ran.

Referring to the region based scenario, five parts of the area are considered separately. These parts are represented in Figure 8. The aim is to compare height information of the generated thermal DSM with visible derived DSM.

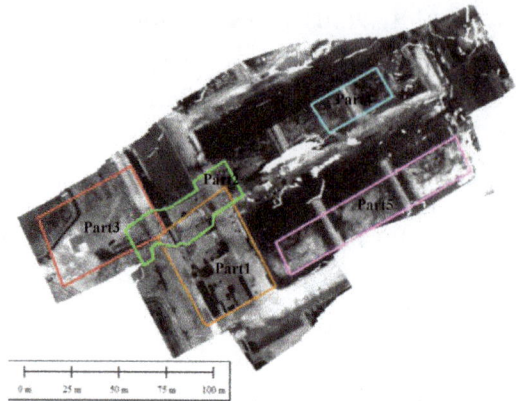

Figure 8. Dividing area to some separated parts.

For each part, the final thermal DSM and the histogram of height differences between visible DSM and thermal one is represented (Figure 9).

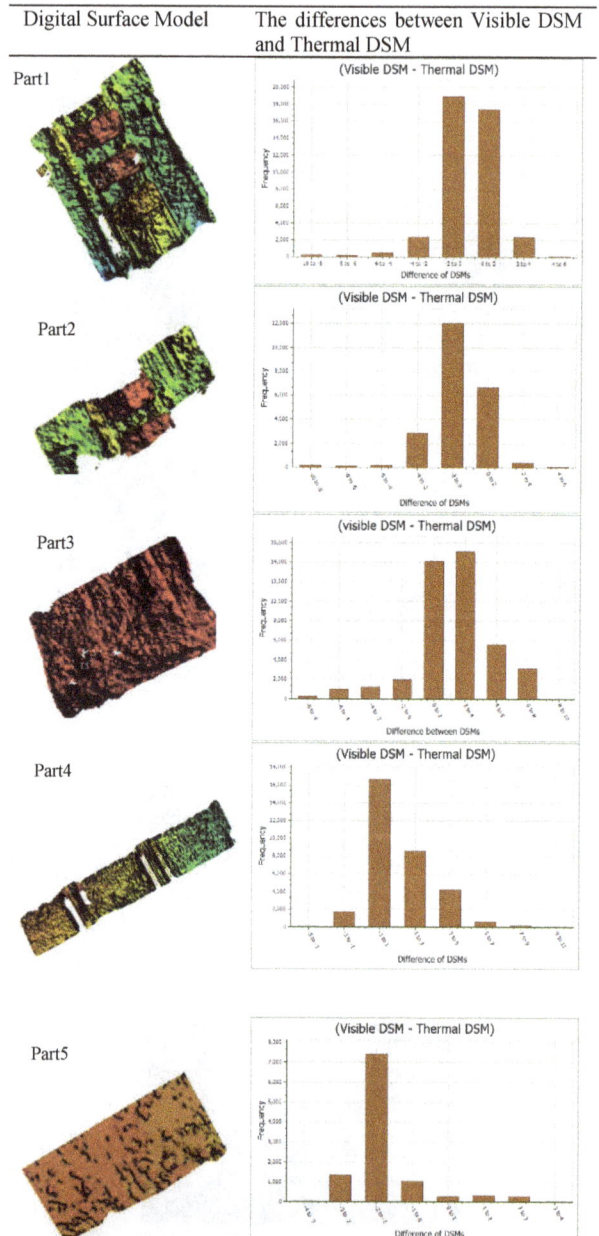

Digital Surface Model	The differences between Visible DSM and Thermal DSM
Part1	
Part2	
Part3	
Part4	
Part5	

Figure 9. The comparison of the thermal DSM with visible DSM.

For point-wise assessment of the thermal derived DSM, 9 check points are considered. Figure 10 shows the planimetric error vector of check points by comparing thermal and visible orthophotos.

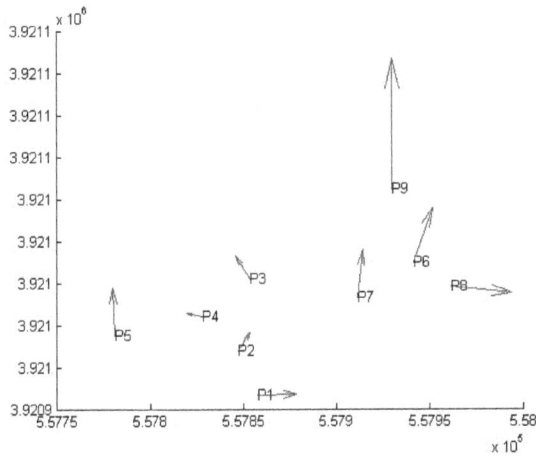

Figure 10. Planimetric error vector of the check points

The obtained results show the comparable accuracy of ortho-photo generated based on thermal images with respect to visible ortho-photo. The statistical analysis on check points coordinates are shown in Table 3. Four statistical parameters are computed. These parameters are the minimum, maximum, MAE and RMSE of check point's error in X, Y and Z directions.

Table 3. Statistical assessment of check point accuracy

Statistics	X	Y	Z
Min (m)	0.001	0.025	0.202
Max (m)	1.018	1.961	3.175
MAE (m)	0.314	0.563	1.433
RMSE (m)	0.442	0.804	1.629

As presented in Table 3, the planimetric position is more accurate than vertical coordinate. There are 2 points with large error values which seems to be the blunders. In order to improve the results, blunder detection should be applied to remove such blunders.

4. CONCLUSION

In this paper the potential of thermal video imagery in 3D surface generation was inspected. For this purpose three steps method consists of geometric camera calibration, flight planning and 3D surface generation presented. Obtained results show the comparable accuracy of 3D model generated based on thermal images with respect to DSM generated from visible images, although thermal based DSM is somehow smoother with lower level of texture. Results show the RMSE of 0.44m, 0.8m and 1.62m in X, Y and Z directions, respectively.

Although the blurry appearance and low resolution of the thermal images cause a bit vagueness in thermal products, further investigations should be carried out in combination of the thermal and visible outputs for mapping applications.

REFERENCES

Berni, J., Zarco-Tejada, P.J., Suarez, L. and Fereres, E., 2009. Thermal and narrowband multispectral remote sensing for vegetation monitoring from an unmanned aerial vehicle. *IEEE Transaction on Geoscience and Remote Sensing*, Vol. 47, No. 3, pp. 722-738.

Deseilligny, M. P., & Clery, I., 2011. Apero, an open source bundle adjusment software for automatic calibration and orientation of set of images. *ISPRS-International Archives of the Photogrammetry, Remote Sensing and Spatial Information Sciences*, 38, 5.

Fraser, C.S., 1997. Digital Camera Self-Calibration. *ISPRS International Journal of Photogrammetry and Remote Sensing*, Vol. 52, pp. 149-159.

Gaszczak, A., Breckon, T. P., & Han, J., 2011, January. Real-time people and vehicle detection from UAV imagery. In IS&T/SPIE Electronic Imaging. International Society for Optics and Photonics, pp. 78780B-78780B.

Hobi, M. L., & Ginzler, C., 2012. Accuracy assessment of digital surface models based on WorldView-2 and ADS80 stereo remote sensing data. *Sensors*, Vol. 12, No. 5, pp. 6347-6368.

Kuenzer, C. and Dech, S., 2013. *Thermal Infrared Remote Sensing: Sensors, Methods, Applications*. Springer Science+Business Media Dordrecht, USA.

Lagüela, S., Díaz-Vilariño, L., Roca, D. and Armesto, J., 2014. Aerial oblique thermographic imagery for the generation of building 3D models to complement Geographic Information Systems. *Proc. of QIRT'14*.

Leira, F.S., Johansen, T.A. and Fossen, T.I., 2015. Automatic detection, classification and tracking of objects in the ocean surface from UAVs using a thermal camera. *IEEE Aerospace Conference*, Big Sky, MT.

Luhmann, T., Piechel, J., and Roelfs, T., 2013. *Thermal Infrared Remote Sensing: Sensors, Methods, Applications*, Springer Netherlands, pp. 27-42.

Mouget, A., Lucet, G., 2014. Photogrammetric Archeological Survey with UAV, *ISPRS Annals of the Photogrammetry, Remote Sensing and Spatial Information Sciences*, Volume II-5.

Pech, K., Stelling, N., Karrasch, P., Maas, H.G., 2013. Generation of multitemporal thermal orthophotos from UAV data. *International Archives of the Photogrammetry, Remote Sensing and Spatial Information Sciences*, Vol. XL-1/W2.

Pless, S., Vollheim, B., Haag, M. and Dammaß, G., 2012. Infrared cameras in airborne remote sensing: IR-Imagery for photogrammetric processing at German Aerospace Center DLR, Berlin. *11th International Conference on Quantitative InfraRed Thermography*.

Roy, S., Cox, I.J., 1998. A Maximum-Flow formulation of the N-camera Stereo Correspondence Problem, Proc. *IEEE Internation Conference on Computer Vision*, pp 492-499, Bombay.

Vidas, S., Lakemond, R., Denman, S., Fookes, C., Sridharan, S. and Wark, T., 2012. A mask-based approach for the geometric calibration of thermal-infrared cameras. *IEEE Transactions on Instrumentation and Measurement*, Vol. 61, No. 6, pp. 1625-1635.

7

MOVING OBJECTS TRAJECTOTY PREDICTION BASED ON ARTIFICIAL NEURAL NETWORK APPROXIMATOR BY CONSIDERING INSTANTANEOUS REACTION TIME, CASE STUDY: CAR FOLLOWING

M. Poor Arab Moghadam [a], P. Pahlavani [b]*

[a] MSc. Student in GIS Division, School of Surveying and Geospatial Eng., College of Eng., University of Tehran, Tehran, Iran
[b] Assistant Professor, Center of Excellence in Geomatics Eng. in Disaster Management, School of Surveying and Geospatial Eng., College of Eng., University of Tehran, Tehran, Iran- Pahlavani@ut.ac.ir
*Corresponding author: Parham Pahlavani

E-mail addresses: pahlavani@ut.ac.ir

Commission

KEY WORDS: Moving Object Trajectory, Artificial Neural Network, Reaction Time, Mean Square Error, Index of Agreement, Microscopic Simulation

ABSTRACT

Car following models as well-known moving objects trajectory problems have been used for more than half a century in all traffic simulation software for describing driving behaviour in traffic flows. However, previous empirical studies and modeling about car following behavior had some important limitations. One of the main and clear defects of the introduced models was the very large number of parameters that made their calibration very time-consuming and costly. Also, any change in these parameters, even slight ones, severely disrupted the output. In this study, an artificial neural network approximator was used to introduce a trajectory model for vehicle movements. In this regard, the Levenberg-Marquardt back propagation function and the hyperbolic tangent sigmoid function were employed as the training and the transfer functions, respectively. One of the important aspects in identifying driver behavior is the reaction time. This parameter shows the period between the time the driver recognizes a stimulus and the time a suitable response is shown to that stimulus. In this paper, the actual data on car following from the NGSIM project was used to determine the performance of the proposed model. This dataset was used for the purpose of expanding behavioral algorithm in micro simulation. Sixty percent of the data was entered into the designed artificial neural network approximator as the training data, twenty percent as the testing data, and twenty percent as the evaluation data. A statistical and a micro simulation method were employed to show the accuracy of the proposed model. Moreover, the two popular Gipps and Helly models were implemented. Finally, it was shown that the accuracy of the proposed model was much higher - and its computational costs were lower - than those of other models when calibration operations were not performed on these models. Therefore, the proposed model can be used for displaying and predicting trajectories of moving objects being followed.

1. INTRODUCTION

Swift social and economic development caused very significant traffic congestion, accidents, and environmental pollution. In this regard, Intelligent transportation systems (ITS) have been used for improving of the motion and the roads system dynamic, enhancing in the traffic immune, as well as increasing in the traffic management exploitation (Wu, et al., 2009). One of the most important issues in ITS is microscopic traffic simulation which indicates the real traffic situation by describing each driver-vehicle-unit (DVU)'s state, including speed, acceleration, position, and route choice decision (Xiaoliang Ma and Ingmar Andréasson, 2007). Micro traffic simulation systems provide an efficient platform to assess the impacts of the different traffic controls and management strategy under virtual road network. These systems can revert drawbacks of conventional traffic analysis approaches and represent solution for it. The core of a micro traffic simulation is driver behavior. Hence, the quality of the driver behavior directly influences in accuracy and reliability of simulation results. Car following models relate to vehicles' space-time modeling and discrete interaction among them in a single lane

route and it is considered as back bone of driver behavior (Hsun-Jung Cho and Yuh-Ting Wu, 2008). Following vehicle driver perceives acceleration/deceleration, distance, and velocity of lead vehicle and after short delay in reaction, (s) he implements changing in acceleration. Car following modeling has been studied more than a half century (Saifuzzaman and Zheng, 2014). At the first time, this model was proposed by Reuschel (Reuschel, A, 1950) and Pipes (Pipes, Louis A, 1953) and had been extensively refined by Herman et al. (Herman and Potts, 1900; Herman et al. 1959; Herman and Rothery, 1965, Herman and Rothery, 1967). Among the most prominent models widely used in many simulation software, several models can be mentioned, including GHR (Gazis, 1959), Collision Avoidance (CA) (Kometani and Sasaki, 1959), Gipps (Gipps, 1981), Helly (Helly, 1961), Fuzzy logic (Chakroborty and Kikuchi, 1999), and neural network (Panwai and Dia, 2007) models. Due to the multi-disciplinary study scope, all of the above mentioned have many parameters to calibrate. This process is time-consuming and costly, and any change in these parameters even slight creates disturbances in output of these models. The paper is composed of four sections. More applicable conventional

---

* Corresponding author

models is briefly introduced in the subsequent section, followed by the description of the proposed methodology based on artificial neural network. In Section 3, the proposed methodology is thoroughly validated. The last section is devoted to the conclusions.

2. LITERATURE REVIEW

In this section, the most notable car-following models would be reviewed.

2.1 Gazis-Herman-Rothery model (GHR)

The GHR model is one of the most famous car motion models proposed in the late 1950s and early 1960s by Gazis and colleagues (Gazis, 1959). The basic equation of the model is as follows:

$$a(t+T) = \alpha \frac{\Delta V(t)}{\Delta X(t)} \tag{1}$$

where a(t+T) is the Acceleration or deceleration at time (t), $\propto$ is the sensitivity coefficient, $\Delta V(t)$ is the speed difference of vehicle at time t, T is the driver's reaction time, and $\Delta X(t)$ is the distance headway at time t.

2.2 Collision Avoidance model (CA)

The CA model, which is known as the safe distance model was introduced in 1959 by Kometani and Sasaki (Kometani and Sasaki, 1959). The equation of this model is as follows:

$$\Delta X(t-T) = \alpha V_{n-1}^2(t-T) + \beta_1 V_n^2(t) + \beta V_n(t) + b_0 \tag{2}$$

Where α, β_1, β, b_0 are the constant coefficients to be determined.

2.3 Gipps model

One of the most significant developments done on the CA model was in 1981 by Gipps (Brackstone and McDonald, 1999). He considered several drivers' behavior factors neglected in the previous model. High computational cost for calibration of parameters is the main disadvantage for this model. Eq. (3) demonstrates the Gipps model used in this paper:

$$V_n(t+T) = \min(\{V_n + 2.5A_fT(1 - \frac{V_n}{\max(V_n)})(0.025 + \frac{V_n}{\max(V_n)})^{0.5}\}$$

$$,\{-B_fT + B_f\{2(\Delta X - S) - TV_n + \frac{V_{n-1}^2}{B_l^\wedge}\}\} \tag{3}$$

Where $A_f = 1.7, B_f = 3, B_l^\wedge = 3.5, S = 6.5$ and ΔX is the space length between the follower and the leader vehicles. Also S is the safety distance that is based on the maximum velocity of vehicles. So, $\Delta X < S$ corresponds to an incident, which may involve the vehicle crashing. This parameters are selected according to (Wilson, 2001).

2.4 Helly model

The Helly linear model was defined in 1959 and includes additional parameters to adjust and tune the acceleration of the car while facing the brake of the leading car and the two front cars (Helly, 1961). The equation of this model is as follows (Brackstone and McDonald, 1999):

$$a_n(t) = C_1\Delta V(t-T) + C_2(\Delta X(t-T) - D_n(t)) \tag{4}$$

$$D_n(t) = \alpha + \beta V_n(t-T) + \gamma a_n(t-T) \tag{5}$$

where D_n(t) is the desired following distance at time t and $C_1, C_2, \alpha, \beta, \gamma$ are the constant coefficients to be determined.

2.5 Fuzzy-logic-based & Artificial Neural Network (ANN) model

Fuzzy logic-based modeling has played a prominent role in the car-following field. Fuzzy logic was first used in car-following in 1992. The first attempt to use fuzzy rules in GHR was conducted (Chakroborty and Kikuchi, 1999). In 2007, Hussein Dia and Panwai (Panwai and Dia, 2007) introduced car-following model, based on ANNs and showed that ANN models outperformed Gipps model.

3. PROPOSED METHODOLOGY

The car-following models are generated mathematically by examining the behavior of the driver following the leading car in traffic flow. When there is a leading car, the driver tries to control his/her own driving behavior by considering the speed of the leading car and the speed difference with the leading car, as well as the distance to the leading car through accelerating or braking. Therefore, the car-following model can be expressed as follows (Rahman, 2013):

$$\frac{dV}{dt} = f(V_{n-1}, \Delta X, \Delta V) \tag{6}$$

Given the numerous parameters of the conventional models and the complexity of the parameters calibration process, a very effective, efficient method was proposed in this paper based on an artificial neural network, which not only has a higher accuracy, but also eliminates the parameters calibration process. In this section, at first, a definition of this feed-forward neural network method is presented, then the dataset used in this study will be introduced.

3.1 Neural networks algorithm

Since 1990s, there has been a gained interest concerning artificial neural networks in variety of disciplines (Dougherty, 1995; Kalogirou, 2000; Kalyoncuoglu and Tigdemir, 2004; Karlaftis and Vlahogianni, 2011). Typically, two merits contribute to the popularity of neural networks. One virtue is that neural networks are capable to handle noisy data and estimate any degree of complexity in non-linear systems (Kalogirou, 2000). The other virtue is that, they do not require any simplifying hypothesis or prior knowledge of problem solving, in comparison with statistical models (Kalyoncuoglu and Tigdemir, 2004; Karlaftis and Vlahogianni, 2011). In terms of the topic discussed in this study, driver–vehicle reaction delay and car following are two complicated concepts. Although there are some findings in previous studies, it is not clear what factors impact the reaction delay and car-following behavior, as well as what their underlying relationships are. Therefore, by virtue of the above mentioned features, we believe that neural networks are more flexible than statistical approaches. In fact, validation results in Section 4 also confirm our judgment. A neural network is a massively parallel distributed processor that has a natural propensity for storing experiential knowledge and making it available for use (Haykin 1999). It stimulates the human brain in two respects: the knowledge is acquired by the network through a learning process, and inter-neuron connection strengths, known as synaptic weights, are used to store the knowledge. A schematic diagram of a typical multilayer neural network is displayed in Figure.1.

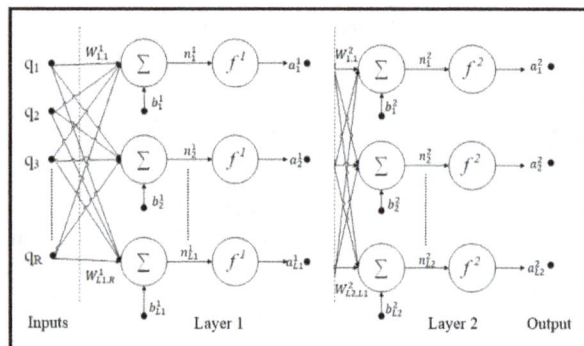

Figure 1.The schematic diagram of a two-layer neural network

In this study the network is composed of inputs and two layers. The last layer of a neural network is also called the output layer. The input vector consists of R elements q_1, q_2... q_R. Each input element q_i is multiplied by a weight $w_{j, i}{}^1$ to form $w_{j, i}{}^1$, one of the terms that is sent to the adder. The other input, the constant 1, is multiplied by a bias $b_j{}^1$ and then passed to the adder. The adder output is:

$$n_j^1 = \sum_{i=1}^{R} w_{j,i}^1 q_i + b_j^1, \qquad (7)$$

Often recourse to as the net input, goes into the transfer function f^1, which produces the neuron output $a_j{}^1$ in the layer one. The bias $b_j{}^1$ has the effect of increasing or decreasing the net input of the transfer function, depending on whether it is positive or negative, respectively. L^1 shows the number of neurons adopted in layer one. If we relate the artificial neural network to a biological neuron, the strength of a synapse is shown by the weights and bias $w_{j, i}{}^1$, $b_j{}^1$. A cell body is represented by the adder and transfer function, and the neuron output $a_j{}^1$ represents the signal on an axon. In the same way, handling the outputs of layer one as the inputs of layer two, the signals from layer one are passed through second layer. The transfer function in layer two f^2 can be totally vary from that in the first layer. The outputs of second layer $a_k^2\, k \in [1, 2, ..., L^2]$, are also the outputs of the discussed neural network. L^2 shows the number of neurons used in layer two. The neuron network is written in the following matrix form:

$$a^1 = f^1(w^1 q + b^1), a^2 = f^2(w^2 a^1 + b^2), \qquad (8)$$

where w1, w2 are the weight matrixes and q, b^1, b^2, a^1, a^2 denote the input, bias and output vectors in layers one and two.

3.2 Dataset

The actual data on car following from the NGSIM project was used to determine the performance of the proposed model. This dataset was used for the purpose of expanding behavioral algorithm in micro simulation. One part of this dataset is related to the Emeryville Highway and includes routes of about 1500 cars including about 1.5 million information records for the first 15 minutes between 5:00 p.m. and 5:15 p.m. on April 13, 2005. Every record contains 18 information fields taken by the sensors embedded for this purpose at 0.1-second intervals, and the most important fields include accurate spatial position, speed, acceleration, and distance to the car in front. From among all the data, a suitable sample was considered that included three important conditions: both cars should be of the same type, none of them should change lanes during car following, and no other car should come between them. Finally, the car following

should last for at least 30 seconds (Kim et al., 2003; Kim and Taehyung, 2005). After the preliminary study of the noise in the acceleration data taken from the dataset (Punzo et al., 2011) it was found that the use of the moving average method reduced this noise as much as possible. Sixty percent of the data was entered into the designed neural network as training data, 20 percent as testing data, and 20 percent as evaluation data.

3.3 Experimental result

In this paper, in order to determine a trajectory model for vehicle movements, a multilayer neural network approximator was used with 10 neurons in one hidden layer, one neuron in output layer, and Levenberg-Marquardt (Vogl et al., 1988) backpropagation training algorithm. The hyperbolic tangent sigmoid function (Vogl et al., 1988) and purelin function (Vogl et al., 1988) were employed as the transfer functions of the neurons of hidden layer and the output layer, respectively.

One of the important aspects in identifying driver behavior is reaction time. This parameter shows the period between the time the driver recognizes a stimulus and the time a suitable response is shown to that stimulus. One of the very popular models that are used for calculating reaction time is the Ozaki model that calculates this parameter separately for the two states of acceleration and deceleration (Ozaki, 1993). In this research, the inputs for determining the trajectory model are the distance between the two cars being followed at any instant, the speed of the leading car, and the difference between the speeds of the two cars. In addition, the reaction time of the driver was calculated using the Ozaki model and was considered as the new input. As will later be shown, the effects of this parameter are very important in modeling trajectories of car movements using the proposed artificial neural network approximator. Figure.2 and Figure.3 demonstrate the fitting curve between the real data and prediction of the proposed model by neglecting the reaction time and by considering it, respectively.

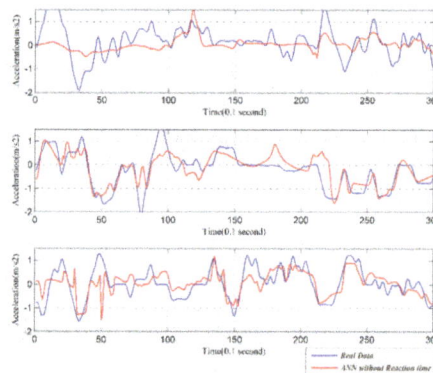

Figure 2. Artificial Neural Network prediction without considering the reaction time

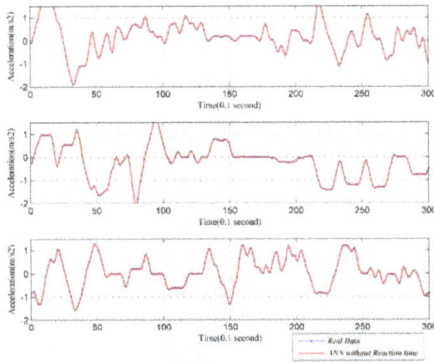

Figure 3. Artificial Neural Network prediction by considering the instantaneous reaction Time

Figure.4 demonstrates histogram error of training, testing, and validation data. The proposed ANN model was stopped after 23 epoch. Figure.5 shows the epoch gained the best accuracy. The reason of training stopping was based on achieving the best accuracy goal. Furthermore, if there is not any better accuracy after 310 epochs, training process of the proposed ANN will be finished.

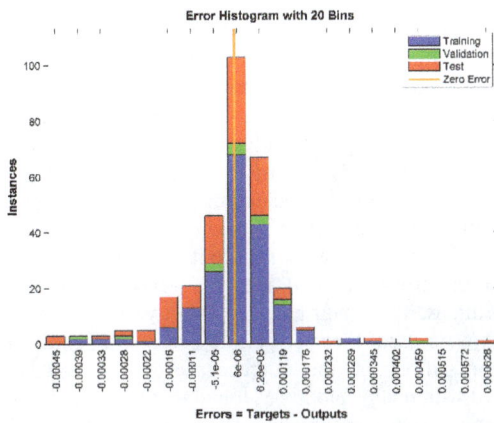

Figure 4. Error Histogram for the ANN outputs (Lane 1)

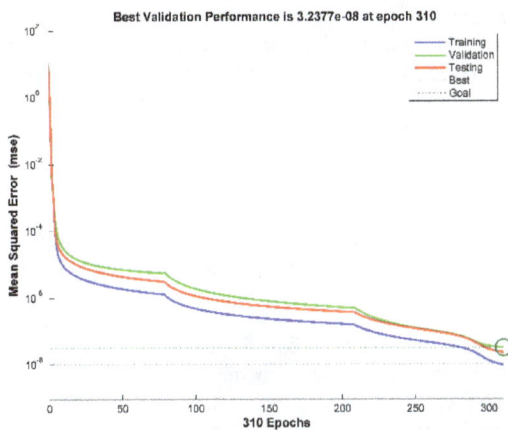

Figure 5. The best validation performance (Lane 1)

The outputs of the conventional models were then compared to the outputs of the proposed method in Figure.6.

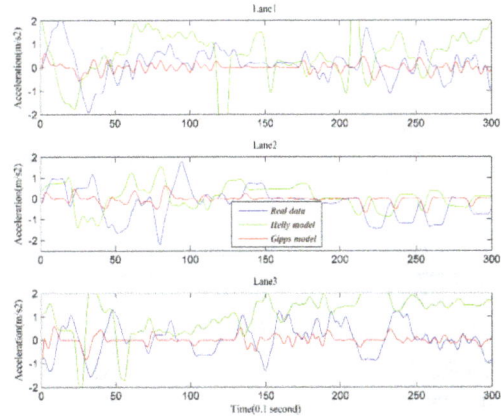

Figure 6. Difference between the real data and two conventional model (Helly and Gipps)

4. VALIDATION OF METHODOLOGY

Three measures were employed to validate the proposed methodology. The first measure is the mean square error, Eq. (9). The second measure is the Index of agreement, Eq. (10). Moreover, micro simulation validation was carried out as the last measure.

4.1 Statistical validation

To demonstrate the accuracy of the proposed model, the following criteria were used: The Mean Square Error (MSE) method was used to determine the accuracy of the model. The following Equation shows the calculations of the measure of accuracy:

$$MSE = \frac{\sum_{i=1}^{n}\left(y(t_i)-\left(y(t_i)'\right)^2\right)}{n}, \tag{9}$$

Index of agreement, i.e., d, indicates the extent that the predicted values are error-free. The closer the value of this parameter becomes to 1, the better our model of prediction will be. The following Equation shows this index formula (Papanastasiou et al., 2007):

$$d = 1 - \frac{\sum_{i=1}^{n}(y_i - y_i')^2}{\sum_{i=1}^{n}(\left|y_i' - \overline{y}\right| + \left|y_i - \overline{y}\right|)^2} \tag{10}$$

where $y(t_i)'$ is the value predicted by the model, $y(t_i)$ is the real value or the test data value and $\overline{y}$ is the mean values of real data. The following table summarizes the statistical results of the model for various states.

Mean Square Error(MSE)			
	Lane 1	Lane 2	Lane 3
ANN without Reaction Time	0.347	0.223	0.150
ANN with Reaction Time	**1.0E-08**	**7.0E-08**	**1.0E-08**
Helly	2.242	0.988	1.500
Gipps	0.423	0.581	0.295

Table 1. The mean squared error (MSE) between the real data and the output of models with reaction time and without it

Index of Agreement			
	Lane1	Lane 2	Lane 3
ANN without Reaction Time	0.55%	0.89%	0.85%
ANN with Reaction Time	**0.99%**	**0.99%**	**0.99%**
Helly	0.17%	0.41%	0.37%
Gipps	0.38%	0.34%	0.34%

Table 2. Statistical summary for each observed and simulated vehicle (*d* is index of agreement)

Based on the above table and by considering the instantaneous reaction, the proposed model yields the best statistical parameters for all lines of the route. This model can be used for the car guidance systems in the non-collision state.

4.2 Microscopic validation

In this section, we show the validity of the proposed model using the micro-simulation technique. After determining the model accuracy using the error test with the least squares error method and the use of index of agreement parameter as the determinant of prediction accuracy in the previous section, we firstly considered Highway Emeryville and three lines, where data was collected. The following figure shows the study area for simulation:

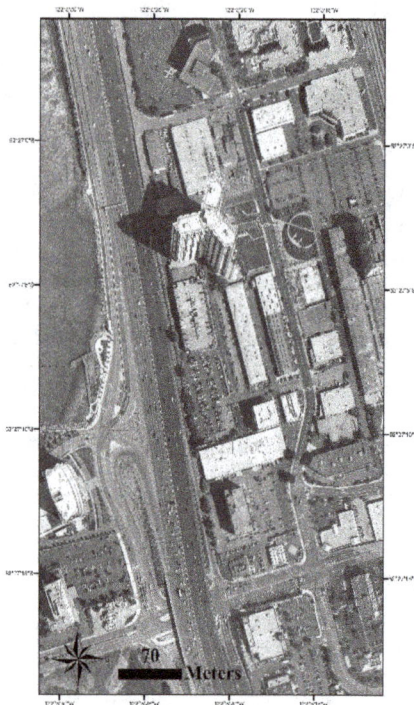

Datum: WGS_84   Projection System: UTM_Zone 10 S

Figure 7. Study area

During the simulation, two cars were assumed for each line: one as the front car and the other as the follower car. The front car started moving with the speed available at the collected data and the follower car instantaneously calculated acceleration based on the proposed model by understanding the speed of the front car, their distance, and relative speed. Based on this acceleration, displacement was calculated for the next moment.

Simulation was performed with 0.1-second steps and it was found that the proposed model has a good validity according to the proximity to the real situation of car-following in its simulation. Figure .18 depicts simulation results in three lane.

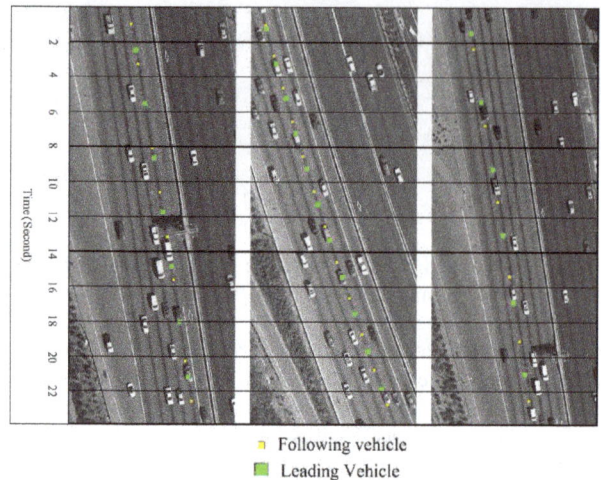

▫ Following vehicle
■ Leading Vehicle

Figure 8. Simulation prototype

5. CONCLUSION

Car-following models are among the most important topics in the field of traffic simulation at the micro level. All famous models that are used today include a set of parameters that require careful calibration, and any changes, even trivial, will severely affect the output. To solve this problem, this paper uses an artificial neural network to propose a car-following model. The model predicts the desired output, i.e. acceleration changes, by allocating four parameters, including the speed of the front car, the relative distance to the front car, the relative speed between the front car, and the follower car with different reaction times. To demonstrate the validity of this model, two methods were used: theory of errors and micro simulation. In the theory of errors, the output data of the models were compared with reality and it was found that the proposed model has a good accuracy in the car-following modeling, considering the instantaneous reaction time as an input parameter. In the simulation, it was shown that the results from the moving vehicles in the actual route follow the actual conditions of the car-following flow, based on the understanding of the follower car's driver about the performance of the front car, including the car speed and distance with it in every 0.1 seconds. This model can also be used in driver support tools, maintaining the safe distance, guiding unmanned vehicles and other applications of intelligent transportation systems.

REFERENCES

Brackstone, M. and M. McDonald. 1999. "Car-following: a historical review." Transportation Research Part F: Traffic Psychology and Behaviour 2(4): 181-196.

Chakroborty, P. and S. Kikuchi., 1999. "Evaluation of the General Motors based car-following models and a proposed fuzzy inference model." Transportation Research Part C: Emerging Technologies 7(4): 209-235.

Cho, H.-J, and Y.-T. Wu., 2008. "Microscopic analysis of desired-speed car-following stability." Applied Mathematics and Computation 196(2): 638-645.

Dougherty, M., 1995. "A review of neural networks applied to transport." Transportation Research Part C: Emerging Technologies 3(4): 247-260.

Gazis, D. C., R. Herman and R. B. Potts., 1959. "Car-following theory of steady-state traffic flow." Operations research 7(4): 499-505.

Gipps, P. G., 1981. "A behavioural car-following model for computer simulation." Transportation Research Part B: Methodological 15(2): 105-111.

Helly, W., 1961. Simulation of bottlenecks in single-lane traffic flow.

Herman, R., E. W. Montroll, R. B. Potts and R. W. Rothery., 1959. "Traffic dynamics: analysis of stability in car following." Operations research 7(1): 86-106.

Herman, R. and R. B. Potts., 1900. "Single lane traffic theory and experiment."

Herman, R. and R. Rothery., 1967. PROPAGATION OF DISTURBANCES IN VEHICULAR PLATOONS. IN VEHICULAR TRAFFIC SCIENCE. Proceedings of the Third International Symposium on the Theory of Traffic Flow.

Herman, R. and R. W. Rothery., 1965. Car following and steady-state flow. Proceedings of the 2nd International Symposium on the Theory of Traffic Flow. Ed J. Almond, OECD, Paris.

Haykin, S., 1999. "Neural Networks–A Comprehensive Foundation." New Jersey: Printice-Hall Inc.

Kalogirou, S. A., 2000. "Applications of artificial neural-networks for energy systems." Applied Energy 67(1): 17-35.

Kalyoncuoglu, S. F. and M. Tigdemir., 2004. "An alternative approach for modelling and simulation of traffic data: artificial neural networks." Simulation Modelling Practice and Theory 12(5): 351-362.

Karlaftis, M. and E. Vlahogianni., 2011. "Statistical methods versus neural networks in transportation research: Differences, similarities and some insights." Transportation Research Part C: Emerging Technologies 19(3): 387-399.

Kim, T., 2005. "Analysis of Variability in Car-Following Behavior over Long-Term Driving Maneuvers."

Kim, T., D. J. Lovell and Y. Park., 2003. Limitations of previous models on car-following behavior and research needs. 82nd Annual Meeting of the Transportation Research Board, Washington, DC.

Kometani, E. and T. Sasaki., 1959. "A safety index for traffic with linear spacing." Operations Research 7(6): 704-720.

Ma, X. and I. Andreasson., 2007. "Behavior measurement, analysis, and regime classification in car following." Intelligent Transportation Systems, IEEE Transactions on 8(1): 144-156.

Ozaki, H., 1993. "REACTION AND ANTICIPATION IN THE CAR-FOLLOWING BEHAVIOR." Transportation and traffic theory.

Panwai, S. and H. Dia., 2007. "Neural agent car-following models." Intelligent Transportation Systems, IEEE Transactions on 8(1): 60-70.

Papanastasiou, D., D. Melas and I. Kioutsioukis., 2007. "Development and assessment of neural network and multiple regression models in order to predict PM10 levels in a medium-sized Mediterranean city." Water, air, and soil pollution 182(1-4): 325-334.

Pipes, L. A., 1953. "An operational analysis of traffic dynamics." Journal of applied physics 24(3): 274-281.

Punzo, V., M. T. Borzacchiello and B. Ciuffo., 2011. "On the assessment of vehicle trajectory data accuracy and application to the Next Generation SIMulation (NGSIM) program data." Transportation Research Part C: Emerging Technologies 19(6): 1243-1262.

Rahman, M. 2013. "Application of parameter estimation and calibration method for car-following models". Master of Science Thesis, Clemson University.

Reuschel, A., 1950. "Vehicle movements in a platoon." Oesterreichisches Ingenieur-Archir 4: 193-215.

Saifuzzaman, M. and Z. Zheng., 2014. "Incorporating human-factors in car-following models: a review of recent developments and research needs." Transportation research part C: emerging technologies 48: 379-403.

Vogl, T. P., J. Mangis, A. Rigler, W. Zink and D. Alkon., 1988. "Accelerating the convergence of the back-propagation method." Biological cybernetics 59(4-5): 257-263.

Wilson, R. E. 2001. "An analysis of Gipps's car-following model of highway traffic". IMA journal of applied mathematics, 66, 509-537.

Wu, J., Y. Sui and T. Wang., 2009. "Intelligent transport systems in China." Proceedings of the ICE-Municipal Engineer 162(1): 25-32.

8

INTRODUCING NOVEL GENERATION OF HIGH ACCURACY CAMERA OPTICAL-TESTING AND CALIBRATION TEST-STANDS FEASIBLE FOR SERIES PRODUCTION OF CAMERAS

M. Nekouei Shahraki [a] [*] , N. Haala [a]

[a] Institute for Photogrammetry (ifp), University of Stuttgart, Geschwister-Scholl-Str. 24D, 70174 Stuttgart, Germany

KEY WORDS: Fisheye Camera Calibration, Single-Shot Camera Calibration, Calibration Test-Stands, Automated Calibration, Test-Stand Design, Camera Alignment-Testing, Camera Optical-Testing

ABSTRACT

The recent advances in the field of computer-vision have opened the doors of many opportunities for taking advantage of these techniques and technologies in many fields and applications. Having a high demand for these systems in today and future vehicles implies a high production volume of video cameras. The above criterions imply that it is critical to design test systems which deliver fast and accurate calibration and optical-testing capabilities. In this paper we introduce new generation of test-stands delivering high calibration quality in single-shot calibration of fisheye surround-view cameras. This incorporates important geometric features from bundle-block calibration, delivers very high (sub-pixel) calibration accuracy, makes possible a very fast calibration procedure (few seconds), and realizes autonomous calibration via machines. We have used the geometrical shape of a Spherical Helix (Type: 3D Spherical Spiral) with special geometrical characteristics, having a uniform radius which corresponds to the uniform motion. This geometrical feature was mechanically realized using three dimensional truncated icosahedrons which practically allow the implementation of a spherical helix on multiple surfaces. Furthermore the test-stand enables us to perform many other important optical tests such as stray-light testing, enabling us to evaluate the certain qualities of the camera optical module.

1. INTRODUCTION

1.1. Video-Based Driver Assistant Systems

There is an ever growing demand for using optical sensors and cameras in different systems and environments such as the driver-assistant (DA) systems. The video-based driver-assistant systems incorporate different types of video cameras such as mono/stereo front-view camera or fisheye surround-view cameras which allow gathering and analysing information of the surroundings of vehicle, creating a state of situational awareness. This makes it possible to perform the vehicle motion, manoeuvre and trajectory control. In computer-vision applications such as the video-based driver-assistant systems there is a need for accurate image to world transformation which presumes accurate camera calibration. This requires designing and developing optical-testing and calibration systems for performing various optical tests on every single camera.

1.2. Test-Stand Single-Shot Calibration

Calibration test-stands are often used when speed is the key requirement in performing camera calibration in a single-shot operation, together with required high (sub-pixel) accuracy. In this paper we introduce new generations of test-stands designed based on the implementation of ideal bundle-block calibration benefitting from state-of-the-art manufacturing techniques.

2. REQUIREMENTS IN TEST-STAND DESIGN

2.1. Implementing Bundle-Block Geometrical Features

The nature of fisheye cameras used in surround-view systems is to have extreme opening angles bigger than 180 degree which means having high spherical projection factor or high radial distortion depending on the interpretation of projection model such as (Mei and Rives, 2004). This implies that the radial distortion is significantly big in comparison with other lens distortion characteristics. Therefore we need to be able to accurately model this distortion. In test-field-based bundle-block calibration we have generally many points that are - because of its multiple view geometry - well distributed over the image and provide us with high calibration quality, and make an accurate distortion modelling possible. Therefore we need to understand the specifications in point distribution of bundle-block calibration solutions and implement it in the test-stand design.

2.2. Automatic Calibration and High Repetitive Accuracy

A fast and automatic calibration (Abraham and Hau, 1997) is required for sustaining certain efficiency and operation capability when dealing with huge number of cameras for driver assistant systems. Therefore it is required to realize the capability of automatic calibration by taking advantage of automation techniques in software and hardware e.g. in image processing algorithms or the use of robots. This brings an operational automation into our process to avoid any direct human interaction. In comparison with solutions in which human interaction is necessary (which brings unpredicted effects and instability in calibration procedure), using automated machines instead would increase the stability and repetitive accuracy of the calibration and thus ensured the quality of the calibration process. This feature (high repetitive accuracy) is very important for example when using the test-stand platform for analysing different environmental effects on the camera or performing hardware and software simulations.

2.3. Maintaining Calibration Accuracy for Complete Image Area

In surround-view systems taking advantage of omnidirectional cameras the calibration sub-pixel accuracy at the edges of image have a special importance in many applications. Therefore we need to make sure that the designed test-stand and the used calibration model will provide us with this accuracy at the edges. One of the most important criterions in this matter is the Runge's Phenomenon (Dahlquist and Björck, 1974). This phenomenon has been discussed in numerical analysis and refers to the increase in errors at the edges of a domain when trying to fit a high-degree polynomial to a function over that domain. Runge's phenomenon is mostly present in conditions in which the high degree of polynomial comes close to the number of equidistant sampled points in the domain. Because the modelling of lens distortion could be considered as a mathematical function fitting or polynomial fitting (Tang et al, 2012a) (Tang et al, 2012b) we need to make sure that this phenomenon does not happen when performing the test-stand calibration. This is also valid when using calibration models such as C.Mei model (Mei and Rives, 2004) or the recent Free-Function model (Nekouei and Haala, 2015) for which different types of polynomials or function series are used with degrees from 6 up to 32 are used for distortion modelling.

2.4. Camera Alignment-Testing Capability

When performing the camera intrinsic calibration, we also require the information about the alignment of the camera internal parts. This information include the alignment of sensor with respect to the optical axis, the camera housing, or the reference camera coordinates system. Therefore we need to consider the capability of performing camera alignment testing in the design of test-stand.

2.5. Additional Demands on Optical Tests

It is required and advantageous to have multiple functionalities available in our test systems. Therefore the test-stand needs to be designed to encompass the capability for other optical tests on our cameras such as Stray-Light testing (Raizner, 2012). These optical tests provide us with further quality measures of our optical module such as the appearance of optical artefacts or unwanted optical effects such as ghosts in image. This helps us determine if a certain level of quality is reached for each camera. Therefore the existence of these extra functionalities should be considered in the test-stand design.

3.   TEST-STAND DESIGN

3.1.   Analysing the Bundle-Block Calibration

As discussed in chapter 2.1 we need to implement certain features of multiple-view test-field-based calibration as point distribution in our test-stand. In order to calculate the point distribution for the bundle-block calibration, we set the image height value (observed point radius) for these points as observations. To calculate distribution factors, we set the radial distance from principal point or correspondingly the sight angle for each consecutive point as new observation and then calculate the PDF based on these values. For this purpose identical radial observations which are image points closer than 1 Pixel were omitted from the dataset (the threshold for which the point distribution is studied in image space). By avoiding the recurrent points, we were able to calculate the distribution parameters for bundle-block calibration of our fisheye cameras.

The distribution of the points (measured as image height) for bundle-block calibration is very close to a uniform distribution up to the image edge with image height of nearly 550 [pixels]. This means that a uniform point distribution is present when performing bundle-block calibration. The geometry of multiple-view test-field-based calibration helps us to have points with different image heights and sight angles similar to uniform spacing or uniformly distributed points. Assuming a normal polynomial-fitting (for simplifications) this would mathematically help us to better fit our model to the projection curve (Guest, 1958) (Benhenni and Degras, 2011) and means that by having points that are uniformly distributed over radius we will be able to estimate and model the lens distortion effectively.

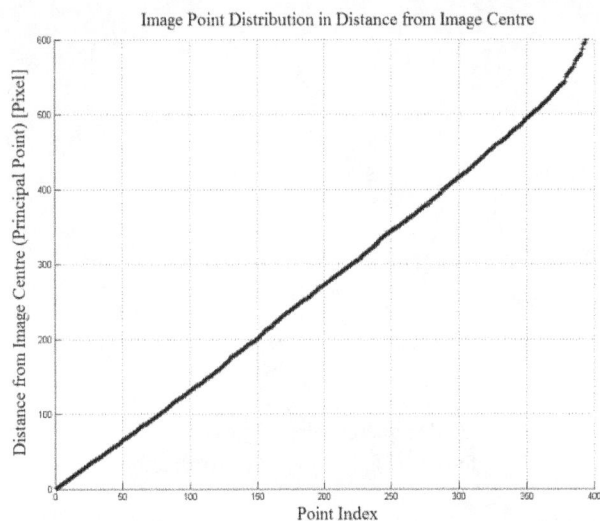

Figure 1. Distribution of point distances in bundle-block calibration for fisheye cameras

In Fig 1 we can see the ideal point distribution is almost a straight line and looks very similar to a uniform distribution. Considering a normal distribution for the distance between consecutive points, we get $\mu = 1.6902$ [pixel], $\sigma = 1.7069$ [pixel]. We can see that it has a standard deviation (which shows the quality of point distribution over the whole image) almost equal or not bigger than the expectation for the point distances (that shows the mean distance of consecutive points) which implies a dense point distribution.

3.2. Test-Stand Geometrical Design

As the first step for the geometrical design of the test-stand we need to study the specifications of the video camera which is going to be calibrated using this device. The video camera is the NRC (Near-Range Camera) used as a surround-view video system in driver-assistant systems. The specifications of the NRC camera are provided in Table 1:

Camera Parameter	Value
Image Sensor	CMOS
Image Height	960
Image Width	1280
Pixel Size	3.75 [μm]
Colour Filter Array (CFA) Pattern	RGGB (Bayer Pattern)
Nominal Focal Length	5.2 [mm]
Spectral Range (SR)	400-700 [nm]
HFOV	≈ 192 [deg]
VFOV	≈ 144 [deg]

Table 1. Specifications of the NRC camera

In order to have high calibration quality in single-shot calibration comparable to the bundle-block solution we have tried to simulate similar geometrical features while designing the test-stand.

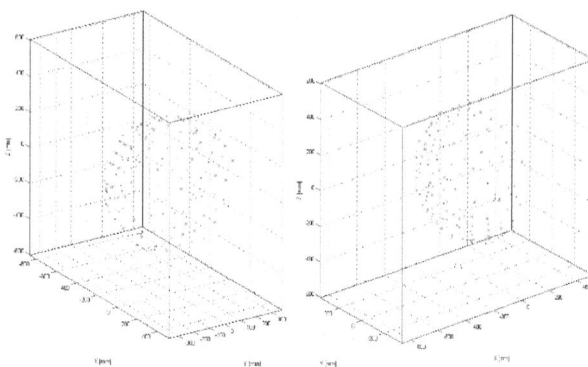

Figure 2. 3D representation of the geomtric design of test-stand for fisheye camera calibration

As mentioned above, the ideal bundle-block calibration has a point distribution characteristic similar to a uniform point distribution which increases the camera calibration quality and validity all over the image. To have this advantageous features from bundle-block calibration in our calibration test-stand, we have used the geometrical shape of a Spherical Helix (Type: 3D Spherical Spiral). One of the special characteristics of this geometrical shape is that it has a uniform radius which corresponds to the uniform motion (Pottmann et al, 2001) (Pottmann et al 2004).

Also in order to realize an automated calibration process in software (as discussed in chapter 2.2), there are certain coded point-groups added to the test-stand control points allowing us to use automated algorithms in the calibration process and thus avoid any direct human interaction. This process is performed by automatically detecting these distinct point-groups and performing target point-matching automatically.

3.3. Test-Stand Mechanical Design and Features

The geometrical feature was realized using three dimensional truncated icosahedrons (Stakhov, 2015) (Hosoya 2011) which practically allows the implementation of a spherical helix on multiple surfaces. To keep the ideal designed distribution in image space, the image points are re-projected analytically on the surfaces of the icosahedrons and the initial three-dimensional space coordinates on the surfaces are calculated.

The design characteristics of the control points were finally achieved by analysing the geometrical features of the test-stand and the stray-light testing as the second necessary optical test (as discussed in chapter 2.5) required to be performed on the camera. The points are designed to be active-illuminated diffuser points with an accurate no-shadow design for which the not-directly-measurable points are measured using the mechanical touch measurement techniques.

All the control point mechanical parts and elements are constructed using precise machinery with accuracy of around 50 [μm]. This accuracy threshold is derived from the pre-analysis calculations in chapter 3.5. Also, the test-stand was finally calibrated mechanically using a measurement arm which provides us point coordinates with the required accuracy available in chapter 3.5. Furthermore, because the control points are actively illuminated we can also use the test-stand for other purposes such as optical stray-light testing.

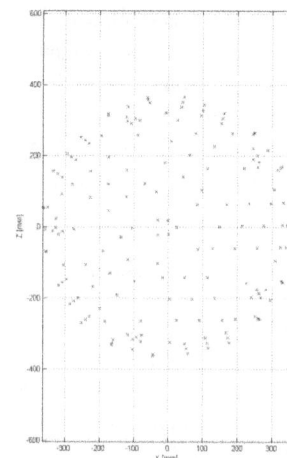

Figure 3. Representation of the geometric design of test-stand for fisheye camera calibration

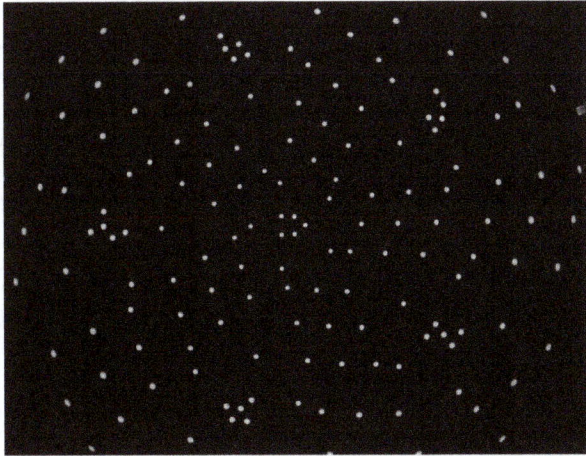

Figure 4. Captured image from test-stand with a NRC Camera

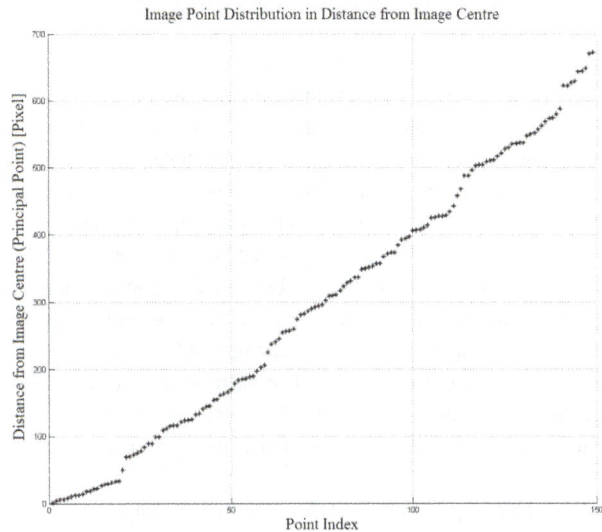

Figure 5. Image point distribution in radius for Test-Stand calibration of fisheye Cameras

The final design and the test-stand is represented in Figure 3 and Figure 4, showing the achieved geometrical characteristics of projected points in the camera. The projection of the (modified) three dimensional spherical helix (spherical spiral) is clearly visible in the image.

As discussed in chapter 2.3 we should guarantee a certain point distribution to avoid Runge's phenomenon from happening. By studying the 3D geometry of test-stand points and image points, it would be clear that by having an (almost) equidistance point distribution (similar to equispaced interpolation points), having a high point density i.e. the number of points far exceeding the degree of the distortion model, and avoiding extremely high degrees of the distortion model the Runge's Phenomenon does not happen (Dahlquist and Björck, 1974). Furthermore, this uniform and dense point distribution would help us to have extrapolation capabilities (Laderman and Laderman, 1982) in the distortion model such as C.Mei further beyond the last detected image point near the image edge.

We can also perform camera alignment testing using this test-stand as discussed in chapter 2.4. This means evaluating the geometrical alignment of the camera housing and the interior elements such as the alignment of sensor surface compared to the defined optical axis and the reference coordinates system. This is realized by designing an accurate and stable (machine-operated) camera fixture defined as origin of the unified test-stand coordinates system, achieved by using reference points on both test-stand and camera fixture.

3.4. Design Statistical Analysis and Point Distribution

Having the test-stand point coordinates on the surface icosahedron, we can calculate the projected coordinates in camera and analyse the point distribution in image space. We analyse the distance of consecutive points from each other and from the image centre (principal point) and generate the point distribution graph comparable to Figure 1. We can then study the histogram of the point distribution over the whole image to get further information about the point distribution in image space.

Figure 5 illustrates the test-stand point distribution in image by considering each single point (point index) and its corresponding image height. We can see that it is very much similar to a uniform distribution. The small deviations from the straight line are because of some practical and mechanical limits of construction techniques and the introduction of geometrically-coded point-groups for performing automatic point matching in the calibration process.

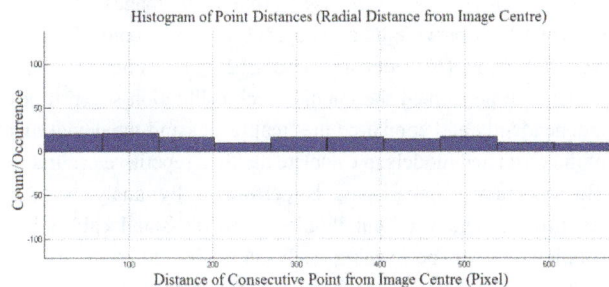

Figure 6. Histogram of point distribution (image height) for Test-Stand calibration of fisheye cameras

Figure 6 again illustrates the point distribution as a histogram chart. We can see that there is an almost constant value in point count/occurrence over the image height which means the point distribution could well be considered a uniform distribution.

After analysing the test-stand point distribution, we get the values μ = 4.5 - 5.4 [pixel], σ = 4.9874 [pixel] for a normal distribution of the distance difference between consecutive points. This shows compliance with the design criterion discussed in chapter 3.1 as the standard deviation is almost equal to the expectation for the point distances i.e. mean distance of consecutive projected points in image. We can finally say that this point geometry potentially helps in estimating a valid projection function and modelling the lens distortion accurately all over the image comparable to bundle-block calibration accuracy.

3.5. TEST-STAND MEASUREMENT ACCURACY

We perform the test-stand pre-analysis in order to determine the required accuracy for test-stand calibration and measurement. We refer to pre-analysis as the operation performed while designing an adjustment network which helps us to make decisions about network geometry and the accuracy of observations to ensure the required accuracy and significance for the unknown parameters. Having an approximate design of the test-stand, and some expectations for the calibration accuracy in image space, we can calculate the required accuracy of the observations and determine accuracy thresholds with a certain factor of safety (e.g. 3-Sigma). The accuracy of the test-stand measurement should be at least within those limits or higher so that the accuracy of estimated parameters meets our expectations.

After performing the pre-analysis using the design geometry of the test-stand and the required image application accuracy of about 0.1 [pixel], the required accuracy of initial 3D point measurement was estimated to be about 50 [μm] and used for measuring the coordinates of the points and calibrating the test-stand itself.

4. TEST-STAND ACCURACY ANALYSIS

4.1. Comparison with bundle-block

In our tests we performed the bundle-block calibration which is referred to the multiple-view test-field-based calibration for fisheye cameras. We have used around 40 images for each camera which provide us around 1500 observed control points per calibration (Nekouei and Haala, 2015). Using the same camera we performed the bundle-block calibration seven times independently and compared the final results and the deviations of the estimated models to calculate the final repetitive accuracy. The accuracy comparisons, is performed by analyzing the repetitive accuracy of bundle-block and test-stand calibration solutions. When performing each calibration for both of the solutions, the camera has been mounted/placed again in the camera fixture.

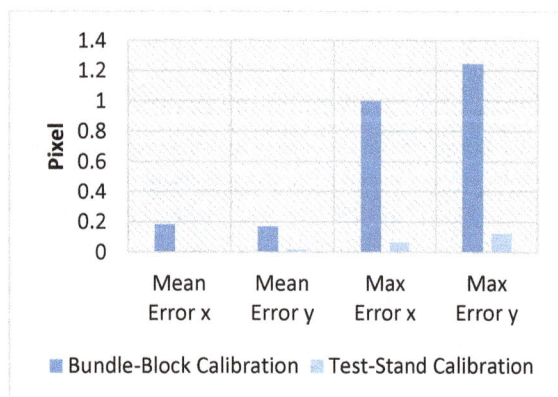

Figure 7. Analysis of repetitive accuracy using test-stand and bundle-block calibration and a single fisheye camera

Fig 7 represents the accuracy factors corresponding to the repetitive accuracy taken from processing of the calibration datasets generated using test-stand and bundle-block calibration. The "Mean Error" and "Max Error" values are calculated in image x and y directions respectively from generated distortion-field of each calibration. The latter is estimated by calculating their deviation from each other on defined projected 3D points i.e. a simulated dense sphere in space.

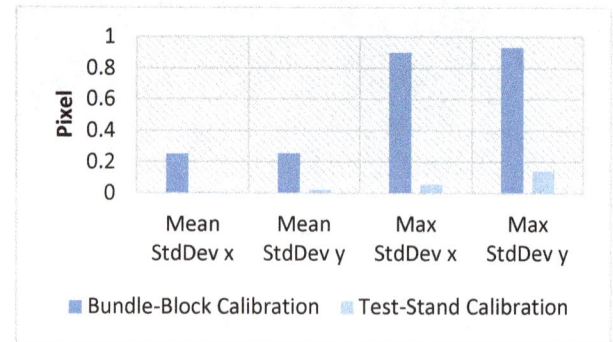

Figure 8. Accuracy analysis of multiple calibrations using test-stand and bundle-block calibration and a single camera

Figure 8 illustrates the standard deviation of re-projection errors for all of the simulated 3D points as further measures calculated using the calibration dataset for both of the calibration solutions. We should also note that these values could be interpreted as the amount of noise or random error in two different calibration processes which again implies a higher repetitive accuracy for test-stand calibration, which is one of the many important aspects in the camera calibration.

We can see in Fig 7 and Fig 8 that the repetitive accuracy of test-stand calibration is higher than bundle-block calibration. The reason behind this accuracy difference is that it is usually not possible to guaranty the realization of the same ideal point distribution in every bundle-block calibration and thus the deviations of point distributions (or the low stability in point distribution) directly affect the estimated camera intrinsic parameters. In contrast to that, in test-stand calibration the presence of a stable ideal point distribution has increased the accuracy of the test-stand calibration. Also the advantage of an automated calibration procedure (i.e. performed with machine) in test-stand has contributed to a more stable calibration procedure which has further increased the calibration accuracy and stability.

5. CAMERA CALIBRTION ACCURACY ANALYSIS

We performed the test-stand camera calibration for a fisheye camera using the Free-Function calibration model (Nekouei and Haala, 2015). The designed test-stand point distribution in image space is especially beneficial in this case because it makes possible a high quality modelling of local lens distortions. Furthermore the high accuracy threshold for test-stand measurement makes it possible to achieve high sub-pixel accuracies in distortion modelling of the optical module.

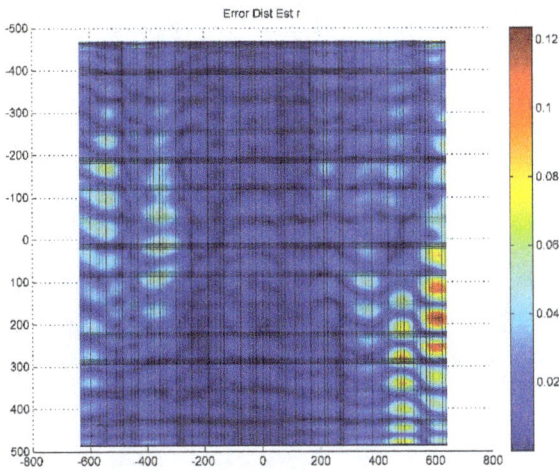

Figure 9. The remaining errors from Free-Function calibration
(of degree 16) for a fisheye image
The colour-bar and axes coordinates are in pixel units.

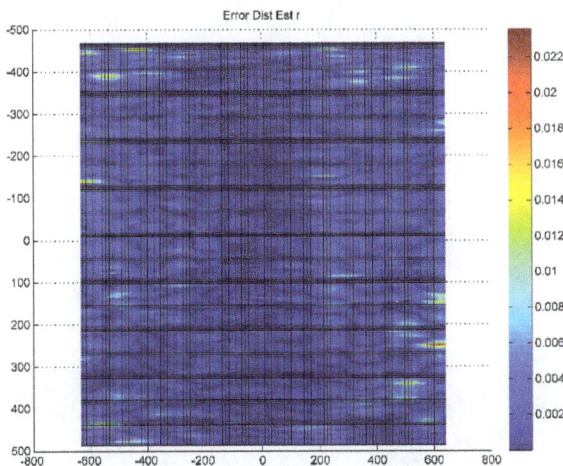

Figure 10. The remaining errors from Free-Function calibration
(of degree 32) for a fisheye image
The colour-bar and axes coordinates are in pixel units.

Figure 9 and Figure 10 show the remaining errors of local lens distortion modelling. As we can see the remaining errors are very small which means the local distortion is very well modelled. In Table 2 the accuracy parameters (after performing camera calibration) are available. The "RMSE" values are calculated using all of the observed control points in image and the "Max Error" values are calculated over the entire image (Nekouei and Haala, 2015).

Parameter Name	Free-Function Model (degree 16)	Free-Function Model (degree 32)
RMSEx	0.012	0.001
RMSEy	0.016	0.001
Max Error x	0.061	0.014
Max Error y	0.123	0.022

Table 2. Accuracy analysis of the Free-Function model used in test-stand calibration (All the parameters are in pixel units)

As we can see in Table 2 using the designed test-stand as the calibration hardware platform, and the Free-Function model we could achieve outstanding sub-pixel accuracy compared to classical calibration solutions.

6. CONCLUSION

We designed a test-stand in order to realize the specifications and geometrical features of multiple-view test-field-based camera calibration referred to as bundle-block calibration. To have the advantageous point distribution features from bundle-block calibration in our calibration test-stand, we have used the geometrical shape of a spherical helix (type: 3D spherical spiral). One of the special characteristics of this geometrical shape is that it has a uniform radius which corresponds to the uniform motion. The geometrical feature was realized using three dimensional truncated icosahedrons which practically allow the implementation of a spherical helix on multiple surfaces.

This uniform and dense point distribution helps us to have extrapolation capabilities in the calibration model (after last-squares adjustment) further beyond the last detected image point at the image edge. Depending on the calibration model such as C.Mei or Free-Function model used to model the optical parts, this point geometry can help in estimating a valid projection function and thus modelling the lens distortion accurately all over the image (comparable to bundle-block calibration accuracy).

Furthermore, such a point distribution is beneficial when using calibration models such as Free-Function model which enable us to model of local lens distortion with high accuracy and quality all over the image.

A very important feature of this test-stand is having the capability of performing camera/sensor alignment-testing, a feature which is very important for testing the geometrical alignment of the internal mechanical elements of each camera.

There is also another special advantage of this test-stand design which is having operational automation i.e. the use of machines to perform the calibration procedure. Using automated machines and algorithms would increase the stability and accuracy of the calibration and thus ensures the quality and speed of the calibration for cameras used in video-based driver assistant systems.

REFERENCES

Abraham, S., and Hau, T., 1997. *Toward Autonomous High-Precision Calibration of Digital Cameras*, Institute of Photogrammetry, Bonn University, SPIE 3174, Videometrics V, 82

Benhenni, K., Degras, D., 2011. Local Polynomial Regression Based on Functional Data, *Journal of Multivariate Analysis*

Dahlquist Germund and Björck Åke, 1974. *Numerical Methods*, Dover Publication Inc., p 101-103

Guest, P. G., 1958. The Spacing of Observations in Polynomial Regression, University of Sydney, Australia, (JSTOR, *The Annuals of Mathematical Statistics*)

Hosoya, H., 2011. *High π-Electronic Stability of Soccer Ball Fullerene C₆₀ and Truncated C₂₄ Among Spherically Polyhedral Networks*, The Mathematics and Topology of Fullerences, Springer, Chapter 13, p 249-252

Laderman, J., and Laderman J.D., 1982. Simplified Forecasting by Polynomial Regression with Equally Spaced Values of the Independent Variable, *Mathematics of Computation*, Volume 38, Number 158

Mei, C., and Rives, P., 2007. *Single View Point Omnidirectional Camera Calibration from Planar Grids*, IEEE International Conference on Robotics and Automation (ICRA) 3945-3950 doi: 10.1109/ROBOT.2007.364084

Nekouei, M., and Haala, N. 2015. *Introducing free-function camera calibration model for central-projection and omni-directional lenses*, Proc. SPIE 9630, Optical Systems Design 2015: Computational Optics, 96300P (September 23, 2015); doi: 10.1117/12.2191121

Pottmann, H., Wallner, J., Leopoldseder, S., 2001. *Kinematical methods for the classification, reconstruction and inspection of surfaces,* Comptes rendus du Congr`es national de math´ematiques appliqu´ees et industrielles, (Corr`eze, France), p 51–60.

Pottmann, H., et al., 2004. *Line Geometry for 3D Shape Understanding and Reconstruction, Classification of Surfaces by Normal Congruences,* Computer Vision – ECCV: 8th European Conference on Computer Vision, (Springer), p 300-301

Raizner, C., 2012. *Objective and Automated Stray Light Insection of High-Dynamic-Range Cameras*, Shaker Verlag, Herzogenrath 2012, ISBN 978-3-8440-1287-3, 240 S

Stakhov, A., 2009. *The Mathematics of Harmony: From Euclid to Contemporary Mathematics and Computer Science*, World Scientific, Vol 22, p 144-146

Tang, R., Fritsch, D., Cramer, M., 2012a. *A novel family of mathematical self-calibration additional parameters for airborne camera systems*, European Calibration and Orientation Workshop (EuroCOW 2012), 7 pages on CD-ROM

Tang, R., Fritsch, D., Cramer, M., 2012b. New rigorous and flexible Fourier self-calibration models for airborne camera calibration, *ISPRS Journal of Photogrammetry and Remote Sensing*, Volume 71, p 76–85

9

AUTOMATIC ROAD EXTRACTION BASED ON INTEGRATION OF HIGH RESOLUTION LIDAR AND AERIAL IMAGERY

Sara Rahimi[a], Hossein Arefi[a], and Reza Bahmanyar[b]

[a]School of Surveying and Geospatial Engineering, University of Tehran, Tehran, Iran -
{rahimi.sara, hossein.arefi}@ut.ac.ir
[b]Institute of Remote Sensing Technology (IMF), German Aerospace Center (DLR), Wessling, Germany -
gholamreza.bahmanyar@dlr.de

KEY WORDS: Automatic Road Extraction, High Resolution Aerial Imagery, Hough Transform, LiDAR, Principal Component Analysis

ABSTRACT

In recent years, the rapid increase in the demand for road information together with the availability of large volumes of high resolution Earth Observation (EO) images, have drawn remarkable interest to the use of EO images for road extraction. Among the proposed methods, the unsupervised fully-automatic ones are more efficient since they do not require human effort. Considering the proposed methods, the focus is usually to improve the road network detection, while the roads' precise delineation has been less attended to. In this paper, we propose a new unsupervised fully-automatic road extraction method, based on the integration of the high resolution LiDAR and aerial images of a scene using Principal Component Analysis (PCA). This method discriminates the existing roads in a scene; and then precisely delineates them. Hough transform is then applied to the integrated information to extract straight lines; which are further used to segment the scene and discriminate the existing roads. The roads' edges are then precisely localized using a projection-based technique, and the round corners are further refined. Experimental results demonstrate that our proposed method extracts and delineates the roads with a high accuracy.

1. INTRODUCTION

In recent years, various tasks such as traffic navigation and urban planning have been affected by the rapid growth of transportation networks, therefore, increasing the interest in road extraction methods in fields such as EO and Geographic Information System. Recent advances in EO imaging have made it possible to obtain high resolution aerial and satellite images which represent land cover, such as roads, in more detail. In order to use this large quantity of data for road extraction while avoiding the costs of the manual processing, developing efficient automatic road extraction methods are in high demand. In order to automatize road extraction, previous works have followed two main directions, namely semi-automatic and fully-automatic methods. While semi-automatic methods require human interaction (Miao et al., 2014, Anil and Natarajan, 2010), most of the proposed fully automatic methods require human supervision as a ground truth either for training or parameter estimation (Shi et al., 2014, Samadzadegan et al., 2009). Since both human interaction and ground truth generation requires considerable human effort, several unsupervised fully-automatic methods have been recently introduced (Bae et al., 2015, Singh and Garg, 2014). While in many recent works, the focus is to improve road network detection, the roads' precise delineation has been studied to a lesser degree. However, in various applications such as road map updating, a precise determination of the roads' components (e.g., edges) is essential.

In this paper, we propose a new method for road extraction and precise delineation, using the integration of a LiDAR image; and the red, green, and blue (RGB) bands of a high resolution aerial image. While the LiDAR image provides objects' height information, the aerial image represents their color properties and textures from a bird's eye view. Figure 1 shows the overview of our proposed method. As a first step, in order to integrate the height and color information, we use the PCA (Pearson, 1901, Hotelling, 1933) method and convert the 4 information bands

(i.e., R, G, B, height) into 4 principal components. Among the resulting components, we select the one which provides the largest distinction between the roads and their neighboring objects. In the next step, a Hough transform (Hough, 1962) is applied to the selected component in order to extract straight lines. The lines are then intersected to segment the scene into different regions. Using the LiDAR and the aerial images, we verify whether the segments contain road parts, and discard the ones which do not. To this end, since the roads' color and height are usually more homogeneous than their surrounding objects (e.g., buildings), we suppose that the pixel value deviation of the segments containing road parts is smaller than those of the other segments. Thus, we discard the segments with the pixel value deviations larger than a certain threshold. Additionally, using the LiDAR image, the segments with average height values larger than a certain threshold are removed.

After extracting the road segments, we merge the segments which belong to the same road. The points within each segment are then projected on a perpendicular plane to the segment's main orientation, so that the majority of the points are able to generate the road's profile. The profile is generated for both the height and the gray value of the aerial image, in which the road edges are supposed to be the points with a significant change in either height or gray value. The road's edges are then detected by integrating the two profiles (height and gray value), and taking the derivative of the result. The maximum points of the resulting derivative localize the roads' edges. As a final step, the edges at the round corners, which usually occur at the road intersections, are refined. Experimental results demonstrate that our proposed method both detects the road network, and delineates the road edges with a high accuracy.

The rest of this paper is organized as follows: Section 2., provides an introduction to Hough transform. Section 3., presents our proposed method in detail. Section 4., discusses our experimental results, and Section 5., concludes this paper.

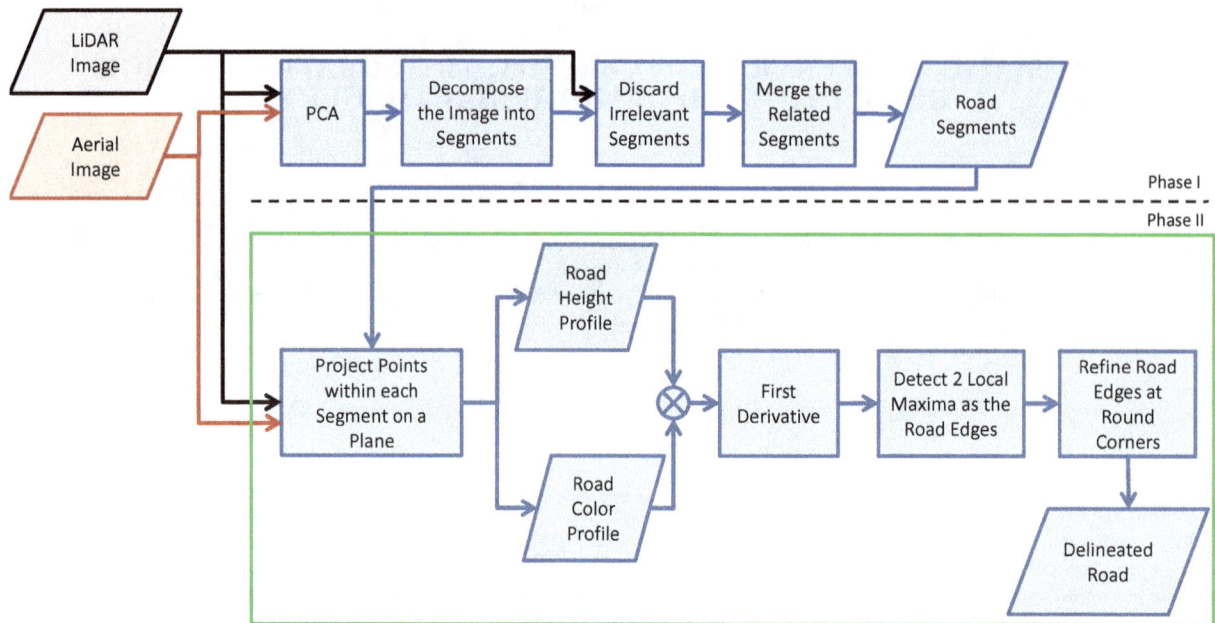

Figure 1: The overview of our proposed method; in which the process within the green rectangle is repeated for every road segment.

2.   HOUGH TRANSFORM

Hough transform is an image feature extraction method usually used to detect simple structures such as straight lines, circles or ellipses (Hough, 1962, Duda and Hart, 1972). The Hough transform is employed when, due to the image properties or the edge detection performance, grouping the extracted edges into a structure (e.g., straight line) is not possible. It considers a set of parametrized structure candidates. It then groups the image edges (for example, the ones extracted by Canny edge detector) into an appropriate structure candidate through a voting procedure. Assume detecting a straight line is required. Each line is defined by the slope parameter α and the y-intercept parameter b, $y = \alpha x + b$. However, in practice, since the infinite value of the slope parameter for the vertical lines causes computational burden, using Hesse normal form of the line is suggested in (Duda and Hart, 1972) as the following,

$$\rho = x\,cos(\theta) + y\,sin(\theta), \tag{1}$$

where ρ is the distance of the line from origin, and θ is the angle between the x-axis and the line's normal vector which passes through origin ($\theta \in [-\frac{\pi}{2}, \frac{\pi}{2}]$). Using this line form, any point in the x-y space is equal to a sinusoidal curve in ρ-θ space. In order to detect the existing straight lines, Hough transform quantizes the two parameters (ρ and θ), and for each quantization considers a bin in a histogram. Then for every pixel, it tries to find evidences of a straight line and assign it to one of the histogram bins. In a next step, the bins with higher values are selected as the parameters of the existing straight lines. Finally, since the line lengths are not determined (as an essential step) the lines are localized on the image.

3.   METHODOLOGY

The main contribution of our paper is the extraction of road edges in urban area by integrating high resolution LiDAR and aerial imagery. Figure 1 shows the overview of our proposed method. In the following, we explain every step in detail.

3.1   Apply PCA to the Images

In order to extract the existing roads in a scene, the first step is to discriminate the roads from the other objects. Thus, we integrate the LiDAR and aerial images in order to make them more discriminative. To this end, we apply PCA to the 4 available information bands (e.g., R, G, B, and height) to convert them into 4 principal components, where each component has a unique property. Among them, we empirically find out that the second component discriminates the roads from the rest of the scene better than the others. Therefore, we use it as the input for the road extraction step.

3.2   Decompose the Image into Segments

In urban areas, roads are usually in the form of straight lines. Using this assumption, we apply Hough transform to the PCA's second component in order to extract the existing straight lines in the scene. We then elongate these lines up to the scene borders in order to segment the scene (Figure 2.a). Among the segments, some are aligned with the roads and contain road parts. To find these segments (which contain part of a road), we assess all the segments in the next step.

3.3   Discard Irrelevant Segments

In this step, we perform two strategies in order to discard the segments which contain no road parts. The first strategy is based on the homogeneity of the segments in the aerial image. In order to measure the homogeneity, we compute the pixel values' average standard deviation. Since roads are usually homogeneous in color, the standard deviation of the segments which contain road parts are supposed to be smaller than those of the other segments. Thus we discard the segments with standard deviation larger that a certain threshold. Figure 2.b presents the remaining segments, and as depicted, most of the irrelevant segments are removed according to their homogeneity. However, there are segments which contain homogeneous objects, except roads, covered almost by grass and partially by a homogeneous house roof. In order to handle these situations, we apply a second strategy which uses the height information taken from the LiDAR image.

Figure 2: Segmentation of the image using the Hough lines. (a) All the segments, (b) Segments after discarding the irrelevant ones based on their homogeneity. (c) The remaining segments after removing the ones with a high average height value, (d) The road segments after merging and discarding ones which are not connected.

According to how we segment the scene, the segments containing road parts usually have a lower average height than the ones containing other homogeneous objects. Thus, we discard the segments with the average height larger that a threshold. Figure 2.c shows the remaining segments after applying the second strategy. According to the figure, although most of the irrelevant segments are removed, there are still segments which contain no road parts. These segments are homogeneous and their average height value is small; and are therefore, not detected by the two strategies. In the next step, we propose a method to find and remove these segments.

3.4 Merge the Related Segments

In order to represent each road as a whole in a single segment, we find and merge the segments containing parts of the same road. To this end, we extract the corner points of every segment using Douglas-Peucker line simplification algorithm (Douglas and Peucker, 2011). The corner points are considered as intersections of any two adjacent segments' edges with dissimilar orientations. Due to the roads' structure, we usually obtain four corner points for each segment. To merge the segments, we randomly select a segment and determine its main orientation as the orientation of its longest adjacent Hough line. Then we randomly take one of its edges (a line between a pair of corner points) which is not parallel to the segment's main orientation. Using the aerial image, the mean gray value of the pixels along the selected edge is then compared to that of the closest edge of its neighboring segment. If the edges' mean gray values are close enough, the segments are considered to be containing the same road; and thus, they are merged. This procedure is repeated until the main orientation of the newly merged segment is changed. The same procedure is repeated, by randomly selecting a segment from the remaining ones, until all the smaller segments are merged into larger ones. This results in connected segments representing the network structure of the roads. The results show that since the remaining irrelevant segments from the previous step do not contain a road part, they cannot connect to the discovered road network. Thus, we can detect and remove them. Figure 2.d shows the obtained segments.

3.5 Project Points within each Segment on a Plane

Although the resulting segments from the previous step localize the roads, the precise locations of the roads' edges are still missing. In order to delineate the roads accurately, in this step, we integrate the points within each segment along the segment's main orientation to project them on a perpendicular plane to the segment's main orientation. The idea behind this projection is that a majority of the points can represent the road's profile within a segment. In our experiments, we generate the roads' profiles using both LiDAR and aerial images. In the profiles, the roads' edges are the points where they experience a significant change

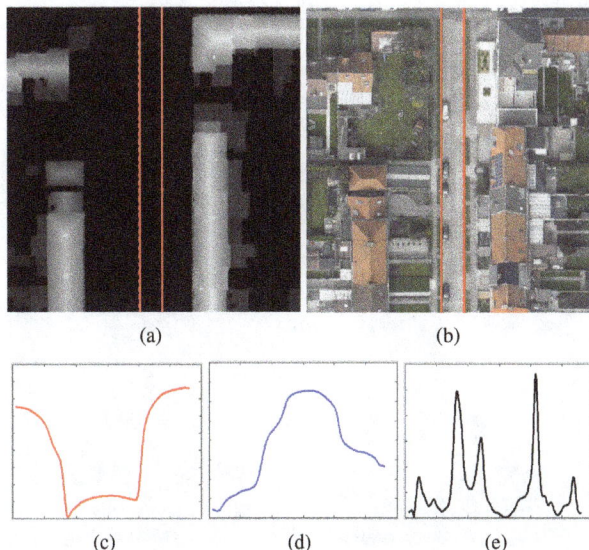

Figure 3: Projection of the points within each segment on a plane perpendicular to the main orientation of the segment. (a) A road depicted on the LiDAR image, (b) The same road depicted on the aerial image, (c) The road profile obtained from the LiDAR image, (d) The road profile obtained from the aerial image, (d) The first derivative of the integrated road profiles.

either in height value (when LiDAR image is used), or the pixels' gray value (when the aerial image is used).

In order to increase the area for searching the road edges, we expand the segments to the sides (i.e., the perpendicular direction to the segment's main orientation) and increase each segment's width by 65%. Considering the LiDAR image (Figure 3.a) and the aerial image (Figure 3.b), we compute the height profile (Figure 3.c) and the gray value profile (Figure 3.d). In order to localize the road edges as the points of significant changes, we integrate the two road profiles (height and gray value), and take the absolute value of the result's first derivative. The local maxima represent the places of significant change. Figure 3.e shows the computed derivative for the two road profiles. Due to the existence of objects with large contrasts either in height (e.g., parked cars) or in color (e.g., artifacts in sidewalks), the resulting derivate usually has more than two local maxima. In order to identify the ones which correspond to the road's edges, we take the 4 larger local maxima indicating the edge candidates' locations. In contrast to the road edges, which continue along the road, the high contrast objects occur only at some points. Therefore, in order to find and discard the irrelevant edge candidates, we randomly sample 50 points along each edge candidate. After that, we compute each point's gradient in an orientation perpendicular to that of its corresponding edge candidate. For each can-

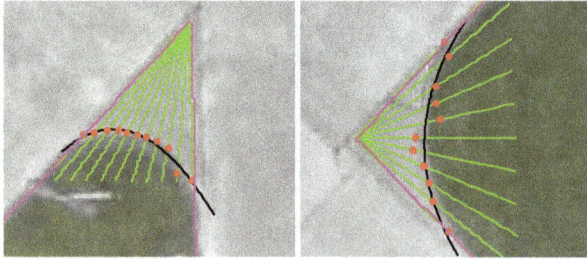

Figure 4: Refinement of the road edges at the round corners. In this figure, the magenta lines are the extracted straight road edges, the green lines are the search lines, the red points are the search results for the round corner edges, and the blue line is the curve fitted to the points determining the road's round corner edge.

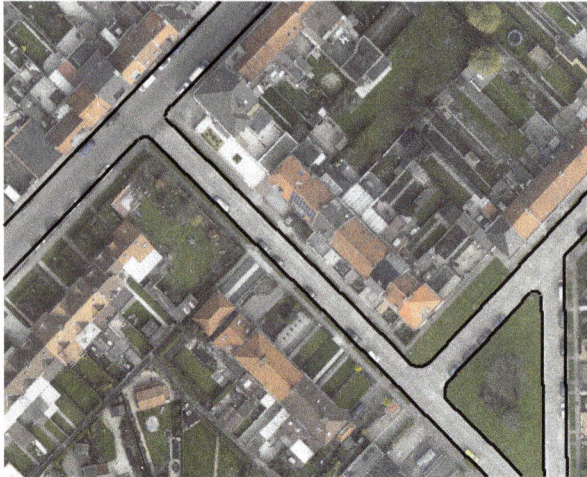

Figure 5: Visualization of the extracted road network and the roads' delineation using our proposed method.

didate, the computed gradients are then averaged. Due to their noncontinuous effect, the high contrast objects' average gradient is usually smaller than that of the road edges. Thus, we localize the road edges by the two edge candidates with the largest average gradient. In order to increase the stability of our method in the presence of random selection, we run the experiments three times; and then consider the average result.

3.6 Refine Road Edges at Round Corners

Since our road extraction method (up to this step) is based on the straight lines, the round corners, especially at the road intersections are missed. As Figure 4 shows, the extracted road edges depicted by magenta lines do not fit the roads' round corners. In this step, in order to improve the road delineation at round corners, we look for local round edges at the roads' intersections. To this end, we split the angle between every two intersected road edges into 10 equal sized angles. Then we compute the gradient of the pixels' gray values along every smaller angles' side (depicted by green lines in Figure 4), up to a certain distance from the intersection point (70 pixels in our experiments). On each angle side, the point with the largest gradient value is then taken as a sample of the round edge (the red points in Figure 4). These points are then used to fit a quadratic polynomial which delineates the round corner (the blue curve in Figure 4).

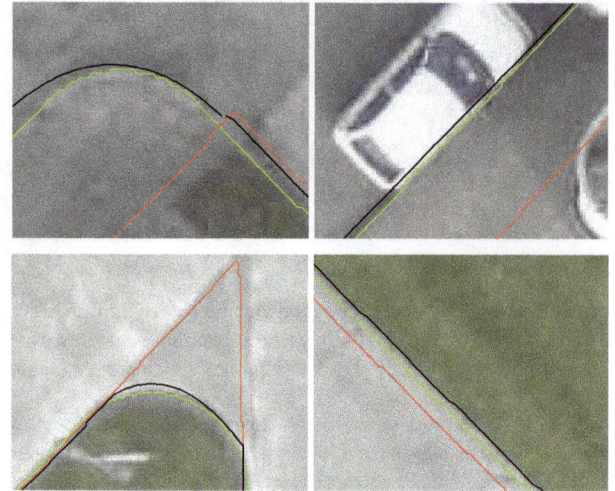

Figure 6: A closer look at some roads' delineations resulted by our proposed method (the blue lines); comparing to the ground truth (the green lines), and the segment boundaries computed by Hough transform (the red lines).

	Road Segments	Delineation & Refinement
RMS	5.75	2.57

Table 1: RMS errors computed for our method's result after road discrimination based on the Hough transform, and after delineation of the roads and refinement of their round corners.

4. EXPERIMENTAL RESULTS AND DISCUSSION

In our experiments, we use images of an area in Zeebruges, Belgium, which were acquired on March 13th, 2011[1]. We use its high resolution LiDAR and RGB aerial images. The LiDAR image has a spatial resolution of 10 cm, while the aerial image's spatial resolution is 5 cm. Therefore, before using them in our experiments, we downsample the aerial image to have the same resolution as the LiDAR image.

Figure 5 shows the extracted road network and the roads' delineation by our proposed method. According to the results, our method is able to detect all the existing roads in the scene. Moreover, Figure 6 demonstrates our proposed method's delineation precision. In this figure, the green lines are the ground truth, the red lines are the segment boundaries obtained by Hough transform, and the blue lines are the road edges determined by our proposed method. The results show that using the projection-based technique for delineation of the roads, and refining the round corners, brings the extracted road edges close to their real location the ground truth.

For a quantitative evaluation, we compute the Root Mean Square (RMS) error of the roads' edges extracted by our method comparing to the ground truth, as proposed in (Heipke et al., 1997). Table 1 shows the average RMS errors after extracting the road segments, and after delineating the roads and refining the round corners. As a further evaluation, we measure the average angular displacement of the extracted road edges from the ground truth, which is only 0.3 degrees. Considering the both evaluations, our method is able to precisely localize and delineate the roads in the scene.

---

[1] This data has been provided by the Belgian Royal Military Academy and presented at 2015 IGRSS Data Fusion Contest.

5. CONCLUSION

In this paper, we propose an unsupervised automatic road extraction method which uses the integration of the LiDAR and RGB aerial image of a scene. As an integration technique, PCA is used on the 4 available information bands, namely R, G, B, and height. The most road discriminating principal component is then used in Hough transform to extract the possible locations of the roads as straight lines. Using these lines, the scene is segmented into various regions. The segments containing a road part are then detected and merged to shape the road network in the scene. The roads are then delineated using a projection-based method, and their round corners are further refined. For a quantitative evaluation, the extracted roads are compared to ground truth data using RMS and angular displacement measures. Experimental results demonstrate that our proposed method detects and delineates the roads precisely in a given scene.

ACKNOWLEDGEMENTS

The authors would like to thank A. Murillo Montes de Oca who assisted in the proofreading of this paper.

REFERENCES

Anil, P. and Natarajan, S., 2010. A novel approach using active contour model for semi-automatic road extraction from high resolution satellite imagery. In: International Conference on Machine Learning and Computing (ICMLC), pp. 263–266.

Bae, Y., Lee, W.-H., Choi, Y., Jeon, Y. W. and Ra, J. B., 2015. Automatic road extraction from remote sensing images based on a normalized second derivative map. IEEE Geoscience and Remote Sensing Letters 12(9), pp. 1858–1862.

Douglas, D. H. and Peucker, T. K., 2011. Algorithms for the reduction of the number of points required to represent a digitized line or its caricature. John Wiley & Sons, Ltd, pp. 15–28.

Duda, R. O. and Hart, P. E., 1972. Use of the hough transformation to detect lines and curves in pictures. Communications of the ACM 15(1), pp. 11–15.

Heipke, C., Mayer, H., Wiedemann, C. and Jamet, O., 1997. Evaluation of automatic road extraction. International Archives of Photogrammetry and Remote Sensing 32(3 SECT 4W2), pp. 151–160.

Hotelling, H., 1933. Analysis of a complex of statistical variables into principal components. Journal of Education Psychics.

Hough, P., 1962. Method and Means for Recognizing Complex Patterns. U.S. Patent 3.069.654.

Miao, Z., Wang, B., Shi, W. and Zhang, H., 2014. A semi-automatic method for road centerline extraction from vhr images. IEEE Geoscience and Remote Sensing Letters 11(11), pp. 1856–1860.

Pearson, K., 1901. On lines and planes of closest fit to systems of points in space. Philosophical Magazine 2(6), pp. 559–572.

Samadzadegan, F., Hahn, M. and Bigdeli, B., 2009. Automatic road extraction from lidar data based on classifier fusion. In: Joint Urban Remote Sensing Event, pp. 1–6.

Shi, W., Miao, Z. and Debayle, J., 2014. An integrated method for urban main-road centerline extraction from optical remotely sensed imagery. IEEE Transactions on Geoscience and Remote Sensing 52(6), pp. 3359–3372.

Singh, P. P. and Garg, R. D., 2014. A two-stage framework for road extraction from high-resolution satellite images by using prominent features of impervious surfaces. International Journal of Remote Sensing 35(24), pp. 8074–8107.

10

A SURVEY OF SMART ELECTRICAL BOARDS IN UBIQUITOUS SENSOR NETWORKS FOR GEOMATICS APPLICATIONS

S.M.R. Moosavi [a, *], A. Sadeghi-Niaraki [b]

[a] Dept. of Geomatic Engineering, Islamic Azad University, Larestan, Iran. - hseyyed@gmail.com
[b] GIS Dept., Geoinformation Technology Center of Excellence, Faculty of Geodesy&Geomatics Eng, K.N.Toosi Univ. of Tech., Tehran, Iran. - a.sadeghi@kntu.ac.ir

KEY WORDS: Smart boards, ubiquitous sensor networks, Arduino, Raspberry Pi, physical computing, ubiquitous GIS

ABSTRACT

Nowadays more advanced sensor networks in various fields are developed. There are lots of online sensors spreading around the world. Sensor networks have been used in Geospatial Information Systems (GIS) since sensor networks have expanded. Health monitoring, environmental monitoring, traffic monitoring, etc, are the examples of its applications in Geomatics. Sensor network is an infrastructure comprised of sensing (measuring), computing, and communication elements that gives an administrator the ability to instrument, observe, and react to events and phenomena in a specified environment. This paper describes about development boards which can be used in sensor networks and their applications in Geomatics and their role in wireless sensor networks and also a comparison between various types of boards. Boards that are discussed in this paper are Arduino, Raspberry Pi, Beagle board, Cubieboard. The Boards because of their great potential are also known as single board computers. This paper is organized in four phases: First, Reviewing on ubiquitous computing and sensor networks. Second, introducing of some electrical boards. Then, defining some criterions for comparison. Finally, comparing the Ubiquitous boards.

1. INTRODUCTION

Nowadays more advanced sensor networks in various fields are developed. There are lots of online sensors spreading around the world. Sensor networks have been used in Geospatial Information Systems (GIS) since sensor networks have expanded. Health monitoring (Kemis et al., 2012), environmental monitoring (Son et al., 2006) (Choi et al., 2015), traffic monitoring (Costanzo, 2013), etc, are the examples of its applications in Geomatics. A sensor is a device that receives a stimulus and responds with an electrical signal (Fraden, 2010). Sensor is a transducer which purpose is to sense some characteristics of its environs. It receives signal or energy from physical environment to make it readable and provide corresponding output information. The readable output of electrical sensors is the variation of output voltage which it has converted to an understandable data to human. Sensors are connected to the Internet or an internal network. They send their location and the recorded or online environmental data. Sensor can be controlled remotely and human interaction with the physical environment can be provided. Sensors' Applications include manufacturing and machinery, airplanes and aerospace, cars, medicine and robotics.it is also included in our day-to-day life.

2. UBIQUITOUS COMPUTING

The phrase "Ubiquitous Computing" has invented by Mark Weiser around 1988, during his tenure as Chief Technologist of the Xerox Palo Alto Research Center (PARC). Weiser wrote some of the earliest papers both alone and with PARC Director and Chief Scientist John Seely Brown on the subject, largely defining it and sketching out its major concerns (Weiser, 1996).

Ubiquitous computing, or ubicomp, is the term given to the third era of modern computing. The first era was defined by the mainframe computer. Second, is the era of the PC, a personal computer used by one person and dedicated to them. The third era, ubiquitous computing, representative of present time, is characterized by explosion of small networked portable computer products in the form of smart phones, PDAs, and embedded computers built into many of the devices, Figure 1 (Krumm, 2010). Ubiquitous computing is the method of enhancing computer used by making many computers available throughout the physical environment, but making them effectively invisible to the user. One of the positive effects from ubiquitous computing is people who do not have skills use the computer and people with the physical lack (the defect) could continue to use the computer for all the needs. Ubiquitous technology means the ability to access to any services and gathering information in any location such as country, city, workplace and even home, any time, by anyone, by any device and in any network (LAN, Wireless etc.). In ubiquitous perspective every element of real world can communicate together.

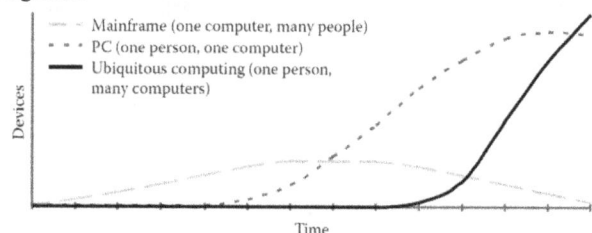

Figure 1. Graph conceptually portraying three eras of modern computing.

---

3. SENSOR NETWORK

Sensor network is an infrastructure comprised of sensing (measuring), computing, and communication elements that gives an administrator the ability to instrument, observe, and react to events and phenomena in a specified environment. The administrator typically is a civil, governmental, commercial, or industrial entity. The environment can be the physical world, a biological system, or an information technology (IT) framework (Sohraby et al., 2007). The technology for sensing and control includes electric and magnetic field sensors; radio-wave frequency sensors; optical-, electro-optic, and infrared sensors; radars; lasers; location/navigation sensors; seismic and pressure wave sensors; environmental parameter sensors (e.g., wind, humidity, heat). Today's sensors can be described as "smart" inexpensive devices equipped with multiple on-board sensing elements; they are low-cost, low-power untethered multifunctional nodes that are logically homed to a central sink node.

4. USN, U-GIS

In order to have Ubiquitous GIS (UBGIS), an integration of Ubiquitous computing and traditional GIS is necessary. By using UBGIS, any user or any system through any communication device can access to geographic information and applications at any time and any place. The dynamic context of the user is playing the major role of UBGI. For achieving the dynamic context in UBGIS, ubiquitous computing concepts should employ Ubiquitous Sensor Networks (USNs) to collect any data on any environmental parameter. Dynamic context is defined as dynamic location and the identity of any object, people and parameter in environment. Ubiquitous sensor networks (USN) consist multiple nodes, each node can independently communicate with a server or using wireless technology that connects nodes together and they all connect to a server through a router node. In addition of sensor, there are hardwares such as electronic board, network modules in order to process data and communicate to the server. Electronic boards have the task of pre-processing of the sensor's output. By developing them, they will have the ability to connect to a network. Boards with the programming ability can analyse the sensor's output and affects its surroundings by controlling lights, motors, and other actuators. Actuators are things like lights and LEDs, speakers, motors, and displays.

5. PHYSICAL COMPUTING

Physical computing is interactive physical systems which can sense and respond to the analog world. It's a creative framework for understanding human beings' relationship to the digital world, Figure 2. Interaction is "A cyclic process in which two actors alternately listen, think, and speak" (Crawford, 2003). "Interactive" is a fuzzy term, and often misused for all kinds of purposes and most physical computing projects can be broken down into these same three stages: listening, thinking, and speaking—or, in computer terms: input, processing, and output (O'Sullican and Igoe, 2004).

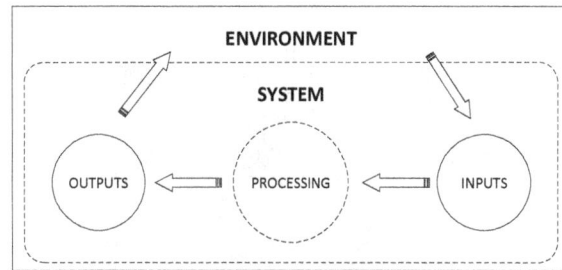

Figure 2. Physical Computing

6. ELECTRONIC BOARDS

For pre-processing sensors' output and communication with the server or a router node, an electronic circuit is necessary. In the past for any type of sensor and application, a circuit should be designed. But by recent development in electronic boards, a lot of boards have seen that can connect to variety of sensors and other equipment such as wireless connectivity modules or GSM modules. These boards have called "Single Board Microcontroller". Single board microcontroller is a microcontroller that built onto the single printed circuit. This board contain all necessary element to do a task such as: microprocessor, RAM, I/O circuits, etc. The intention is that the board is immediately useful to an application developer, without spending much time and effort in developing the controller hardware. Some of single board microcontrollers because of their great potential, computing and connectivity options also known as single board computers. There are types of boards some of them kind of open source computing hardwares that follow the open community to let the users to re-design and develop them. This paper discuss about some boards such as: Arduino, Raspberry Pi, Beagle board and Cubieboard.

6.1 Arduino

Arduino is a tool for making computers that can sense and control more of the physical world than the desktop computer. It's an open-source physical computing platform based on a simple microcontroller board, and a development environment for writing software for the board. The project is based on a family of microcontroller board designs manufactured primarily by SmartProjects in Italy and also have cloned by several other vendors. An Arduino board consists an 8, 16bit or 32bit *Atmel* AVR microcontroller with other components to facilitate the incorporation with other circuits. One of the advantages of Arduino is its standard connectors that let user to connect board to variety of modules known as *shields*. The communication of some shields with Arduino board can be done by directly over various pins, there are also some shield that individually are addressable via serial bus. So many shields can be used in parallel. Early Arduino boards used to be programmed via RS232 ports, however current boards can be programmed via USB and also some unofficial models programmed over a Bluetooth connection. The Arduino board is where the code has written and executed. The board can only control and respond to electricity, so specific components are attached to it to enable it to interact with the real world. These components can be sensors, which convert some aspect of the physical world to electricity so that the board can sense it, or actuators, which get electricity from the board and convert it into something that changes the world. Examples of sensors include switches, accelerometers, and ultrasound distance sensors. Arduino board is real-time which means no operating system installed on them, this capability is useful for fast responses (Margolis, 2012).

Figure 3. Arduino Uno

Figure 4. Arduino Gemma (wearable)

6.2 Raspberry Pi

The Raspberry Pi is a credit-card-sized computer (Single-Board Computer) created by the non-profit Raspberry Pi Foundation in the UK. It all started when a chap named Eben Upton (now an employee at Broadcom) got together with his colleagues at the University of Cambridge's computer laboratory, to discuss how they could bring back the kind of simple programming and experimentation that was widespread among kids in the 1980s on home computers such as the BBC Micro, ZX Spectrum, and Commodore 64 (Sjogelid, 2013). The Raspberry Pi is built off the back of the Broadcom BCM2835 and BCM2836 SoCs. These SoCs are multimedia application processors geared towards mobile and embedded devices which includes *ARM* processor, GPU, 512Mb to 1 GB RAM, and MicroSD socket for boot media and persistent storage (Dennis, 2013). Operating systems like Linux and Windows 10 can be installed on Raspberry Pi. Raspberry Pi main programming language is python and also supports: Java, C, C++ and Ruby.

Figure 5. Raspberry Pi A+

Figure 6. Raspberry Pi 2 B

6.3 BeagleBoard

The BeagleBoard is an open-source single-board computer produced by Texas Instruments in association with some other companies. The BeagleBoard also designed with open source software. The board base on ARM architecture and it's included GPU to provide 2D and 3D rendering that supports OpenGL. Some models have both video outs which provided through S-Video and HDMI connections. A single SD card also included that supports USB On-The-Go.

Figure 7. BeagleBone Black

6.4 CubieBoard

CubieBoard is a Single-Board-Computer which made in china. It can run Android 4 ICS, Ubuntu 12.04 desktop, Fedora 19 ARM Remix desktop, Archlinux ARM, a basic Debian server via the Cubian distribution, or OpenBSD.

Figure 8. CubieBoard 2

Figure 9. CubieBoard 4 (CC A80)

7. CRITERIONS

To choose an appropriate board some criterions are necessary such as price, power consumption, expandability and compatibility, ease of use and development, information resource availability for beginners. For better understanding of power consumption related to Table. 1: Voltage is measured in units of volts (V). With the symbol V, it is the measure of potential in a circuit. The oft used analogy is water - voltage then becomes the height from which the water is flowing or falling. Greater height, more potential energy from the water flow, similarly greater voltage, more potential energy. Current is measured in units of amperes (A), usually abbreviated to amps, and is the rate of flow of electric charge past a point. The symbol used for current is I. To continue the water analogy, current might be considered the width/depth of the water flow. Power is the amount of energy in a system, and is measured in units of watts (W). With the symbol P, in quantitative terms for an electrical circuit, it is equal to current × voltage. Hence, $P = I \times V$. To round out the water analogy, there is a lot more power in Niagara Falls than the downpipe on the side of a house (Oxer and Blemings, 2010). Price is an important criterion because ubiquitous networks are implemented in a large scale. Another important criterion is power consumption, need to know the places that sensor should be installed. There is lack of power in some places, so if power consumption of hardware were low, the boards can be work by small solar cells which solar cells charge batteries in day to save energy for board in night. More power consumption needs bigger solar cells and bigger solar cells get more expensive. Expandability in electronic boards means the board has the ability to support more add-on shields and modules (e.g. GSM shield (Costanzo, 2013), Display controllers, Motor controllers, Wi-Fi module such as Xbee (Harikrishnan, 2015)) boards can do more tasks with shields and modules and compatibility with them. Generally open source boards are easy to use and develop and in case of open source, the hardware and software can be modified by project goal.

Table 1. Board Comparison

Brand	Model	Size (mm)	Cost (US$)	CPU		Flash Memory	Operating Voltage	Power Consumption	Operating System
				Core	Speed (MHz)				
Arduino	Uno	68.6 x 53.4	24.95	1	16	32 KB	5V	< 0.5 W	-
	Pro	52.07x53.34	14.95	1	8	16 KB	3.3V	< 0.5 W	-
	Gemma	⌀ 27.98 Dia	9.95	1	8	8 KB	3.3V	< 0.5 W	-
	LilyPad	⌀ 51 Dia	19.95	1	8	16 KB	2.7V	< 0.5 W	-
	Nano	45x18	4.25	1	16	16 KB	5V	< 1 W	-
Raspberry Pi	Pi 1 A+	65x56.5	25	1	700	MicroSD	5V	1 W	Linux
	Pi 1 B+	86x56	39.95	1	700	MicroSD	5V	3 W	Linux
	Pi 2 B	86x56	41.95	4	700	MicroSD	5V	4 W	Linux/Windows 10
BeagleBoard	Beaglebone Black	86.40x53.3	50	1	1000	4 GB	5V	2.3 W	Linux/Android/WinCE
Cubie Board	Ver. 2	100x60	70	2	1000	4 GB	5V	10-15 W	Linux/Android
	CC A80	111x111	135	8	1300	8 GB	5V	10-15 W	Linux/Android

8. CONCLUSION

Based on comparison table boards with lowest prices are Arduino, Arduino boards also use low energy consumption, when the device energy consumption is low there is room to add more sensor and add-ons. Also lack of operating system on Arduino boards is a benefit for real-time computing and responses. In case of small volume of data that received from sensors, boards do not need much memory and CPU speed. BeagleBoard and Cubie boards have great performances but due to high price and high power consumption are not good choices for Ubiquitous sensor networks, but can be used as a server. Raspberry Pi boards because they host an operating system are good for large computing and data logging and in case of factory ready Ethernet module, better communication to network is available. Raspberry Pi also can connect to multiple Arduino boards (with Wi-Fi or other methods) and receive and analyse the data. In case of existence of good information resources, ease of software and hardware development, Arduino and Raspberry Pi as smart boards are considered as a good choice to build Ubiquitous sensor networks (Ferdoush and Li, 2014). Most researches in ubiquitous GIS have not been tested in the real world (especially in Iran), due to the lack of electrical knowledge of Geomatic students. These boards called DIY (Do-It-Yourself) boards, the operation is simple and lots of resources are available.

REFERENCES

Choi, Y., Hong, S., Joe, I., 2015. A Fire Evacuation Guidance System Based on Ubiquitous Sensor Networks, in: Advanced Multimedia and Ubiquitous Engineering. Springer, pp. 217–222.

Costanzo, A., 2013. An arduino based system provided with GPS/GPRS shield for real time monitoring of traffic flows. 2013 7th Int. Conf. Appl. Inf. Commun. Technol. pp. 1–5.

Crawford, C., 2003. The Art of Interactive Design, No Starch Press. San Francisco, pp. 5.

Dennis, A.K., 2013. Raspberry Pi Home Automation with Arduino. Packt Publishing, Brimingham, pp. 9-11.

Ferdoush, S., Li, X., 2014. Wireless Sensor Network System Design Using Raspberry Pi and Arduino for Environmental Monitoring Applications. Procedia Comput. Sci. 34, pp. 103–110.

Fraden, J., 2010. Handbook of Modern Sensors Physics, Designs, and Applications, Fourth Edition, Springer. New York, pp. 2.

Harikrishnan, R., 2015. An Integrated Xbee Arduino and Differential Evolution Approach for Localization in Wireless Sensor Networks. Procedia Comput. Sci. 48, pp. 447–453.

Kemis, H., Bruce, N., Ping, W., Antonio, T., Gook, L.B., Lee, H.J., 2012. Healthcare monitoring application in ubiquitous sensor network: Design and implementation based on pulse sensor with arduino, pp.34–38.

Krumm, J., 2010. Ubiquitous Computing Fundamentals. Taylor & Francis, New York, pp. 2-3.

Margolis, M., 2012. Arduino Cookbook, O'Reilly. Cambridge. pp. 2.

O'Sullican, D., Igoe, T., 2004. Physical Computing, Thomson. Boston, pp. xx.

Oxer, J., Blemings, H., 2010. Practical Arduino: Cool Projects for Open Source Hardwar. Apress, New York, pp. 2

Sjogelid, S., 2013. Raspberry Pi for Secret Agents. Packt Publishing, Brimingham, pp. 7-8.

Sohraby, K., Minoli, D., Znati, T., 2007. Wireless Sensor Networks: Technology, Protocols, and Applications. John Wiley & Sons, New Jersey, pp. 1-2.

Son, B., Her, Y., Kim, J., 2006. A design and implementation of forest-fires surveillance system based on wireless sensor networks for South Korea mountains. Int. J. Comput. Sci. Netw. Secur. 6, pp. 124–130.

Weiser, M., 1996. Ubiquitous Computing [WWW Document]. URL http://www.ubiq.com (accessed 4.9.15).

11

DEVELOPMENT OF AN OPEN-SOURCE AUTOMATIC DEFORMATION MONITORING SYSTEM FOR GEODETICAL AND GEOTECHNICAL MEASUREMENTS

Ph. Engel[a], B. Schweimler[a]*

[a] Faculty of Landscape Sciences and Geomatics, Hochschule Neubrandenburg, University of Applied Sciences,
(pengel, schweimler)@hs-nb.de

KEY WORDS: automatic deformation monitoring, engineering surveying, software development, open source

ABSTRACT

The deformation monitoring of structures and buildings is an important task field of modern engineering surveying, ensuring the standing and reliability of supervised objects over a long period. Several commercial hardware and software solutions for the realization of such monitoring measurements are available on the market. In addition to them, a research team at the Neubrandenburg University of Applied Sciences (NUAS) is actively developing a software package for monitoring purposes in geodesy and geotechnics, which is distributed under an open source licence and free of charge. The task of managing an open source project is well-known in computer science, but it is fairly new in a geodetic context. This paper contributes to that issue by detailing applications, frameworks, and interfaces for the design and implementation of open hardware and software solutions for sensor control, sensor networks, and data management in automatic deformation monitoring. It will be discussed how the development effort of networked applications can be reduced by using free programming tools, cloud computing technologies, and rapid prototyping methods

1. INTRODUCTION

The automation of deformation measurements by means of computer systems started early. In the 1980s first efforts where made concerning the remote control of geodetical and geotechnical sensors, using personal computers and analog modems (Pelzer, 1988). Since then, many software and hardware systems for deformation monitoring have been developed in universities and corporations. Several proprietary deformation monitoring systems exist, either made by the manufacturers of sensors or by engineering offices for their own businesses. These solutions are often designed as isolated applications. The supported sensors are limited to specific types or producers, and interfaces for data exchange and remote control are not open to third party products to ensure a vendor lock-in.

Automated deformation measurements gain importance due to the increasing building density in public space and the advanced age of existing infrastructure. At the same time, the costs of sensors and information technology in general decrease, which may also facilitate their widespread usage in geodesy and geotechnics.

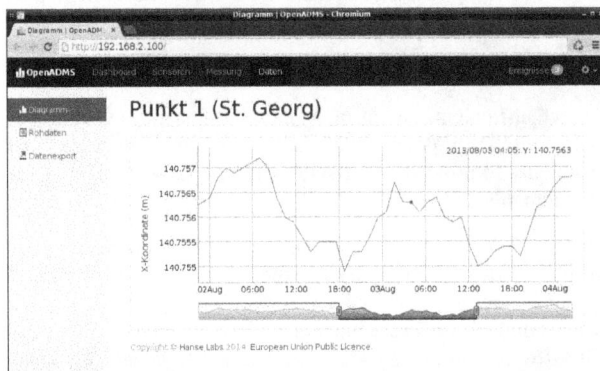

Figure 1: Web-based user interface of the upcoming version of OpenADMS

*Corresponding author

Despite the progress in terms of sensors and hardware appliances for remote control, an open software platform for deformation monitoring is still not available. For this reason, efforts are being made at NUAS to develop a universal open source monitoring solution with platform-independence, secure remote control functions using standardised interfaces for data exchange, and compatibility to cloud computing environments.

2. THE DABAMOS MONITORING SYSTEM

The development of an automatic deformation monitoring system at the Faculty of Landscape Sciences and Geomatics started in 2009 as a students project to simply control geodetical sensors remotely. The project was later called *Datenbank-orientiertes Monitoring- und Analyse-System* (DABAMOS). The first version of the software has been written entirely in Java SE and could be run on conventional personal computers without any special requirements, except the Java Runtime Environment. The integrated remote control interface allows users to start and stop the monitoring process through a TCP/IP network. All sensor data is stores in an object-oriented *db4o* database. Later on, the first steps towards a client–server architecture were taken with the development of an enterprise application written in Java EE for the storage and visualisation of monitoring data on Internet servers. But this application has never left prototype stage.

Coinciding with a rewrite of the software base in the Go programming language in 2013 the decision was made to publish the whole project under an open source licence. The new software was designed to be operated through a Web front-end with modern browsers (fig. 1) and to support ARM-based single-board computers, like the Raspberry Pi or the BeagleBone. These single-board computers can be used to control connected sensors.

The DABAMOS project consists of three parts (fig. 2):

OpenADMS: The *Open Automatic Deformation Monitoring System* is a platform-independent software for sensor control. The measured values of geodetical or geotechnical sensors are stored locally in a NoSQL database and then transmitted to an FTP server or an OpenSDMS instance.

OpenSDMS: The *Open Spatial Data Management Service* is used for the storage, analysis, and visualisation of sensor data in cloud computing environments. OpenSDMS collects and manages data from an arbitrary number of OpenADMS clients. The service is still in conceptual phase.

Middleware: In most cases, sensors have to be connected to a local computer system to be controlled remotely. Beside personal or industrial computers also embedded computers can be used to run OpenADMS. Such a "middleware" has to provide Internet access via Ethernet, WiFi, or wireless networks (3G/4G) and is possibly equipped with an uninterruptible power supply unit.

3. THE POTENTIAL OF OPEN SOURCE LICENCES

Free and open source development models sustain various software and hardware projects and make it easier for groups of people, who may not be acquainted, to work together in a collaborative way (Laurent, 2004).

The open source model leads to many advantages for users and developers of such projects. The fact that open source software is made available free of charge or at a low cost is thereby only the first perceived one, while even not exclusive to open source. More important is the availability of the source code and the right to modify it. It makes it feasible to improve the software by anyone and to port the application to new hardware or to adapt it to further conditions, which were not considered by the original authors.

Another advantage of the open source model is the possibility of splitting a project into subsequent ones ("forking"). The original project serves as a basis for ones with different aims. In case of an automatic deformation monitoring system this means that users can take the source code to develop software for their own purposes or to integrate it into a second, already existing project.

The DABAMOS software is published under the European Union Public Licence (EUPL) v. 1.1. The EUPL has been created on the initiative of the European Commission and is the first European Free/Open Source Software (F/OSS) licence, available in 22 linguistic versions. The EUPL is compatible to the popular GNU GPLv2 and shares its "copyleft" clause: derivated works of EUPL-licenced software must be published under the same or a compatible licence.

The decision to publish the DABAMOS software under the EUPL was pursued by the aim to establish a widely accepted and used platform for all kind of geodetical and geotechnical monitoring tasks. This is why all future work will be made freely available on the project website.

4. SOFTWARE DEVELOPMENT METHODS

Development teams often face several problems with changing and increasing requirements. This is even more of an issue when developing open source software. Traditional methods like the waterfall model are too rigid and prescriptive to handle small distributed teams and many different use cases. Software developed this way is only released when it is feature complete. This means, all of the functionality has to be developed when deploying the software. This is not an option for an open source project, as it is self-organised and often confronted with changing requirements, depending on the current user or contributor. Agile development methods can solve this problem.

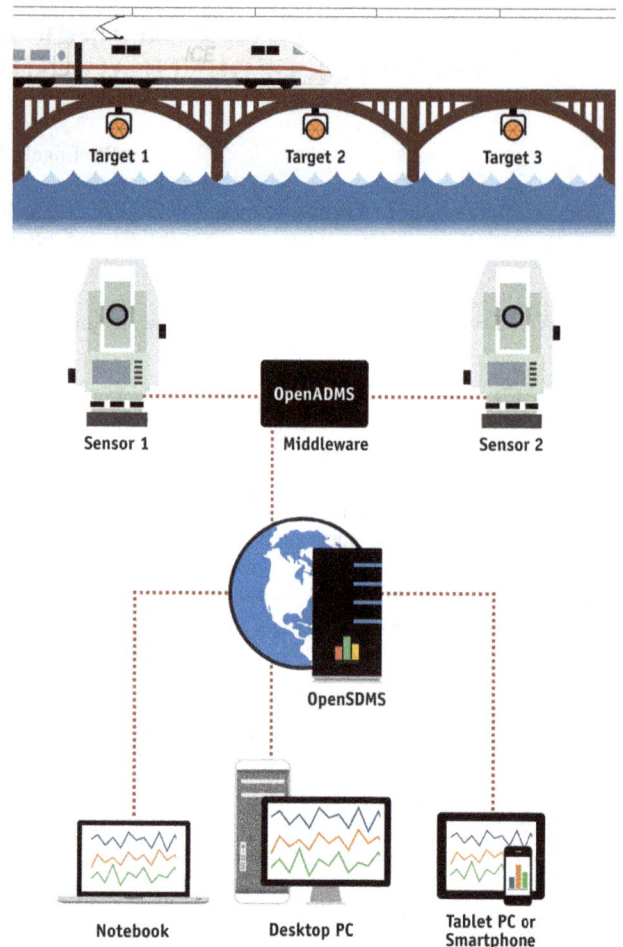

Figure 2: Schematic illustration of the DABAMOS platform

4.1 Agile development

Agile development offers a lightweight framework to feature evolving solutions. Those can grow with the problem and can adapt quickly to new situations. Agile development teams are generally self-organised and cross-functional. Many of the implemented methods focus on the user, forcing teams to deliver software early and continuous in frequent intervals. For measuring the work process, working software is almost all that matters. The principles are formulated in the Agile Manifesto (Agile, 2001).

4.2 Continuous Delivery

An actual implementation of Agile development is Continuous Delivery (CD). It describes a collection of tools, processes, and techniques to provide high quality software using a high degree of automation.

The actual development practice is the integration of source code of an application into a shared repository of a version control system at least daily. The code is provided to a processing chain, which builds the software immediately for all targeted platforms. All software tests are included and executed automatically by the processing chain.

By convention, every commit has to build on every platform. If a build fails, the developer gets an immediate feedback. The occurring error has to be fixed as soon as possible, which forces developers to commit small changes to the repository. This reduces the time spend for debugging and backtracking significantly, as problems are detected early and can be located more easily. The outcome of the processing chain is the built bug-free software. The executables for the targeted platforms should be easy accessible and automatically be deployed to clients. (Humble/Farley, 2010)

4.3 DevOps

To build reliable software, developers have to understand what the software is used for and more importantly how it is used. Agile development describes procedures between developers. It does not overcome breaking points between the development and the actual operation of software.

DevOps, a clipped compound of Development and Operations, is an approach to build a bridge between developers and system administrators who run an application in a live environment. It solely aims for the delivery of reliable software and has to be implemented in a CD context. DevOps teams can fulfil all tasks needed, from developing to testing and finally administrating the software.

The communication is crucial for all team members to understand the underlying business logic of the software and to exchange information with actual users. To be part of a DevOps team, developers and operators need a wide skill set, as DevOps follows a multi-disciplinary approach. (Swartout, 2012)

The described paradigms and techniques are a small cut off of open source development methods. As the development includes many different persons, it is vital for the success of an open source project to react flexible to changing requirements and new situations. The following chapter describes the actual implementation of the described methods to manage and build DABAMOS.

5. MANAGING AND BUILDING AN OPEN SOURCE SOFTWARE PROJECT

A range of hardware and software tools for the management and development of open source engineering projects is available. The following section describes some of the tools used for the implementation of applications and appliances in the DABAMOS project.

5.1 Hardware

The control of sensors and the management of their collected data do require not only specialised software products but also hardware appliances for data transmission, persistence, and interaction. New technologies for rapid prototyping can shorten the development time of hardware solutions.

The analog or digital data of sensors can be read by several secondary devices, like personal computers or microcontrollers. For further processing, analog signals have to be turned into digital values first by using analog-to-digital converters (ADCs). Industrial-grade ADC modules are equipped with a serial port for data transmission to a connected computer, whereas geodetical sensors often have an internal digital interface for data exchange and remote control. Conventional personal computers are adequate for the communication with ADC modules and sensors but are often not suitable for a usage in harsh environmental conditions, such as dirt, heat, cold, wet, or vibration. For a higher level of resistance against environmental impact industrial or embedded computers are used.

In addition, so called single-board computers can also act as sensor control instances. The low-priced systems combine all required components (i. a., processor, memory, I/O, and permanent storage) of a computer on a single small-sized circuit board. Most single-board computers are based on the ARM or the MIPS processor architecture and can be run with GNU/Linux or open source Unix operating systems. Popular single-board computers are, beside others, Rasperry Pi (fig. 3), BeagleBone, and Cubieboard. With prices under 100 Euros per board they are a low-cost alternative to industrial and conventional embedded computers and a convenient basis for the rapid prototyping of sensor control units.

Figure 3: Low-cost single-board computer "Raspberry Pi"

For many applications custom-built cases, moutings, and adapters are required. 3D printers can be used to manufacture single units and small series of such physical components. The work pieces are constructed by 3D printers in layers of liquid or solid material, commonly plastics. These pieces are produced from a three-dimensional CAD model and can be of almost any shape. The Faculty of Landscape Sciences and Geomatics of the Neubrandenburg University of Applied Sciences owns a MakerBot Replicator Z18 3D printer for research and educational purposes (fig. 4). The printing is done in fused deposition modeling technique, where heated thermo-plastics are deposited on a movable table (Grimm, 2004). The printing resolution of 0.1 mm allows the precise construction of fine structures. Depending on the size of the work piece, the printing process can take several hours. Because the used polylactic acid filament is not weatherproof, the printed work pieces should be used indoor only and not be exposed to water and excessive UV radiation. The design of the models can be done in most CAD/CAAD software capable of exporting files in STL format, like Autodesk Inventor, SketchUp, or the open source programs FreeCAD and Blender.

5.2 Software

Modern programming languages and their tools are often published under an open source licence and therefore also suited to manage and build open source projects. This reduces the dependency to single vendors, as it is, for instance, the case for the .NET platform of Microsoft.

5.3 Programming language

The software of the DABAMOS project is mostly written in the Go programming language, which is published under a BSD-style licence. The source code of Go is hosted publicly on GitHub and all tools to build and test the software are available for free. The language enables developers to construct flexible and modular programs. It provides concurrency mechanisms, garbage collection, and run-time reflections. (Go, 2015)

Figure 4: 3D printer "MakerBot Replicator Z18"

The language supports common computer architectures, including ARM, and most modern operating systems. It is not necessary to build the code directly on the specified computer platform, as it can easily be compiled by setting up flags in the development environment (cross-compiling). The languages concurrency mechanisms support techniques that are actively used in the DABAMOS software. It provides so called "Go routines", which are similar to threads in other programming languages. Additionally, there are tools to synchronise and to handle the communication between Go routines.

Many processes are executed simultaneously in a monitoring system. The Go programming language makes it possible to manage those processes. As DABAMOS is a distributed platform, the development also benefits from integrated networking libraries, which are related to open standards, like Transport Layer Security (TLS) for encrypted communication, as well as Extensible Markup Language (XML) and JavaScript Object Notation (JSON) for data exchange.

5.4 Version control system

The source code of the DABAMOS software is managed with Git, an open source version control system. Git can handle software projects of any size and supports Continuous Delivery methods, especially by providing tools for tagging the source code. The typical workflow starts with the addition of a source code file to the local workspace. The change is described by a message and finally committed to the local repository. The commit is afterwards pushed to a central remote repository. As there is often more than one contributor, it is common to use branches, isolated from each other, for the development of features. After the implementation is finished, a branch can be merged into the master branch, which is the default branch of a Git repository.

This way, many developers can evolve complex software without interfering each other. (Preiel, Stachmann, 2013)

GitHub, a repository hosting service, is used as the remote repository for the DABAMOS software which is free of charge for open source projects. The service adds additional so called "social coding" features to Git. Developers can set up several collaboration functions to manage their project and its contributors. (GitHub, 2015)

5.5 Networks

For the management and organisation of the DABAMOS project the team uses an Intel Xeon server with five hard disks, running FreeBSD 10, an open source Unix operating system. The collocation of the server is provided by the University, while the Internet access is available through the X-WiN backbone of the national research and education network DFN. The server is mainly used for serving the project website and the project documentation, for file exchange, project management, continuous integration, and instant messaging, but also as a SSH gateway for remote sensor nodes.

The project network infrastructure has also been equipped with a virtual private network (VPN) hardware gateway to establish encrypted tunnel connections based on IPsec/L2TP. A VPN can be used to integrate remote sensor nodes into a local network, which makes it possible to establish serial data connections through the Internet. To access sensors remotely through an existing TCP/IP network so called "serial device servers" are an easy to use option. The serial device servers are connected to a VPN hardware client which establishes an encrypted tunnel connection to the remote VPN gateway. In this way the local sensors become part of the whole VPN network. They can be accessed and controlled by any privileged user inside the VPN network, even if the user is far away from the sensor location.

The establishment of encrypted tunnel connections is not limited to the operation of hardware appliances. Software-based alternatives, like SSH and OpenVPN, also exist and do not require the acquisition of hardware solutions.

5.6 Documentation

An important task in software development is the writing of a comprehensive documentation for end users and developers. The documentation of the source code is often done by the respective programmer using special comments and tags within the source code. Depending on the programming language, the developer can then create a HyperText-based source code documentation with documentation generators, like "godoc" for Go, or "Javadoc" for Java.

The end user documentation can be created in several ways. In the past, user manuals were often available as a printed work only. But this kind of offline documentation is no more appropriate, since printing and distribution of user handbooks is both time-consuming and expensive. A further disadvantage of printed manuals is that the user has to receive an updated copy for each new software version, which makes the update process more complex.

Therefore, a user manual is often available twice: as a printed hard copy as well as a digital copy in the Portable Document Format (PDF). In comparison with HyperText-based online manuals PDF files are still geared to be printed on paper, which is why they are based on a fixed format while being less interactive for the user.

For a modest online publication of user manuals the HyperText Markup Language (HTML) is more suitable. The manual is simply written or transformed to HTML and uploaded to a web server, so that all users have access to it. Apart from HTML, further markup languages exist, some particularly for the creation of online manuals. These languages are less complex than HTML and therefore easier to learn (e. g., Markdown, reStructured-Text, Textile, and AsciiDoc).

A further alternative to printed books are wikis. They are Web-based applications which allow users to work on content in a collaborative way. The content is structured by a markup language and later exported to HTML. All changes on the content are tracked by the internal version control system of the wiki and can be set back anytime. Popular open source wiki systems are MediaWiki, which is also used for the Wikipedia encyclopedia, and Doku Wiki. Within the DABAMOS project a wiki, based on the DokuWiki software, is used for end user and developer documentation.

Figure 5: Monitoring station with Leica TM30 totalstation and embedded sensor control unit, based on a Raspberry Pi

6.   A PLATFORM FOR GEO-SENSOR NETWORKS IN CLOUD-COMPUTING ENVIRONMENTS

A geo-sensor network consists of sensors with a spatial reference, connected to wired or wireless transmission technologies. A single sensor node is generally formed by the sensor itself and a computer to request, store, and forward collected measurement data. The sensor can transmit the data on its own without caching it on a computer. Usually, information from all sensor nodes is collected in a central storage and processed afterwards.

The central storage of DABAMOS is called OpenSDMS. It is integrated into geo-sensor network and connects all system components. Conceptualised as a platform, OpenSDMS is used to manage monitoring projects and configure sensor nodes centrally. The platform has an expandable toolset to analyse the collected data and to create reports summarising the results.

Core features of such a platform are accessibility and scalability. As the grade of utilisation is unknown before operating, it is important for the system to scale dynamically with changing requirements. This is a typical scenario for cloud computing environments. The cloud represents an abstract model of configurable computing resources, which are accessed over a network.

The management of the cloud infrastructure is handled by a service provider. For DABAMOS the service model "Platform as a Service" (PaaS) is used. PaaS provides the cloud infrastructure and additional tools and services, supporting the operation of the software. The user does not manage nor control the underlying resources (e. g., network, servers). The application installed in such an environment can acquire resources to match higher utilisation as well as release resources when being confronted with lesser load. This process is automated.

The actual application, deployed in the cloud, consists of two components: the application programming interface (API) and the graphical user interface (GUI). The API provides the general functionality to execute specific tasks, the so called business logic. It is based on "Representational State Transfer" (REST), a software architecture model for distributed systems. REST is designed for machine-to-machine communication, which makes it applicable for the automatic communication between sensor nodes and OpenSDMS. The architecture uses the HyperText Transfer Protocol (HTTP) and its methods to access resources via a uniform resource identifier (URI).

When implementing REST with the constraint of "Hypermedia as the Engine of Application State" (HATEOAS) any client is able to navigate the interface via URIs that are exclusively provided by the server. This technique reduces dependencies between the server and the clients. It allows developers to evolve the server API without having to worry about client incompatibilities. (Fielding, 2000)

7.   CURRENT TASKS AND UPCOMING FEATURES

The development of the monitoring system DABAMOS is still work in progress. There is likely to be incompatibilities with different platforms or certain sensors, which will be fixed as soon as possible. However, in the future, the DABAMOS development team has two priorities to improve the project.

One future task is the integration of OpenSDMS into consisting geo-sensor networks. A live system is running on the roof top of the faculty. The system consists of a totalstation, which monitors points in the surrounding area. For the detection of temperature and air pressure corresponding sensors are integrated as well. The current set-up will be replaced by a new prototype of the embedded sensor control unit (fig. 5). At the moment, the station is configured locally. In the future, the station will be connected to the cloud service OpenSDMS, to allow extensive remote control.

The main purpose to operate such a system is to test new features under real conditions. Another intention of the team is to use the project as a showcase for possible users and contributors. The developed solution will allow interested to view and change the configuration settings, after OpenSDMS is integrated. The data will be shown in interactive charts and additionally be provided to the public for download.

As the monitoring with total stations is rather common, it is another priority to extend the feature list by integrating low-cost GNSS receivers. These improvements are part of an ongoing research project at NUAS.

8. REFERENCES

Agile, 2015. Manifesto for Agile Software Development. http://www.agilemanifesto.org/. Visited in September 2015.

Fielding, R., 2000. Architectural Styles and the Design of Network-based Software Architectures. University of California, Irvine.

Humble, J.; Farley, D., 2010. Continuous Delivery: Reliable Software Releases through Build, Test, and Deployment Automation. Addison-Wesley. Mnchen.

GitHub, 2015. GitHub source code hosting service. https://github.com/. Visited in September 2015.

Go (2015). Go programming language. https://golang.org/. Visited in September 2015.

Grimm, T., 2004. Users Guide to Rapid Prototyping. Society of Manufacturing Engineers, Dearborn, MI.

Laurent, A. M. St., 2004. Understanding Open Source and Free Software Licensing. OReilly Media, Sebastopol, CA.

Pelzer, H., 1988. Ingenieurvermessung. Deformationsmessungen. Massenberechnung. Ergebnisse des Arbeitskreises 6 des Deutschen Vereins fr Vermessungswesen (DVW) e. V. Herausgegeben von Hans Pelzer. Wittwer, Stuttgart.

Preiel, R.; Stachmann, B., 2013. Git: Dezentrale Versionsverwaltung im Team Grundlagen und Workflows. dpunkt.verlag, Heidelberg.

Swartout, P., 2012. Continuous delivery and DevOps: A Quickstart Guide. Packt Publishing, Birmingham.

12

DECISION LEVEL FUSION OF LIDAR DATA AND AERIAL COLOR IMAGERY BASED ON BAYESIAN THEORY FOR URBAN AREA CLASSIFICATION

H. Rastiveis*

[a] School of Surveying and Geospatial Engineering, Faculty of Engineering, University of Tehran, Tehran, Iran, hrasti@ut.ac.ir

KEY WORDS: High Resolution LiDAR Data, Naïve Bayes Classifier, Decision Level Fusion, Classification

ABSTRACT

Airborne Light Detection and Ranging (LiDAR) generates high-density 3D point clouds to provide a comprehensive information from object surfaces. Combining this data with aerial/satellite imagery is quite promising for improving land cover classification. In this study, fusion of LiDAR data and aerial imagery based on Bayesian theory in a three-level fusion algorithm is presented. In the first level, pixel-level fusion, the proper descriptors for both LiDAR and image data are extracted. In the next level of fusion, feature-level, using extracted features the area are classified into six classes of "Buildings", "Trees", "Asphalt Roads", "Concrete roads", "Grass" and "Cars" using Naïve Bayes classification algorithm. This classification is performed in three different strategies: (1) using merely LiDAR data, (2) using merely image data, and (3) using all extracted features from LiDAR and image. The results of three classifiers are integrated in the last phase, decision level fusion, based on Naïve Bayes algorithm. To evaluate the proposed algorithm, a high resolution color orthophoto and LiDAR data over the urban areas of Zeebruges, Belgium were applied. Obtained results from the decision level fusion phase revealed an improvement in overall accuracy and kappa coefficient.

1. INTRODUCTION

Airborne Light Detection and Ranging (LiDAR) generates high-density 3D point clouds to provide a comprehensive information of object surfaces. Recently, the use of LiDAR data has increased in many applications, such as 3D city modeling and urban planning. Although the spatial resolution of this data has intensely improved, however, the lack of spectral and textural information is still a big problem of LiDAR technology. On the other hand, high resolution aerial/satellite imageries offer very detailed spectral and textural information but poor structural information. Therefore, LiDAR data and aerial/satellite imagery are complementary to each other and, combining them is quite promising for improving land cover classification (Lee et al., 2008; Li et al., 2007; Pedergnana et al., 2012; Rottensteiner et al., 2005; Schenk and CsathA, 2002).

Many methods for fusion of LiDAR data and multispectral aerial/satellite image have been proposed by researchers, in recent years(Li et al., 2013; Malpica et al., 2013; Schenk and CsathA, 2002; Sohn and Dowman, 2007; Trinder and Salah, 2012; Yousef and Iftekharuddin, 2014). The majority of these studies have applied aerial image instead of satellite image as a complementary of LiDAR data. Moreover, there are a number of researches fused hayper-spectral image and LiDAR data for different applications(Bigdeli et al., 2014; Dalponte et al., 2008). Here, a few number of these studies are briefly discussed.

Bigdeli et al. (2014) addressed the use of a decision fusion methodology for the combination of hyperspectral and LIDAR data in land cover classification. The proposed method applied a support vector machine (SVM)-based classifier fusion system for fusion of hyperspectral and LIDAR data in the decision level. First, feature spaces are extracted from LIDAR and hyperspectral data. Then, SVM classifiers are applied on each feature data. After producing several of classifiers, Naive Bayes as a classifier fusion method combines the results of SVM classifiers from two data sets. The results discovered that the overall accuracies of

SVM classification on hyperspectral and LIDAR data separately were 88% and 58% while the decision fusion methodology receive the accuracy up to 91%(Bigdeli et al., 2014).

An analysis on the joint effect of hyperspectral and light detection and ranging (LIDAR) data for the classification of complex forest area based on SVM algorithm is proposed in (Dalponte et al., 2008).

Hong et al. (2009) proposed a fusion method by fusing the LiDAR points with the extracted points from image matching through three steps: (1) registration of the image and LiDAR data using the LiDAR data as control information; (2) image matching using the LiDAR data as the initial ground approximation and (3) robust interpolation of the LiDAR points and the object points resulted from image matching into a grid(Hong, 2009).

Zabuawala *et al.* (2009) proposed an automated and accurate method for building footprint extraction based on the fusion of aerial images and LiDAR. In the proposed algorithm, first initial building footprint was extracted from a LiDAR point cloud based on an iterative morphological filtering. This initial segmentation result was refined by fusing LiDAR data and the corresponding colour aerial images, and then applying the watershed algorithm initialised by the LiDAR segmentation ridge lines on the surface were founded(Zabuawala et al., 2009).

The fusion of aerial imagery and LiDAR data has been proposed to improve the geometrical quality of the building outlines (Rottensteiner et al., 2005). They are also applied to improve planar segmentation due to the complementary of these data sources (Khoshelham et al., 2008).

In this paper, a multi-level fusion technique is proposed for land cover classification using LiDAR data and aerial imagery. This method is performed through four consecutive phases: pre-processing, pixel-level fusion, feature-level fusion and decision level fusion.

---

* Corresponding author

2. NAÏVE BAYES FUSION METHOD

The Bayesian algorithm combines training data with a priori information to calculate a posteriori probability of a hypothesis. So, the most probable hypothesis according to the training data is possible to figure out. The basis for all Bayesian Learning Algorithms is the Bayes Rule which is Equation 1.

$$P(h \mid D) = \frac{P(D \mid h)P(h)}{P(D)} \tag{1}$$

Where,

$P(h)$ and $P(D)$ are prior probabilities of hypothesis h and D, and $P(h|D)$ and $P(D|h)$ are probability of h given D and D given h, respectively.

Here the conditional independence of the attributes of the instances is required for the use of Naïve Bayesian Classifiers. To brought it into formula, let X be a set of instances $x_i = (x_1, x_2, ..., x_n)$ and w be a set of classifications w_j

$$
\begin{aligned}
w &= \max_{wj \in w} P(w_j \mid x_1, x_2 ..., x_n) \\
&= \max_{wj \in w} \frac{P(x_1, x_2 ..., x_n \mid w_j)P(w_j)}{P(x_1, x_2 ..., x_n)} \\
&= \max_{vj \in w} P(x_1, x_2 ..., x_n \mid w_j)P(w_j)
\end{aligned}
\tag{2}
$$

Where

$$P(x_1, x_2 ..., x_n \mid w_j) = \prod_i P(x_i \mid w_j) \tag{3}$$

And $P(w_j)$ is a priori probability of class w_j.

These formulae can be used in both feature and decision level data fusion. In the feature level case, extracted descriptors/features from a training data set are applied to calculate $P(x_i \mid w_j)$. A priori probability may also be calculated based on the size of the training data in each class. Estimated posterior probability in each classification may also be applied in decision level fusion. In this case, one can calculate a priori information of each classification using its resulted confusion matrix.

3. HIGH RESOLUTION LIDAR AND IMAGE FUSION

The proposed method for fusion of high resolution LiDAR data and aerial orthophoto is based on the flowchart shown in Fig. 1. First, in a pre-processing step, conversion of LiDAR point cloud into grid form and contrast enhancement of the orthophoto are performed. Then, classification of the area is executed in three sequential fusion levels: pixel level fusion, feature level fusion, and decision level fusion.

As seen in Fig. 1, in pixel level fusion useful descriptors are calculated from both LiDAR and image. After that, extracted features are applied for classification of the area in three different strategies: (1) using merely LiDAR data, (2) using merely image data, (3) using all extracted features from LiDAR and image. Finally, all the classification results from these implementations are used in decision level fusion for making last decision about the pixels. Further details of the proposed method are described in the following sections.

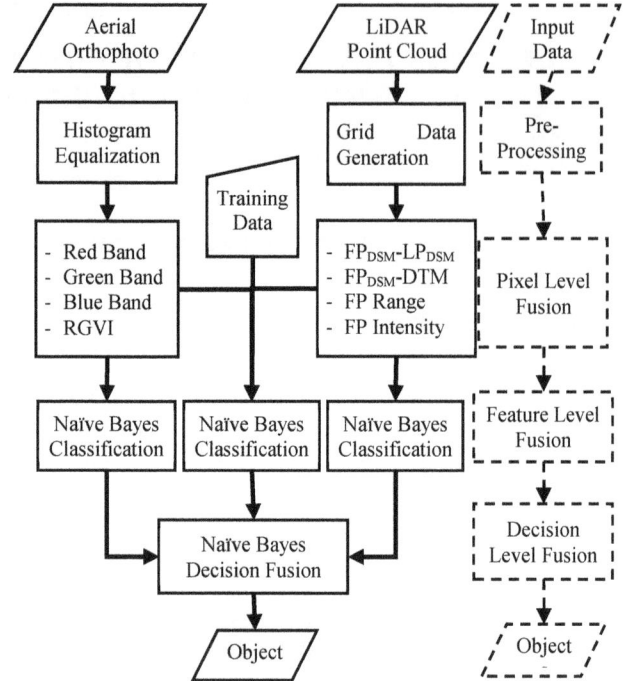

Figure 1. Flowchart of the proposed method

3.1 Pre-Processing

In the pre-processing phase, to simplify the process and, ability to deal with the LiDAR data as an image, irregular 3D point cloud is converted into regular form using interpolation techniques. Although, the interpolation process may cause the loss of information, however, it is negligible in this paper. Note, it is assumed that LiDAR data and aerial image are accurately registered. Moreover, histogram equalization of the color orthophoto is performed due to its effectiveness on contrast enhancement.

3.2 Pixel Level Fusion

The aim of pixel-level phase is to generate the proper descriptors for both data. In this step, eight descriptors are extracted on LiDAR data (four features) and aerial image (four features). These features are selected based on the previous literatures of LiDAR or aerial image classification (Bigdeli et al., 2014; Li et al., 2007). For example, "the height differences between the first pulse range and DTM "to distinguish buildings and trees from other objects and, also, "the height differences between the first pulse and the last pulse" to distinguish tree class from other classes can be seen in several studies (Bigdeli et al., 2014; Li et al., 2007; Rottensteiner et al., 2005). These descriptors can be calculated using equations 4 and 5. "First pulse range" and "First pulse Intensity" are the other descriptors which are extracted from LiDAR data.

$$nDSM = Last\ pulse\ range - DTM \tag{4}$$

$$NDSI = \frac{First\ Pulse\ Range - Last\ pulse\ Range}{First\ Pulse\ Range + Last\ pulse\ Range} \tag{5}$$

From the orthoimage four descriptors of "Red band", "Green band", "Blue band" and "Green-Red Vegetation Index" are considered for classification. Here, therefore, only GRVI feature is calculated through pixel level fusion of Red and Green channels. Same as NDVI in remote sensing data analyses, GRVI may help to distinguish vegetation area from other objects. Equation 6 shows the fusion formula for calculating this feature.

$$GRVI = \frac{G - R}{G + R} \qquad (6)$$

3.3 Feature Level Fusion

In the feature-level fusion phase, the area are classified into six classes of "Buildings", "Trees", "Asphalt Roads", "Concrete roads", "Grass" and "Cars" using aforementioned extracted features from LiDAR and image data. Training and check data set of each class are manually selected for the classification.

Three different classifiers are implemented in this level. (1) a classifier which merely used LiDAR data, (2) a classifier which merely used image data, (3) a classifier that applied all extracted features from LiDAR and image data.

Although in Naïve Bayes classifier as a soft classifier membership degrees of each pixel to the classes are calculated, here, only one class with higher degree of membership (maximum probability), which is shown in Equation 2, is selected. However, the degree of membership to the classes are kept to be used in the next phase of the algorithm, which will be described in the following section.

3.4 Decision Level Fusion

In this step, three previous classification results are integrated for making final decision about a pixel. There are different decision fusion techniques in pattern recognition literature such as Simple Voting, Weighted Voting, Rule Based Fuzzy System, Dempster-Shafer and Naïve Bayes. Voting and Weighted Voting algorithms can be applied for fusing crisp classification results while other methods would be able to integrate soft classification results.

In this paper, the decision level fusion is implemented based on Bayesian theorem. For this purpose, Naïve Bayes provides a method for computing the *a posteriori* degree of membership, based on previous estimated degrees. In the resulted *a posteriori* degrees of membership of each pixel to the classes, the maximum degree can be considered as the final class label. Here, resulted confusion matrix for each classifier can be applied to estimate *a priori* probability of each class.

4. EXPERIMENT AND RESULTS

4.1 Dataset

To evaluate the proposed algorithm, high resolution color orthophoto and LiDAR data over the urban areas of Zeebruges, Belgium were applied. The point density for the LiDAR sensor is approximately 65 points/m² and the color orthophoto were taken at nadir and have a spatial resolution of approximately 5 cm. From this data set a building block which included 1.03 million points was cropped as sample data. Selected area as a test data is depicted in Figure 2.

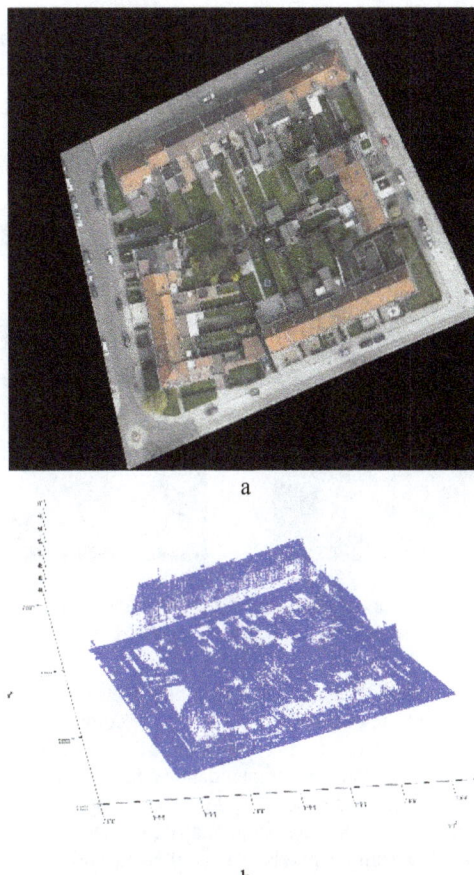

Figure 2. Study area. a. High resolution aerial orthophoto. b. High resolution LiDAR point cloud.

4.2 Results

After generating regular LiDAR data with 5 cm spatial resolution from point cloud, and contrast enhancing of the orthophoto, in the pre-processing step, the features for both LiDAR and orthophoto were extracted. These features from LiDAR data and orthoimage are displayed in Figure 3 and 4, respectively.

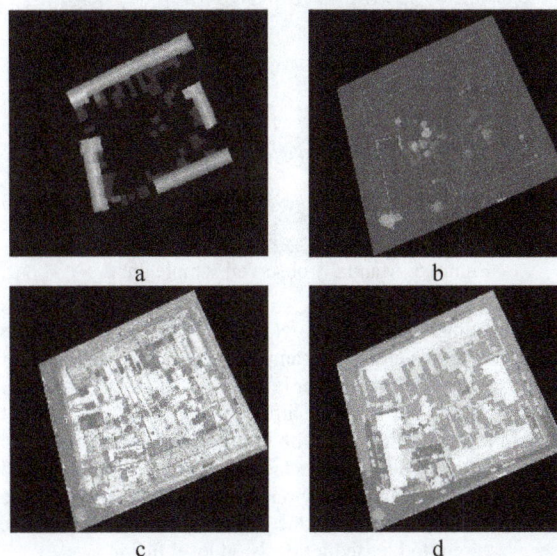

Figure 3. Extracted Features for LiDAR data. a. difference between last pulse and DTM. b. difference between first and last pulse range. c. First pulse intensity. d. First pulse range.

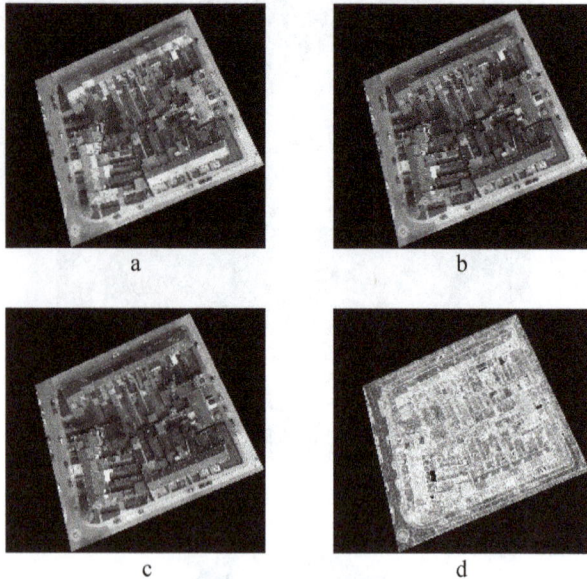

Figure 4. Extracted Features of orthophoto. a. Red channel. b. Green channel. c. Blue channel. d. NDGI.

Sample data collection is the next step after feature extraction. In this case, as all the evaluation parameters are computed based on these samples, here, a huge number of sample data were collected. The collected sample data were divided into two groups of training and check data for calculating the probability density functions and confusion matrix, respectively. In this study, 421515 sample pixels were collected and 266156 and 155359 sample pixels were collected as training and check data, respectively. Selected samples are shown in Figure 5.

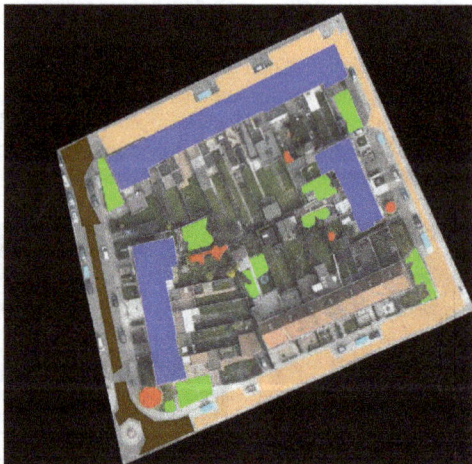

C-Matrix	Trees	Grass	Buildings	Concrete Roads	Asphalt Roads	Cars	Sum (Rows)	Ommision Error	Producer Accuracy
Trees	2347	199	5357	3377	109	148	11537	0.80	0.20
Grass	4512	68764	17945	44607	73	1931	137832	0.50	0.50
Buildings	4854	918	197031	3087	135	702	206727	0.05	0.95
Concrete Roads	1642	784	14671	117923	4393	2105	141518	0.17	0.83
Asphalt Roads	23	1	923	39334	77401	24	117706	0.34	0.66
Cars	0	1	109	0	0	418	528	0.21	0.79
Sum(cols.)	13378	70667	236036	208328	82112	5328	615849		
Commision Error	0.96	0.01	0.19	0.64	0.04	9.30			
User Accuracy	0.04	0.99	0.81	0.36	0.96	-8.30			

Figure 6. Obtained results from Bayesian classifiers based on merely orthophoto.

Figure 5. Manually observed sample data.

After collecting sample data, Naïve Bayes classifiers were designed and executed on the sample data set in three different strategies. Naïve Bayes classifier is a soft classification technique and results degree of memberships for each pixel in different classes. In this case, the class with higher degree of membership (maximum probability) is selected for a pixel. The obtained classification results and corresponding confusion matrix can be seen in Figures 6-8. However, the degree of membership to the classes were kept to be used for decision level fusion.

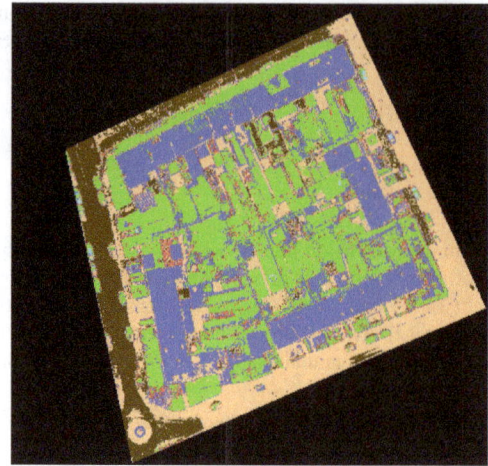

C-Matrix	Trees	Grass	Buildings	Concrete Roads	Asphalt Roads	Cars	Sum (Rows)	Ommision Error	Producer Accuracy
Trees	12734	509	4355	2216	0	1	19815	0.36	0.64
Grass	231	67355	1252	11780	0	568	81186	0.17	0.83
Buildings	138	0	226873	2712	0	0	229723	0.01	0.99
Concrete Roads	175	1954	3242	148952	263	600	155186	0.04	0.96
Asphalt Roads	0	311	61	42489	81846	13	124720	0.34	0.66
Cars	100	538	253	179	0	4146	5216	0.21	0.79
Sum(cols.)	13378	70667	236036	208328	82112	5328	615849		
Commision Error	0.03	0.04	0.04	0.38	0.00	0.23			
User Accuracy	0.97	0.96	0.96	0.62	1.00	0.77			

Figure 7. Obtained results from Bayesian classifiers based on merely LiDAR feature space.

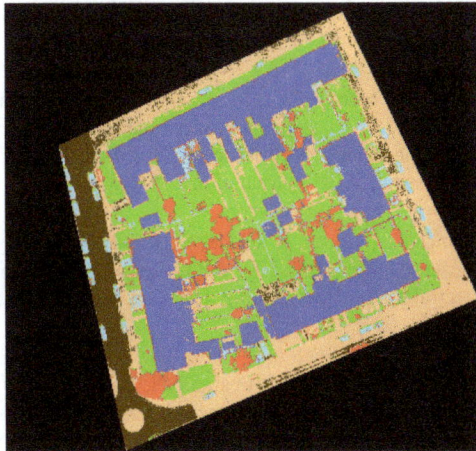

C-Matrix	Trees	Grass	Buildings	Concrete Roads	Asphalt Roads	Cars	Sum (Rows)	Omission Error	Producer Accuracy
Trees	12927	679	4668	2399	1	119	20793	0.38	0.62
Grass	354	69431	1044	37053	26	165	108073	0.36	0.64
Buildings	35	0	227333	2722	0	0	230090	0.01	0.99
Concrete Roads	12	256	2074	149129	163	620	152254	0.02	0.98
Asphalt Roads	0	3	0	16939	81919	10	98871	0.17	0.83
Cars	50	298	917	86	0	4414	5765	0.23	0.77
Sum(cols.)	13378	70667	236036	208328	82112	5328	615849		
Commission Error	0.02	0.01	0.04	0.39	0.00	0.16			
User Accuracy	0.98	0.99	0.96	0.61	1.00	0.84			

Figure 8. Obtained results from Bayesian classifiers based on LiDAR and image feature space.

The results of three classification algorithm were finally integrated based on Naïve Bayes data fusion algorithm. Final classification results is shown in Figure 9. As can be seen from the figure, the confusion matrix obtained from fusion of previous classification presents more promising results.

C-Matrix	Trees	Grass	Buildings	Concrete Roads	Asphalt Roads	Cars	Sum (Rows)	Omission Error	Producer Accuracy
Trees	13329	370	339	4320	6	2	18366	0.27	0.73
Grass	29	69867	21	10050	0	116	80083	0.13	0.87
Buildings	13	0	235577	4899	0	1	240490	0.02	0.98
Concrete Roads	1	366	92	179820	79	320	180678	0.00	1.00
Asphalt Roads	0	0	0	9237	82027	0	91264	0.10	0.90
Cars	6	64	7	2	0	4889	4968	0.02	0.98
Sum(cols.)	13378	70667	236036	208328	82112	5328	615849		
Commission Error	0.00	0.01	0.00	0.16	0.00	0.09			
User Accuracy	1.00	0.99	1.00	0.84	1.00	0.91			

Figure 9. Final obtained classification from decision level fusion.

4.3 Discussion

Among the obtained three classifier in feature level fusion, the one which simultaneously used both LiDAR and Image features presented better results. Furthermore, fusing the three classification through Naïve Bayes classifier fusion method proved an improvement in classification results. Figure 10 shows resulted overall accuracy and kappa coefficient for all the classifications. As can be seen from the figure, the best performance was achieved for final decision level fusion while obtained results from merely image features was dispiriting. The performance of the classifier which used merely image features in detecting trees and cars classes was very disappointing. Low values of user accuracies for these classes in Figure 6 proved this.

The overall accuracy and kappa coefficient of the classifier which used merely LiDAR features is approximately the same as the one which used all LiDAR and Image features. However, it is seen in Figure 7 that lack of spectral information in LiDAR data causes mixing grass, asphalt roads and concrete roads classes, especially in south-eastern part of the test area.

Moreover, as previously reported in researches multiple data resources obtained more promising results in comparison with each data resource individually(Bigdeli et al., 2014).

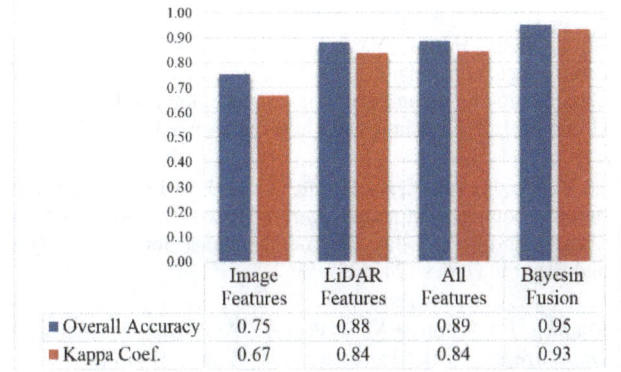

	Image Features	LiDAR Features	All Features	Bayesin Fusion
Overall Accuracy	0.75	0.88	0.89	0.95
Kappa Coef.	0.67	0.84	0.84	0.93

Figure 10. Comparison between classification results.

5. CONCLUSION

In this paper, fusion of high resolution aerial orthophoto and LiDAR data based on Naïve Bayesian algorithm were discussed. Three different classification were designed using training data set: (1) using merely LiDAR data, (2) using merely image data, (3) using all extracted features from LiDAR and image. The results of these classification were integrated using Naïve Bayes algorithm. Among all the classification results, the results of final decision fusion were the best.

Although the features and number of classes have important roles in classification, it is theoretically expected that the same results would be achieved for different feature spaces and number of classes. However, it is recommended to test the algorithm for other case studies. It is also suggested to test other decision fusion algorithm such as fuzzy inference system.

REFERENCES

Bigdeli, B., Samadzadegan, F., Reinartz, P., 2014. A decision fusion method based on multiple support vector machine system for fusion of hyperspectral and LIDAR data. International Journal of Image and Data Fusion 5, 196-209.

Dalponte, M., Bruzzone, L., Gianelle, D., 2008. Fusion of hyperspectral and LIDAR remote sensing data for classification of complex forest areas. Geoscience and Remote Sensing, IEEE Transactions on 46, 1416-1427.

GRSS_DFC, 2015. 2015 IEEE GRSS Data Fusion, 2015 IEEE GRSS Data Fusion Contest.   , in: IEEE (Ed.), Online: http://www.grss-ieee.org/community/technical-committees/data-fusion" .

Hong, J., 2009. Data fusion of LiDAR and image data for generation of a high-quality urban DSM, Proceedings of the joint urban remote sensing event. IEEE., Shanghai, China.

Khoshelham, K., Nedkov, S., Nardinocchi, C., 2008. A comparison of Bayesian and evidence-based fusion methods for automated building detection in aerial data. International Archives of the Photogrammetry, Remote Sensing and Spatial Information Sciences 37, 1183-1188.

Lee, D.H., Lee, K.M., Lee, S.U., 2008. Fusion of lidar and imagery for reliable building extraction. Photogrammetric Engineering & Remote Sensing 74, 215-225.

Li, H., Gu, H., Han, Y., Yang, J., 2007. Fusion of high-resolution aerial imagery and lidar data for object-oriented urban land-cover classification based on svm. Proceedings of the ISPRS Working Group IV/1:â€œDynamic and Multi-dimensional GIS, 179-184.

Li, Y., Wu, H., An, R., Xu, H., He, Q., Xu, J., 2013. An improved building boundary extraction algorithm based on fusion of optical imagery and LiDAR data. Optik-International Journal for Light and Electron Optics 124, 5357-5362.

Malpica, J.A., Alonso, M.C., Papí, F., Arozarena, A., Martínez De Agirre, A., 2013. Change detection of buildings from satellite imagery and lidar data. International journal of remote sensing 34, 1652-1675.

Pedergnana, M., Marpu, P.R., Mura, M.D., Benediktsson, J.A., Bruzzone, L., 2012. Classification of remote sensing optical and lidar data using extended attribute profiles. Selected Topics in Signal Processing, IEEE Journal of 6, 856-865.

Rottensteiner, F., Trinder, J., Clode, S., Kubik, K., 2005. Using the Dempster-Shafer method for the fusion of LIDAR data and multi-spectral images for building detection. Information fusion 6, 283-300.

Schenk, T., CsathA, B., 2002. Fusion of LIDAR data and aerial imagery for a more complete surface description. International Archives of Photogrammetry Remote Sensing and Spatial Information Sciences 34, 310-317.

Sohn, G., Dowman, I., 2007. Data fusion of high-resolution satellite imagery and LiDAR data for automatic building extraction. ISPRS Journal of Photogrammetry and Remote Sensing 62, 43-63.

Trinder, J., Salah, M., 2012. Aerial images and LiDAR data fusion for disaster change detection. ISPRS Annals of Photogrammetry, Remote Sensing and Spatial Information Sciences 1, 227-232.

Yousef, A., Iftekharuddin, K., 2014. Shoreline extraction from the fusion of LiDAR DEM data and aerial images using mutual information and genetic algrithms, Neural Networks (IJCNN), 2014 International Joint Conference on. IEEE, pp. 1007-1014.

Zabuawala, S., Nguyen, H., Wei, H., Yadegar, J., 2009. Fusion of LiDAR and aerial imagery for accurate building footprint extraction, IS&T/SPIE Electronic Imaging. International Society for Optics and Photonics, pp. 72510Z-72510Z-72511.

13

A GIS-BASED MODEL FOR POST-EARTHQUAKE PERSONALIZED ROUTE PLANNING USING THE INTEGRATION OF EVOLUTIONARY ALGORITHM AND OWA

M. Moradi [a]*, M. R. Delavar [b], A. Moradi [c]

[a] School of Surveying and Geospatial Engineering, College of Engineering, University of Tehran, Iran – milad.moradi@ut.ac.ir
[b] Center of Excellence in Geomatic Engineering in Disaster Management, School of Surveying and Geospatial Engineering, College of Engineering, University of Tehran, Iran – mdelavar@ut.ac.ir
[c] Department of Social Sciences, Farhangian University, Resalat Branch, Zahedan, Iran - Asad.moradi1343@gmail.com

KEY WORDS: GIS, Route Planning, Genetic Algorithm, Ordered Weighted Averaging, Earthquake.

ABSTRACT

Being one of the natural disasters, earthquake can seriously damage buildings, urban facilities and cause road blockage. Post-earthquake route planning is problem that has been addressed in frequent researches. The main aim of this research is to present a route planning model for after earthquake. It is assumed in this research that no damage data is available. The presented model tries to find the optimum route based on a number of contributing factors which mainly indicate the length, width and safety of the road. The safety of the road is represented by a number of criteria such as distance to faults, percentage of non-standard buildings and percentage of high buildings around the route. An integration of genetic algorithm and ordered weighted averaging operator is employed in the model. The former searches the problem space among all alternatives, while the latter aggregates the scores of road segments to compute an overall score for each alternative. Ordered weighted averaging operator enables the users of the system to evaluate the alternative routes based on their decision strategy. Based on the proposed model, an optimistic user tries to find the shortest path between the two points, whereas a pessimistic user tends to pay more attention to safety parameters even if it enforces a longer route. The results depicts that decision strategy can considerably alter the optimum route. Moreover, post-earthquake route planning is a function of not only the length of the route but also the probability of the road blockage.

1. INTRODUCTION

The problem of finding the optimum path which can be the shortest path could be regarded as one of the most controversial problems in network analysis (Pahlavani & Delavar, 2014). This problem is in the focus of other researches such as finding the service area and vehicle routing problem (Pourrahmani, Delavar, Pahlavani, & Mostafavi, 2015). The important issue is that the shortest path is not the optimum path necessarily. Sometimes a number of criteria should be considered in a way finding problem. For example, the length of the route, the number of traffic lights and the travel time all can be important criteria in a path finding problem (Karimi, Zhang, & Benner, 2014; Nadi & Delavar, 2011). Therefore, route planning can regarded as a multi-criteria evaluation problem and multi-criteria decision making methods can be used to address this problem (Nadi & Delavar, 2011). Frequent researches have been undertaken on designing and developing a spatial decision support system for route planning problem (Jankowski & Richard, 1994; Niaraki & Kim, 2009; Papinski, Scott, & Doherty, 2009; Torrens, 2014; Xie, Li, Wan, & Li, 2014; Yao, Loo, & Yang, 2015).

A personalized route is a proper route with respect to the combination of the preferences of a decision maker. Criteria such as travel distance, travel time, width of the route, scenery, number of high rises around the route and the number of traffic lights could be considered in a personalized route planning

system. For example, a tourist may have a tendency to visit more landscapes, even if it takes longer hours (Nadi & Delavar, 2011). In after earthquake route planning, road segments which are wider and have lower number of high rises are more preferred (Peiman & Clarke, 2014; Pourrahmani et al., 2015).

Nadi and Delavar (2011) proposed a spatial decision support system for personalized route planning. They designed and developed a web-based GIS which is able to input user preferences for each criterion. The system is able to find the optimum path with respect to user preferences using ordered weighted averaging operator. Jankowski and Richard (1994) was one of the first researchers that worked on the integration of GIS and decision making methods for implementing spatial decision support systems. In (Jankowski & Richard, 1994) they proposed an SDSS for route planning according to some decision rules stated by the user of the system. Pahlavani and Delavar (2014) proposed a route planning model based on invasive weeds optimizations which can search the problem space by simulating the colonizing of weeds. The superiority of the model is that it is very fast in comparison to other optimization methods. The problem of post-earthquake route planning is addressed in work carried out by Pourrahmani et al. (2015). They proposed a route planning algorithm from local shelters to regional ones for long term safety. They proposed a route finding algorithm based on simulated annealing which is able to manage the traffic demand dynamically.

---

* Corresponding author

In this research a GIS-based model is proposed for personalized route planning. The model consists of two main parts. The first part searches the problem space for the optimum path, while the second part tries to find an overall score based on the scores of the route in different criteria. In other words, a multi-criteria decision making method is employed to calculate the degree to which a route is preferred (objective function of the optimization method). In this paper, OWA algorithm is used to find an aggregated score for each route based on a number of criteria associated with earthquake. Then, GA is employed to find the optimum route with respect to the predefined objective function. In other words, the fitness function of the GA algorithm is the aggregation function of OWA. The criteria of route planning are defined based on vulnerability against earthquake. Vulnerability of a route is the probability of blockage after a large earthquake. Thus, the proposed route planning model tries to make a balance between the length of a route and the probability of the blockage of road segments. Obviously, in this paper, it is assumed that there is no real damage data available and the system tries to avoid vulnerable roads and very long ones simultaneously based on the optimism degree of users.

2. METHODOLOGY

2.1 OWA operator

The OWA operator which was introduced by Yager (1993) is an aggregating operator from a class of mean-like operators. The aggregation operator of ordered weighted averaging method is as follows (Yager, 1993):

$$F(a_1,...,a_n) = \sum_{i=1}^{n} w_i * b_i =$$
$$w_1 b_1 + w_2 b_2 + ... + w_n b_n \tag{1}$$

where w_i = the weight associated to ith criteria
             b_i = the ith attribute value.

The OWA operator is able to provide a wide range of answers from the most optimistic to the least optimistic solution. In fact, weight vector (w) is a key element for describing the characteristics of the operator (Milad Moradi, Delavar, & Moshiri, 2015). Frequent methods have been proposed in order to calculate the optimum set of weights for OWA. The most famous one is based on employing linguistic quantifiers (Zarghami, Szidarovszky, & Ardakanian, 2008). Linguistic quantifiers are expressions from natural language which indicates the preferences of user on each criterion. Furthermore, linguistic quantifiers imply the degree to which the decision maker is optimistic or pessimistic (Yager, 1992). The level of optimism can be determined using a linguistic quantifier. Let Q be a linguistic quantifier, the corresponding weight vector could be calculated as (Yager, 1992):

$$W_i = Q_{RIM}(\frac{i}{n}) - Q_{RIM}(\frac{i-1}{n}), \quad i = 1,2,...,n \tag{2}$$

where w_i = the weight of ith criterion
             Q = the linguistic quantifier.

The quantifier can be expressions such as All, Most, Half, Few and At least one. Half means that satisfaction (great score) of

half of the criteria is enough for an acceptable result, while most means that satisfaction of most of criteria is obligatory for an acceptable alternative (Malczewski & Liu, 2014; M Moradi, Delavar, & Moshiri, 2013). OWA weights do not represent importance of criteria. They indicate order weights which represent the degree to which the decision maker is optimistic (Jelokhani-Niaraki & Malczewski, 2015). When the relative weight of the criteria is important, the weight vector is obtained using Eq. 3 (Malczewski & Rinner, 2015):

$$w_j = \left(\frac{\sum_{k=1}^{j} v_k}{\sum_{k=1}^{n} v_k}\right)^{\alpha} - \left(\frac{\sum_{k=1}^{j-1} v_k}{\sum_{k=1}^{n} v_k}\right)^{\alpha} \tag{3}$$

where V_k = the relative importance of the kth criteria.

2.2 Genetic Algorithm

Genetic algorithm is a search algorithm in artificial intelligence which simulates the process of natural selection. This algorithm could be used in shortest path problems because it is able to simulate each route as a chromosome according to which the algorithm can analyse different possible solutions and find the optimum path. Each route is represented by a chromosome (see Figure 1). The problem of route finding is simulated by a weighted digraph $G = (N, E)$ where $N = \{1, 2, ..., n\}$ indicating nodes of the network and $E = \{e_1, e_2, ..., e_n\}$ indicating communication links between them (Abbaspour & Samadzadegan, 2011). Then, a path P is the shortest path if the bandwidth of it is minimum value (Zhang, Wu, Wei, & Wang, 2011):

$$Band(P) = \min(Band(e), \quad e \in E_p) \tag{4}$$

Thus, the problem of finding the shortest path is equal to the problem of finding the path with minimum bandwidth.

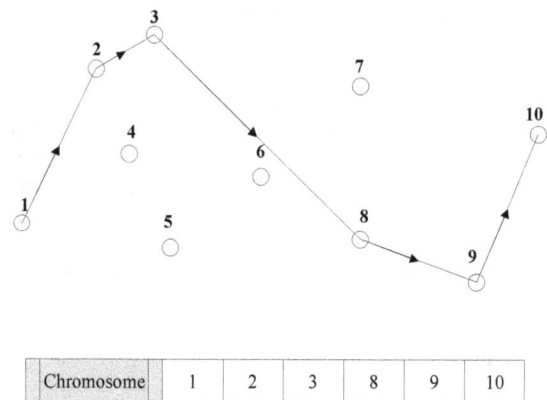

Chromosome	1	2	3	8	9	10

Figure 1. Representation of a route as a chromosome

2.3 Crossover

In order to find the best solution, genetic algorithm make new generation based on the past ones. In other words, the best members of the current generation are selected and the next generation is produced from crossover operation between them

(Vidal, Crainic, Gendreau, & Prins, 2013). Crossover operation represents the natural selection in which the more strong parents produce the next generation. For route finding, crossover is similar to Figure 2.

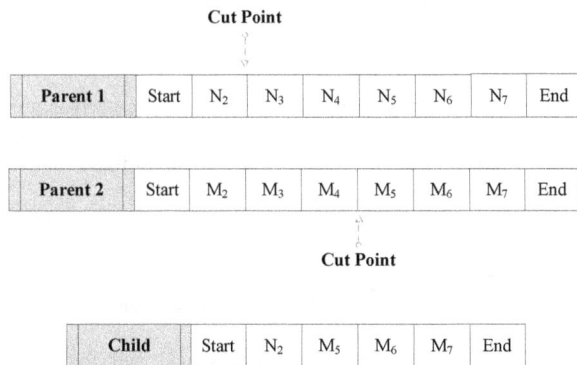

Cut Point

Parent 1	Start	N_2	N_3	N_4	N_5	N_6	N_7	End

Parent 2	Start	M_2	M_3	M_4	M_5	M_6	M_7	End

Cut Point

Child	Start	N_2	M_5	M_6	M_7	End

Figure 2. The procedure of crossover

It means that two valid paths between the start and end points are combined to produce a new valid path between them. In a simple way crossover can be done by fragment of two parent chromosomes and reunion of them. The very important point is that the new path should be valid. In implementation, invalid paths are removed from the set of parents who produce the next generation (Heidari & Abbaspour, 2014).

2.4 Mutation

Mutation is a genetic process which aims to provide the population with diversity (Vidal et al., 2013). Mutation alters a number of genes in a chromosome and produces a new one. In fact, mutation facilitates global search by producing random valid members in each generation. In route finding one or two nodes are changed in a chromosome and a new route will be produced which may not necessarily be valid according to the connection matrix of the problem. Figure 3 illustrates a sample mutation process.

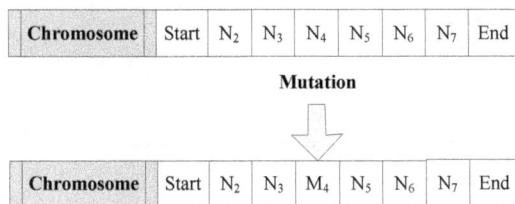

Chromosome	Start	N_2	N_3	N_4	N_5	N_6	N_7	End

Mutation

Chromosome	Start	N_2	N_3	M_4	N_5	N_6	N_7	End

Figure 3. The procedure of mutation

Mutation cause relatively smaller changes in the parent routes in comparison to crossover. Crossover makes substantial changes in the length and characteristics of a route. For example, crossover can cause a route to pass from a road with high percentage of vulnerable building which considerably reduces the preferably of that route for this problem. However, mutation only changes one node and two edges (road segments). Thus, the overall score of the route does not alter dramatically (Rajabi-Bahaabadi, Shariat-Mohaymany, Babaei, & Ahn, 2015).

3. IMPLEMENTATION

3.1 Study Area

This study is undertaken in Tehran metropolitan area where the infrastructure is in a very poor condition due to the rapid urban growth which resulted in urban sprawl (M Moradi, Delavar, & Moshiri). The width of many streets is insufficient and can cause blockage in a natural disaster like earthquake.

3.2 Contributing Criteria

Although the proposed model is a general model for route planning, a case study of after earthquake route planning has been undertaken in this paper. Figure 4 illustrates a number of contributing factors which were employed in previous researches. In this research only 4 criteria are used including length of route, width of road, percentage of high rises around the road and percentage of non-standard buildings around the road. Considering the fact that in this research it is assumed that there is no real damage data, the data layers which indicates the probability of road blockage are employed. The higher percentage of non-standard building and high rises decrease the suitability of this road.

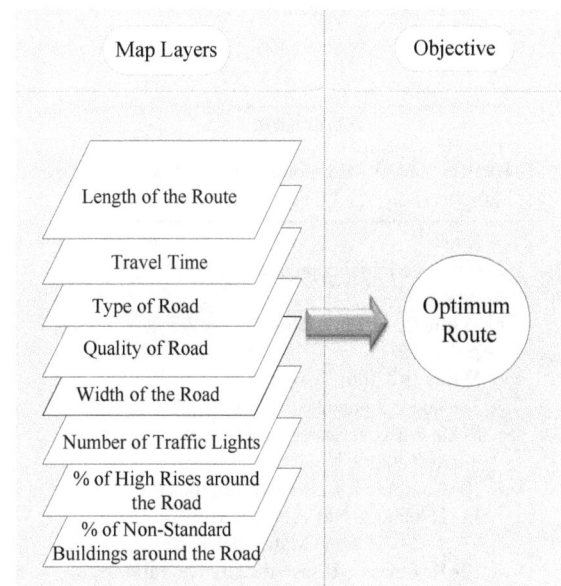

Figure 4. Contributing factors used in this paper

The steps of the proposed model are shown in Figure 5. Firstly, the contributing criteria are determined. Then, relative weights are determined by the expert. These weights indicate the relative importance of criteria in comparison to each other. Following that, genetic algorithm is employed to find the best route from the set of alternative routes. A chromosome is made for each possible route using the nodes and edges. Next, the fitness function of these routes is calculated using the aggregation function of OWA operator. In this step, an overall score in calculated for each route based on the four contributing criteria. After that, crossover and mutation search all the alternative routes and find the best one through an iterative process. The impact of optimism degree is on combination of scores where an overall score for the fitness function is calculated with respect to optimism degree.

1 • Defining contributing criteria

2 • Obtaining relative weights

3 • Selecting a sample route

4 • Calculating fitness function by OWA operator

5 • Crossover

6 • Mutation

7 • Repeating step 3-6 till find the optimum path

8 • Output the path with respect to optimism degree

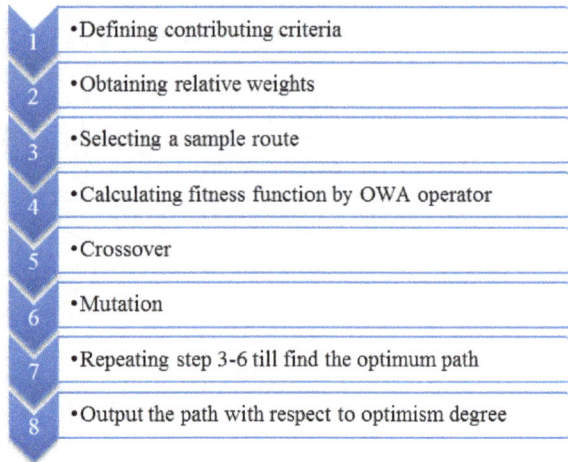

Figure 5. Steps that are done in this research

Figure 6 illustrates the pseudo code of the proposed evolutionary model for route planning. In fact, evolutionary algorithm is employed in research as it is used in finding the shortest path. However, the fitness function is not the length of the route. The fitness function of this model is calculated by OWA which can take into account the optimism degree of expert.

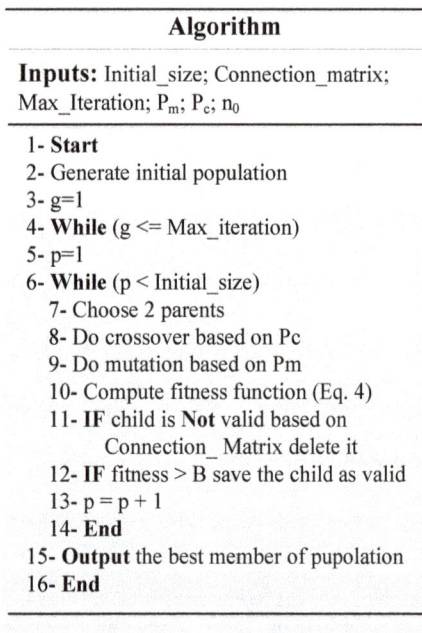

Algorithm

Inputs: Initial_size; Connection_matrix; Max_Iteration; P_m; P_c; n_0

1- **Start**
2- Generate initial population
3- g=1
4- **While** (g <= Max_iteration)
5- p=1
6- **While** (p < Initial_size)
  7- Choose 2 parents
  8- Do crossover based on Pc
  9- Do mutation based on Pm
  10- Compute fitness function (Eq. 4)
  11- **IF** child is **Not** valid based on
       Connection_ Matrix delete it
  12- **IF** fitness > B save the child as valid
  13- p = p + 1
  14- **End**
15- **Output** the best member of pupolation
16- **End**

Figure 6. Pseudo code of the proposed model

One of the main differences of the shortest path problem and route finding problem is that in route planning the shortest path is not the most preferable one necessarily. In this research, the objective is to develop a model for post-earthquake route planning. Hence, the user wants to avoid routes which are surrounded with non-standard buildings and the roads which are near the faults because they are more likely to become blocked. On the other hand, the user does not want to select a route which is considerably long. Therefore, there should be a trade-off between the length of the path and the suitability of it with respecting the earthquake vulnerability criteria. Two main

problems are combined here, the first one is th problem of finding the best route between a number of alternative and the second problem is to find a good fitness function that indicates the degree to which a route is preferable based on user's opinion. The former problem is addressed using genetic algorithm, while the latter is addressed by OWA operator.

3.3 Results and Discussion

In order to evaluate the proposed personalized route planning model, the preferences of one user is asked over contributing criteria. The user assigned normalized weights to the 4 criteria which are as follows:
The length of route: 0.5
The width of road: 0.2
The percentage of high rises around the road: 0.1
The percentage of non-standard buildings around the road: 0.2

The weights are included into Equation (3) and the order weights are calculated. The order weights then are used to run the model. Following that, the optimum route based on the preferences of the user is selected for three different optimism degrees. Figure 8 illustrates the three different routes. The optimistic route is more similar to the shortest path because the user wants to take the risk and go through the shortest path available. On the other hand, the route selected based on the pessimistic strategy tends to select road segments which are high ways because the user tends to avoid narrow roads which are surrounded by non-standard buildings. Therefore, the length of the route increases. Quantifier-guided OWA can make a trade-off between the length of the route and other factors.

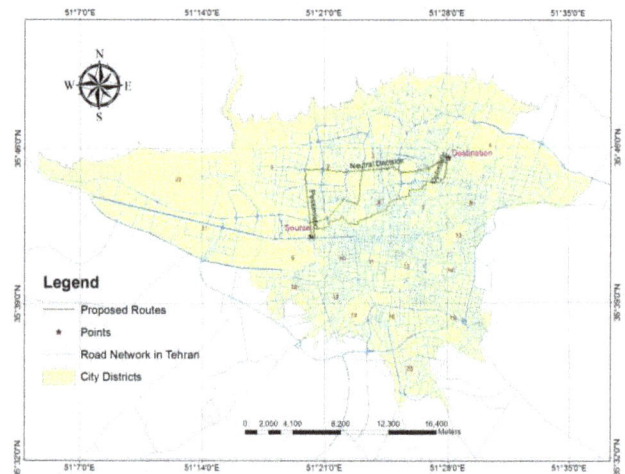

Figure 7. Routes based on optimistic and pessimistic views

Figure 8 indicates 3 different routes for after earthquake based on different point of views. An optimistic decision maker pays no attention to the risk of road blockage and selects the shortest path, whereas a pessimistic one selects the route that is away from those non-standard buildings even if this route is longer (pessimistic users select the safest route).

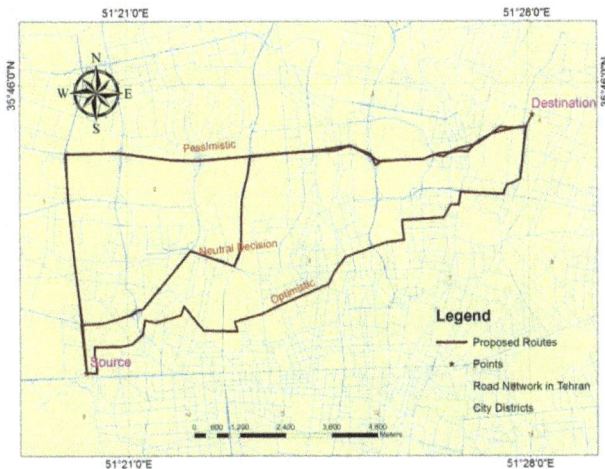

Figure 8. The influence of optimism degree on route finding

Although the proposed route planning model is efficient, due to the computational complexity of the model it is not suitable for large number of road segments. In that case, other network analysis techniques may be used to reduce the complexity of the model.

4. CONCLUSION

This paper proposes a novel personalized route planning model for after earthquake based on the integration of evolutionary algorithm and ordered weighted averaging. The issue of personalization is addressed in this paper. Previous route finding models tend to find the shortest path, while this model tends to find a route based on not only the length of the path but also the safety measures. Furthermore, this model makes a trade-off between the length of the route and other factors such as the width of it. Thus, the output of this model is not a shortest path but also the optimum path with respect the user's preferences. OWA operator, which is employed to calculate the fitness of each route, enables users to include a wide range of decision strategies instead of only neutral decision making. This model can be used in other route finding problems with minor modifications.

REFERENCES

Abbaspour, R. A., & Samadzadegan, F. (2011). Time-dependent personal tour planning and scheduling in metropolises. *Expert Systems with Applications, 38*(10), 12439-12452.

Heidari, A., & Abbaspour, R. (2014). A gravitational black hole algorithm for autonomous UCAV mission planning in 3D realistic environments. *International Journal of Computer Applications, 95*(9).

Jankowski, P., & Richard, L. (1994). Integration of GIS-based suitability analysis and multicriteria evaluation in a spatial decision support system for route selection. *Environment and Planning B, 21*, 323-323.

Jelokhani-Niaraki, M., & Malczewski, J. (2015). Decision complexity and consensus in Web-based spatial decision making: A case study of site selection problem using GIS and multicriteria analysis. *Cities, 45*, 60-70.

Karimi, H. A., Zhang, L., & Benner, J. G. (2014). Personalized accessibility map (PAM): A novel assisted wayfinding approach for people with disabilities. *Annals of GIS, 20*(2), 99-108.

Malczewski, J., & Liu, X. (2014). Local ordered weighted averaging in GIS-based multicriteria analysis. *Annals of GIS, 20*(2), 117-129.

Malczewski, J., & Rinner, C. (2015). Multiattribute Decision Analysis Methods *Multicriteria Decision Analysis in Geographic Information Science* (pp. 81-121): Springer.

Moradi, M., Delavar, M. R., Moshiri, B., & Khamespanah, F. (2014). a Novel Approach to Support Majority Voting in Spatial Group Mcdm Using Density Induced Owa Operator for Seismic Vulnerability Assessment. ISPRS-International Archives of the Photogrammetry, Remote Sensing and Spatial Information Sciences, 1, 209-214.

Moradi, M., Delavar, M. R., & Moshiri, B. (2013). Sensitivity Analysis of Ordered Weighted Averaging Operator in Earthquake Vulnerability Assessment. ISPRS-International Archives of the Photogrammetry, Remote Sensing and Spatial Information Sciences, 1(3), 277-282.

Moradi, M., Delavar, M. R., & Moshiri, B. (2015). A GIS-based multi-criteria decision-making approach for seismic vulnerability assessment using quantifier-guided OWA operator: a case study of Tehran, Iran. *Annals of GIS, 21*(3), 209-222.

Nadi, S., & Delavar, M. R. (2011). Multi-criteria, personalized route planning using quantifier-guided ordered weighted averaging operators. *International Journal of Applied Earth Observation and Geoinformation, 13*(3), 322-335.

Niaraki, A. S., & Kim, K. (2009). Ontology based personalized route planning system using a multi-criteria decision making approach. *Expert Systems with Applications, 36*(2), 2250-2259.

Pahlavani, P., & Delavar, M. R. (2014). Multi-criteria route planning based on a driver's preferences in multi-criteria route selection. *Transportation Research Part C: Emerging Technologies, 40*, 14-35.

Papinski, D., Scott, D. M., & Doherty, S. T. (2009). Exploring the route choice decision-making process: A comparison of planned and observed routes obtained using person-based GPS. *Transportation research part F: traffic psychology and behaviour, 12*(4), 347-358.

Peiman, R., & Clarke, K. (2014). The Impact of Data Time Span on Forecast Accuracy through Calibrating the SLEUTH Urban Growth Model. *International Journal of Applied Geospatial Research (IJAGR), 5*(3), 21-35.

Pourrahmani, E., Delavar, M. R., Pahlavani, P., & Mostafavi, M. A. (2015). Dynamic Evacuation Routing Plan after an Earthquake. *Natural Hazards Review*, 04015006.

Rajabi-Bahaabadi, M., Shariat-Mohaymany, A., Babaei, M., & Ahn, C. W. (2015). Multi-objective path finding in stochastic time-dependent road networks using non-dominated sorting genetic algorithm. *Expert Systems with Applications, 42*(12), 5056-5064.

Torrens, P. M. (2014). High-fidelity behaviours for model people on model streetscapes. *Annals of GIS, 20*(3), 139-157.

Vidal, T., Crainic, T. G., Gendreau, M., & Prins, C. (2013). A hybrid genetic algorithm with adaptive diversity management for a large class of vehicle routing problems with time-windows. *Computers & Operations Research, 40*(1), 475-489.

Xie, J., Li, Q., Wan, Q., & Li, X. (2014). Near optimal allocation strategy for making a staged evacuation plan with multiple exits. *Annals of GIS, 20*(3), 159-168.

Yager, R. R. (1992). Applications and extensions of OWA aggregations. *International Journal of Man-Machine Studies, 37*(1), 103-122.

Yager, R. R. (1993). Families of OWA operators. *Fuzzy sets and systems, 59*(2), 125-148.

Yao, S., Loo, B. P., & Yang, B. Z. (2015). Traffic collisions in space: four decades of advancement in applied GIS. *Annals of GIS*, 1-14.

Zarghami, M., Szidarovszky, F., & Ardakanian, R. (2008). A fuzzy-stochastic OWA model for robust multi-criteria decision making. *Fuzzy Optimization and Decision Making, 7*(1), 1-15.

Zhang, Y., Wu, L., Wei, G., & Wang, S. (2011). A novel algorithm for all pairs shortest path problem based on matrix multiplication and pulse coupled neural network. *Digital Signal Processing, 21*(4), 517-521.

14

THE ROLE OF ASTRO-GEODETIC IN PRECISE GUIDANCE OF LONG TUNNELS

Mirahmad Mirghasempour[a], Ali Yaser Jafari[b]

[a] Dept. of Civil Engineering, Shahid Rajaee Teacher Training University, Tehran, Iran – m.a.ghasempour@srttu.edu
[b] Dept. of Architecture and Urbanism, Shahid Rajaee Teacher Training University, Tehran, Iran – eminemcent23@yahoo.com

KEY WORDS: Astro-geodetic, TZK2-D, Underground surveying, vertical deflection.

ABSTRACT

One of prime aspects of surveying projects is guidance of paths of a long tunnel from different directions and finally ending all paths in a specific place. This kind of underground surveying, because of particular condition, has some different points in relation to the ground surveying, including Improper geometry in underground transverse, low precise measurement in direction and length due to condition such as refraction, distinct gravity between underground point and corresponding point on the ground (both value and direction of gravity) and etc. To solve this problems, astro-geodetic that is part of geodesy science, can help surveying engineers. In this article, the role of astronomy is defined in two subjects:

1- Azimuth determination of directions from entrance and exit nets of tunnel and also calibration of gyro-theodolite to use them in Underground transvers: By astronomical methods, azimuth of directions can be determine with an accuracy of 0.5 arcsecond, whereas, nowadays, no gyroscope can measure the azimuth in this accuracy; For instance, accuracy of the most precise gyroscope (Gyromat 5000) is 1.2 cm over a distance of one kilometre (2.4 arcsecond). Furthermore, the calibration methods that will be mention in this article, have significance effects on underground transverse.

2- Height relation between entrance point and exit point is problematic and time consuming; For example, in a 3 km long tunnel (in Arak- Khoram Abad freeway), to relate entrance point to exit point, it is necessary to perform levelling about 90 km. Other example of this boring and time consuming levelling is in Kerman tunnel. This tunnel is 36 km length, but to transfer the entrance point height to exit point, 150 km levelling is needed. According to this paper, The solution for this difficulty is application of astro-geodetic and determination of vertical deflection by digital zenith camera system TZK2-D. These two elements make possible to define geoid profile in terms of tunnel azimuth in entrance and exit of tunnel; So by doing this, surveying engineers are able to transfer entrance point height to exit point of tunnels in easiest way.

1.1 Introduction

Tunnel construction for transport and other usage have existed for centuries. They have been developed both in urban environments for mass traffic transports and in interurban environments. Tunnels are long and deep, especially in mountainous regions. Surveying represents an important role within these tunnels' lifecycles by applying different technologies and methodologies, for different purposes, from the guidance of new tunnels to the monitoring of old ones *(Boavida et al, 2012)*. Astro-geodetic technique is one of the oldest and the most fundamental technique can be used for this application.

The complete astro-geodetic works have a significant influence on the tunnel construction expenses, starting with the preparation of project documentation, tunnel cutting, staking out the route axis, control of work performance and surveying the completed situation *(Zrinjski, 2006)*.

Until the middle of the last century, exclusively astro-geodetic methods allowed the absolute determination of longitude and latitude related to the global terrestrial coordinate system. Essential early applications were positioning (e.g. on expeditions), orientation of geodetic networks or reference ellipsoids, determination of geoid profiles using the method of astronomical leveling *(Hirt and Bürki, 2006)*. Major improvements of astro-geodetic observation techniques could be achieved since the 1970's when transportable photographic zenith cameras were successfully designed and constructed at the University of Hannover to determine vertical deflection component *(Hirt et al, 2010)*.

Moreover, tunneling projects frequently involve the construction of long tunnels whose azimuths are to be determined very accurately, particularly prior to holing. Although conventional traverse methods may be employed, generally, these cannot guarantee the accuracy required and contractual conditions may then specify that independent gyro-theodolite bearings must be obtained *(Whetherelt and Hunt, 2002)*. So to do this operation exactly, the gyro-theodolite must be calibrated.

This research has yielded that astro-geodetic methods provide a fast result in controlling and for guidance of tunnel excavation.

1.2 Theory and Concept

In this research the role of astro-geodetic in precise guidance of long tunnels have been dealt in two main subjects:

1.2.1 ΔH determination

Initial relative positioning results using the satellites of the Global Position system (GPS) encourage users to compute orthometric height differences, $\Delta H = H_2 - H_1$, by the use of well-known relation *(Hein, 1984)*:

$$H_2 - H_1: (h_2 - h_1) - (N_2 - N_1) \qquad (1)$$

Or $\quad \Delta H_{12} = \Delta h_{12} - \Delta N_{12}$

Where $\Delta h = h_2 - h_1$ is difference in ellipsoidal heights and $\Delta N = (N_2 - N_1)$ is the difference in geoid heights. Whereas Δh can be derived by GPS with an accuracy of 0.1 ppm, ΔN has to be determined using other data sources and formulas that will be mentioned. But the main problem is $\Delta H_{1,2}$ determination in long tunnel, because when entrance point height is known, to have exit point height, surveyors must do geometry levelling several kilometre more than the tunnel length especially in mountainous areas. So in this way, astro-geodtic can help surveyors to determine $\Delta H_{1,2}$ and after that engineers are able to calculate exit point height without long time consuming. According to this research steps of $\Delta H_{1,2}$ determination will be as the below diagram:

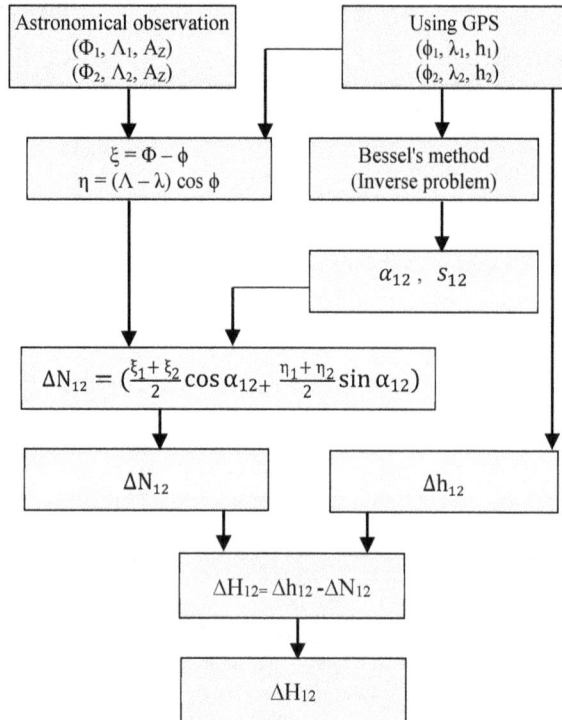

Diagram 1. Basic principle of ΔH determination

1.2.2 Calibration and Correction in gyro-theodolite

As known gyro-theodolite measure horizontal angles from the astronomical North (astronomical Azimuth) with an accuracy of ±3″ (*Lewén, 2006; Lambrou and Pantazis, 2004*). The gyroscope theodolites used to be calibrated before and after their use in an arranged time period frame for their proper function. This is a very important check, which ought to be done carefully and repeatedly. So in this research 5 correction will be introduced to in order to assure the proper function of the instrument and the correct value of the measured astronomical azimuth.

After calibration of gyroscope and applying required correction to examine whether determined azimuth is right or wrong, that azimuth will be compare to the azimuth determined by astronomical (with an accuracy of 0.5″) method.

1.3 Data processing

1.3.1 Vertical defelection components

The astronomical coordinates (Φ, Λ) is obtained by means of *direction measurements* to celestial objects, primarily stars, whose equatorial coordinates right ascension α and declination δ are given in the International Celestial Reference System ICRS. Longitude Λ and latitude Φ define the spatial direction of the plumb line with respect to the International Terrestrial Reference System ITRS (Fig. 1). ITRS and ICRS are linked by Greenwich Sidereal Time GAST being a measure for Earth's rotation phase angle. Astro-geodetic methods use the equivalence of astronomical coordinates (Φ,Λ) and equatorial coordinates (α,δ) for a star exactly located in zenith *(Farzaneh, 2009)* or other directions. When we observe star in zenith direction the equation will be: $\quad \Phi = \delta \quad , \quad \Lambda = \alpha - GAST$ (2)
But for stars in other than the zenith direction, the geodetic coordination can be calculate by reading star height and time.

Vertical deflections (ξ, η) are directly obtained by calculating the difference between astronomical coordinates and geodetic coordinates (ϕ, λ) to be determined with GPS. In linear

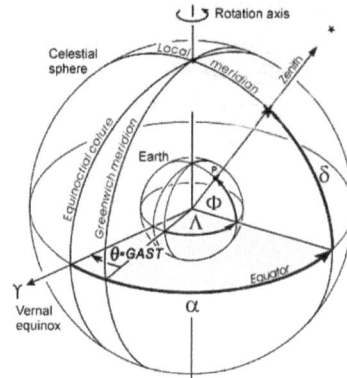

Figure 1. Astronomical coordinate and observation

approximation, the components (ξ, η) are usually computed as *(Hirt et al, 2010)*: $CT=$ Correction Term

$$\xi = \Phi - \phi + CT \quad , \quad \eta = (\Lambda - \lambda) \cos \phi + CT \qquad (3)$$

Nowadays these components can be determined with GPS and Digital zenith camera *(Abedini, 2015)*.

1.3.2 Geodetic azimuth (α_{12}) and distance (s_{12})

We can compute geodetic azimuth by using inverse problem equation that could be called Bessel Bessel's method and have a history dating back to F. W. Bessel's original paper on the topic titled: 'On the computation of geographical longitude and latitude from geodetic measurements.

Inverse problem

In this problem we are given $P_1(\varphi_1, \lambda_1)$ and $P_2(\varphi_2, \lambda_2)$ With the ellipsoid constants a, f, b= a (1-f), e²= f (2-f) and $e'^2 = \frac{e^2}{1-e^2}$ (Fig 2) and (Fig 3).

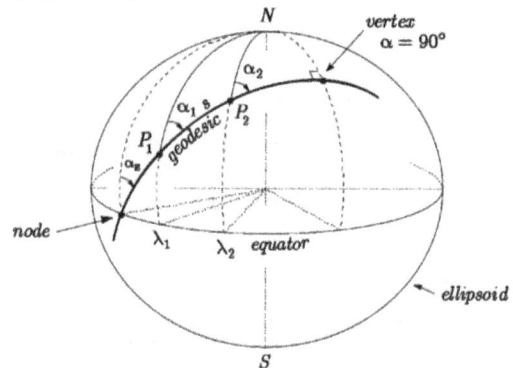

Figure 2. Geodesic on ellipsoid

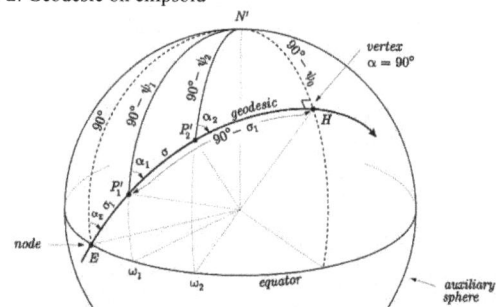

Figure 3. Geodesic on auxiliary sphere

A. Compute reduced latitude ψ_1 and ψ_2 of P_1 and P_2 from

$$\text{Tan } \psi = (1 - e^2)^{\frac{1}{2}} \tan \varphi \tag{4}$$

B. Compute the longitude difference $\Delta\lambda$ on the ellipsoid

$$\Delta\lambda = \lambda_2 - \lambda_1 \tag{5}$$

C. Compute the longitude difference $\Delta\omega$ on the auxiliary sphere between P_1' to P_2' by iteration using the following sequence of equations until there is negligible change in $\Delta\omega$.

$\text{Sin}\,\sigma = \sqrt{sin^2\ \sigma}$ and $\cos\sigma$. This will give $-180° < \sigma \leq 180°$.

$$sin^2\sigma = (\cos\psi_2 \sin\Delta\omega)^2 + (\cos\psi_1 \sin\psi_2 - \sin\psi_1 \cos\psi_2 \cos\Delta\omega)^2 \tag{6}$$

$$\text{Cos } \sigma = sin\,\psi_1 \sin\psi_2 + \cos\psi_1 \cos\psi_2 \cos\Delta\omega \tag{7}$$

$$\text{Tan } \sigma = \frac{sin\sigma}{\cos\sigma}$$

$$\text{Sin } a_E = \frac{cos\,\psi_1 \cos\psi_2 \sin\Delta\omega}{\sin\sigma} \tag{8}$$

$$\text{Cos } 2\,\sigma_m = \cos\sigma - \frac{2\sin\psi_1 \sin\psi_2}{cos^2\,a_E} \tag{9}$$

$$\Delta\omega = \Delta\lambda + (1-C)\, f \sin a_E \{\sigma + C \sin\sigma[\cos 2\sigma_m +$$

$$C \cos\sigma(-1 + 2\cos^2 2\sigma_m)]\} \tag{11}$$

Where:

$$C = \frac{f}{16} cos^2\, a_E\, (4 + f(4 - 3cos^2 a_E) \tag{12}$$

The first approximation for $\Delta\omega$ in this iterative solution can be taken as $\Delta\omega \simeq \Delta\lambda$

D. Compute the reduced latitude of the geodesic vertex ψ_0 from

$$\text{Cos } \psi_0 = \sin a_E \tag{13}$$

E. Compute the geodesic constant u^2 from

$$u^2 = e'^2 sin^2\, \psi_0 \tag{14}$$

F. Compute Vincenity's constants A' and B' from

$$A' = 1 + \frac{u^2}{16384} (4096 + u^2(-768 + u^2(320 - 175u^2))) \tag{15}$$

$$B' = \frac{u^2}{1024} (256 + u^2(-128 + u^2(74 - 47u^2))) \tag{16}$$

G. Compute geodesic distances s from

$$\Delta\sigma = B' \sin\sigma \{\cos 2\sigma_m + \frac{1}{4}B' [\cos\sigma (2cos^2 2\sigma_m - 1) - \frac{1}{6}B' \cos 2\sigma_m (-3 + 4sin^2\sigma)(-3 + 44cos^2 2\sigma_m)]\} \tag{17}$$

$$s_{12} = bA\ (\sigma - \Delta\sigma) \tag{18}$$

H. So finally the geodetic azimuth will be:

$$\text{Tan } \alpha_{12} = \frac{cos\psi_2 \sin\Delta\omega}{\cos\psi_1 \sin\psi_2 - \sin\psi_1 \cos\psi_2 \cos\Delta\omega} \tag{19}$$

I. Compute azimuth α_2 from

$$\text{Tan } \alpha_2 = \frac{cos\psi_1 \sin\Delta\omega}{-\sin\psi_1 \cos\psi_2 + \cos\psi_1 \sin\psi_2 \cos\Delta\omega} \tag{20}$$

So reverse azimuth α_{21} will be

$$\alpha_{21} = \alpha_2 \pm 180° \tag{21}$$

1.3.3 Geoid undulation (ΔN)

The basic principle of astronomical levelling gives us a definite mathematical relationship between geoid undulations and vertical deflection *(Völgyesi, 2005; Tse and Bâki Iz, 2006, Ceylan, 2009)*. According to the notations of Figure 4 we get:

$$dN = \vartheta\, ds \tag{22}$$

where ϑ is the Pizzetti-type deflection of the vertical in the

azimuth α. Between any points P_1 and P_2 the geoid height change is:

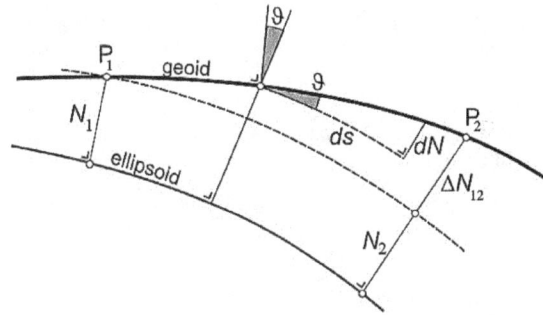

Figure 4. Basic principle of astronomical levelling

$$\Delta N_{12} = \int_{p_1}^{p_2} \vartheta(s)\, ds \tag{23}$$

If P_i and P_k are close together and $\vartheta(s)$ is a linear function between these points the integral (23) can be evaluated by a numerical integration (Völgyesi 1998; Tóth, Völgyesi 2002):

$$N_{p_2} - N_{p_1} = (\frac{\xi_1 + \xi_2}{2} \cos\alpha_{12} + \frac{\eta_1 + \eta_2}{2} \sin\alpha_{12})\, S_{12} \tag{24}$$

According to equation (3)

$$\Delta N_{p_1 p_2} = (\frac{\Phi_1 - \phi_1 + \Phi_2 - \phi_2}{2} \cos\alpha_{12} + \frac{(\Lambda_1 - \lambda_1)\cos\phi_1 + (\Lambda_2 - \lambda_2)\cos\phi_2}{2} \sin\alpha_{12})\, S_{12} \tag{25}$$

To estimate the accuracy we assume that: $\phi_m = \frac{\phi_1 + \phi_2}{2}$

$$\Delta N_{p_1 p_2} = (\frac{\Phi_1 - \phi_1 + \Phi_2 - \phi_2}{2} \cos\alpha_{12} + \frac{(\Lambda_1 - \lambda_1)\cos\phi_m + (\Lambda_2 - \lambda_2)\cos\phi_m}{2} \sin\alpha_{12})\, S_{12} \tag{26}$$

As $\quad \frac{\phi_1}{2} + \frac{\phi_2}{2} = \phi_1 + \frac{\phi_2 - \phi_1}{2}$

$$= \phi_1 + \frac{\Delta\phi_{12}}{2} \tag{27}$$

And

$$\frac{\lambda_1}{2} + \frac{\lambda_2}{2} = \lambda_1 + \frac{\Delta\lambda_{12}}{2} \tag{28}$$

$$\Delta N_{p_1 p_2} = ((\frac{\Phi_1 + \Phi_2}{2} - \frac{\phi_1 + \phi_2}{2})\cos\alpha_{12} + (\frac{(\Lambda_1 + \Lambda_2)}{2} - \frac{\lambda_1 + \lambda_2}{2})\cos\phi_m \sin\alpha_{12})\, S_{12} \tag{29}$$

As there is only point positioning in astronomy, accuracy of error of $\frac{\Phi_1 + \Phi_2}{2}$ depends only on accuracy of Φ_1 and Φ_2, but in GPS accuracy of relative positioning is more than point positioning. So from equations (27), (28) and (29):

$$\Delta N_{p_1 p_2} = ((\frac{\Phi_1 + \Phi_2}{2} - (\phi_1 + \frac{\Delta\phi_{12}}{2}))\cos\alpha_{12} + (\frac{(\Lambda_1 + \Lambda_2)}{2} - (\lambda_1 + \frac{\Delta\lambda_{12}}{2}))\cos\phi_m \sin\alpha_{12})\, S_{12} \tag{30}$$

So from determined geodetic azimuth and geodesic distance from equations (18) and (19), ΔN_{12} can be computed. After that from the equation $\Delta H_{12} = \Delta h_{12} - \Delta N_{12}$, surveyors can determine ΔH_{12} and due to known entrance point height, by using the equation $\Delta H_{12} = H_2 - H_1$, exit point height is computable.

As an example for mentioned equations, in Kerman tunnel according to table 1, estimated ΔN was 2.13 m. Baesd on accuracy of astronomical point positioning and Gps about $0.2''$, if we assume that the maximum error for ξ and η is $0.4''$, when we change this value for ξ and η, the ΔN value change below 5 cm, that is acceptable for administrative project. Morever, Length between entrance and exit point is the important factor that cause error, For instance, in above example for Kerman tunnel, if we assume 3 km instead of 38 km for tunnel length, the error will change in under 1 cm.

	ϕ	λ	Φ	Λ	ξ	η
A	$29°20'$ $24.17169''$	$56°57'$ $10.07054''$	$29°20'34''$	$56°57'12''$	$10''$	$2''$
B	$29°41'$ $05.32196''$	$56°58'$ $57.17907''$	$29°41'18.3''$	$56°59'03''$	$13''$	$5''$
	$S_{12}=38.3238$km		$\alpha_{12}=04°18'33.8''$			
ΔN_{ab}	2.13 m					

Table 1. Kerman tunnel parameters

1.4 Correction in gyro-theodolite

1.4.1 Geodetic correction

The quality of geodetic networks for guiding Tunnel inside long tunnels depends largely on the correct use of a gyroscope. The gyroscope theodolite or gyro-theodolite has a built-in free swing and fast rotation gyroscope that vacillates automatically provides astronomical azimuth (*Lambrou and Pantazis, 2004*). But the determined azimuth with gyro-theodolite is less accurate than the determined azimuth by astronomical methods.

So to control and check the accuracy of azimuth determined by gyro-theodolite, astronomical observation and gyro-theodolite observations should be compare. By the comparison, the necessary geodetic correction can be applied to gyro-theodolite results. the following corrections must be applied to reduce such observations to grid bearings:

  A. Correction for polar motion (actual pole to CIO pole)
  B. Arc to chord correction
  C. Convergence of the meridian
  D. Instrumental corrections
  E. Correction for the deflection of the vertical

That astronomical observation in relation to items (A), (D) and (E) can help surveyor engineers to apply the correction.
The following details can be added concerning these corrections:

1.4.1.1 Polar motion

The earth rotation axis will not remain fixed with respect to the earth body, rather move periodically around a certain mean axis. Such movement of the earth rotation axis is called Polar motion. Due to polar motion, it is important to reduce all observations so they refer to a certain mean pole. The most widely used mean pole is the International Convention of Origin (CIO), which is defined as the mean position of the instantaneous pole during the period 1900 to 1905.However, the deviation between an arbitrary rotation axis and the rotation axis corresponding to CIO is less than 0.1 mgon and therefore this correction is neglected *(Lewén, 2006)*.

1.4.1.2 Arc to chord correction

The arc to chord corrections is applied by reference to formulae of the projection concerned. This correction is negligible for short distances.

1.4.1.3 Convergence of the meridians

Bearing of theodolite telescope will vary from point to point *(Lewén, 2006)*. Thus as one proceeds along a straight line set out by a theodolite on the earth's surface, the bearing of the line will not remain constant but will gradually alter. In latitudes in the neighbourhood of 60°, the alternation amounts to almost a minute of arc in a line of one kilometre in lengths, and in higher latitudes the alteration is even bigger.

When one is using a gyro, the above stated problem is reversed. A gyro will in fact seek out and eventually settle in a meridian (true north) but when one wants to implement the gyro observations on a predefined map grid one has to keep in mind that the observed meridian only coincides with the map grid along the middle meridian of the map grid. The further east or west one gets from the middle meridian the larger the deviation between direction of North of the map grid and the meridian of longitude that the gyro shows.

This deviation (c) may be calculated using the following formula:

$$\text{Tan}\,c = \tan(\lambda - \lambda) * \sin\phi \qquad (31)$$

Where:
ϕ = latitude of gyro position, λ = longitude of gyro position
λ_0= longitude of the middle meridian of the map grid system

1.4.1.4 Instrumental corrections

An alignment error can exist between the indicated heading of the gyroscope and the horizontal optical axis of the theodolite. This constant error can be determined at a measuring range where the azimuth is known.

1.4.1.5 Correction for the deflection of the vertical

The influence of the irregularity of the earth's gravitational field (deflection of the vertical) thus merits special attention in regions where the deviation of the vertical is suspected to be large.

A study of this problem has been carried out for the St Gotthard and Lötschberg tunnels (Carosio et al., 1997). Because of the length of the tunnels, gyroscopic observations are needed in addition to conventional methods. However, in a mountainous area such as the St Gotthard range, the effects of the variation of the earth's gravitational field are not negligible. Experiments have thus been carried out on the effects of such variations on gyroscopic azimuths. The instrument that was used in these experiments was the Gyromat 2000 supplied by Deutsche Montan Technologie (DMT) of Bochum. This instrument has a measuring time of 8 minutes, with a nominal precision of 0.7 mgon.

The application of this correction allows an astronomical azimuth to be converted to geodetic azimuth, as follows *(Heiskanen and Moritz, 1967)*:

$$A = \alpha - \eta \tan\varphi - (\xi \sin\alpha - \eta \cos\alpha) \cot z \qquad (32)$$

Where:
A is the astronomical azimuth
α is the geodetic azimuth
η is the east-west component of the deflection of the vertical
ξ is the north-south component of the deflection of the vertical
φ is the geographical latitude
z is the zenith distance to the observed point

In the case of a tunnel, where the lines of sight are approximately horizontal, cot z = 0, will only the η component of the deflection account to the correction.

Conclusion

This study presented functions for the determination of ∆H between entrance and exit point of tunnel and opened new perspective for guiding long tunnel and geodetic correction and calibration of gyro-theodolite based on astro-geodetic method. In spite of fact that the geometric levelling is time consuming, astronomical can be accomplished in a much shorter time interval. Using astrogeodetic cost in guidance of tunnel with a higher and more valuable amount information and quality. Th shortened surveying can be, by itself, a major advantage in most cases and can also be a decision issue. Furthermore By astronomical methods, azimuth of directions can be determine with an accuracy of 0.5 arcsecond, whereas, nowadays, no gyroscope can measure the azimuth in this accuracy.

Morever, If gyro observation are to be used in an adjustment to improve the network, it is very important that the observations are checked within themselves, i.e.that all corrections are applied and that the surveying and computation methods are such that the influence of gross and systematic errors are minimized.

REFERENCES

Abedini, A., Farzaneh, S., 2014. New approach for vertical deflection determination using digital Zenith cameras. *Iran Geophysic*. 9(1), pp. 29-15.

Baovida, J., Oliveira, A., Santos, B., 2012. Precise long tunnel survey using the Riegl VMX-250 mobile laser ranging system. Conference of riegl lidar 2012.

Ceylan, A., 2009. Determination of the deflection of vertical components via GPS and leveling measurement: A case study of a GPS test network in Konya, Turkey. Scientific Research and Essay. 4 (12), pp. 1438-1444.

Deakin, R. E., Hunter, M. N., 2009. Geodesics on an ellipsoid - Bessel′ method. School of Mathematical & Geospatial Sciences, RMIT University publication, Melbourne, Australia.

Hein, G. w., 1985. Orthometric height determination using GPS observation and integrated geodesy adjustment model. *NOAA technical report NOS 110 NGS 32*, NOAA technical publication, USA.

Hirt, C., Bürki, B., Somieski, A., Seeber., G., 2010. Modern determination of vertical deflections using digital zenith cameras. *Surveying Engineeri*ng. 136(1), PP. 1-12.

Krakiwsky , E.J., 1995. Thomson, D.B., Geodetic position computation. Department of Geodesy and Geomatics Engineering University of New Brunswick publication, Canada.

Lewén, I., 2006. Use of gyrotheodolite in underground control network. *Master's of Science Thesis in Geodesy*. Royal Institute of Technology, Stockholm, Sweden.

Lambrou, E., Pantazis, 2004. Accurate Orientation of the Gyroscope's Calibration System. *FIG Working Week meeting*, Athen, Greek.

Tse, C. M., Bâki Iz, H., 2006. Deflection of the Vertical Components from GPS and Precise Leveling Measurements in Hong Kong. *Surveying Engineering*. 132(3), pp. 97-100.

Völgyesi, L., 2005. Deflections of the vertical and geoid heights from gravity gradients. *Acta Geodaetica et Geophysica Hungarica*. 40(2), pp. 147-157.

Velasco- Gómez, j., Prieto, J. F., Molina, L., Herreo, T., Fábrega, j., Pérez-Martin, E., Use of the gyrotheodolite in underground networks of long high-speed railway tunnels. http://www.maneyonline.com/doi/abs/10.1179/1752270615Y.0000000043?journalCode=sre.

15

COMPARISON OF TARGET- AND MUTUAL INFORMATON BASED CALIBRATION OF TERRESTRIAL LASER SCANNER AND DIGITAL CAMERA FOR DEFORMATION MONITORING

Mohammad Omidalizarandi* and Ingo Neumann

Geodetic Institute, Leibniz Universität Hannover, Germany – (zarandi, neumann)@gih.uni-hannover.de

Commission I, WG I/4

KEY WORDS: Terrestrial Laser Scanner, Digital Camera, Extrinsic Calibration, Bundle Adjustment, Mutual Information

ABSTRACT

In the current state-of-the-art, geodetic deformation analysis of natural and artificial objects (e.g. dams, bridges,...) is an ongoing research in both static and kinematic mode and has received considerable interest by researchers and geodetic engineers. In this work, due to increasing the accuracy of geodetic deformation analysis, a terrestrial laser scanner (TLS; here the Zoller+Fröhlich IMAGER 5006) and a high resolution digital camera (Nikon D750) are integrated to complementarily benefit from each other. In order to optimally combine the acquired data of the hybrid sensor system, a highly accurate estimation of the extrinsic calibration parameters between TLS and digital camera is a vital preliminary step. Thus, the calibration of the aforementioned hybrid sensor system can be separated into three single calibrations: calibration of the camera, calibration of the TLS and extrinsic calibration between TLS and digital camera. In this research, we focus on highly accurate estimating extrinsic parameters between fused sensors and target- and targetless (mutual information) based methods are applied. In target-based calibration, different types of observations (image coordinates, TLS measurements and laser tracker measurements for validation) are utilized and variance component estimation is applied to optimally assign adequate weights to the observations. Space resection bundle adjustment based on the collinearity equations is solved using Gauss-Markov and Gauss-Helmert model. Statistical tests are performed to discard outliers and large residuals in the adjustment procedure. At the end, the two aforementioned approaches are compared and advantages and disadvantages of them are investigated and numerical results are presented and discussed.

1. INTRODUCTION

In the current state-of-the-art, geodetic deformation analysis of natural and artificial objects (e.g. dams, bridges, towers, railroads, landslides,...) is an ongoing research in both static and kinematic mode. In this research, due to increasing the accuracy of geodetic deformation analysis, TLS and a high resolution digital camera are integrated to complementarily benefit from each other. On the one hand, TLS can provide high resolution 3D data in the sub-millimetre range in combination with reflectivity values. Consequently, a reflectance image can be generated using central perspective representation to project the 3D point clouds to a virtual image plane. On the other hand, digital camera can acquire rich and high quality colour images. In the integrated sensor system, high resolution cameras are advantageous due to having high angular accuracy of sub-pixel accuracy image measurements which would improve the lateral accuracy of laser scanners (Schneider & Maas, 2007). In addition, this integration focuses at filling gaps in TLS data to compensate modeling errors and to reconstruct more details in higher resolution (Moussa et al., 2012).

The main purpose of this research is to high accurately estimate extrinsic parameters between TLS and digital camera to compensate the deficiency of the TLS measurements for deformation monitoring of the objects, e.g. in case of large incidence angle, by using high resolution camera. Furthermore, digital images would assist us to detect deformation analysis in both direction of laser beam and perpendicular to laser beam. Moreover, this integration leads to increasing redundancy in the adjustment procedure.

In order to ideally relate digital camera coordinate frame to the TLS coordinate frame, the digital camera is mounted on top of the TLS using clamping system (figure 1, right). To avoid any vibration of digital camera and blurring of images, Nikon wireless mobile utility application is setup on the cell phone and by the usage of Wi-Fi connection, photographs are captured indirectly.

Figure 1. The employed sensors. Laser tracker (left), A D750 digital camera and Z+F Imager 5006 TLS and their corresponding coordinate systems (right)

The calibration of the aforementioned hybrid sensor system can be separated into three single calibrations: calibration of the camera, calibration of the TLS and extrinsic calibration between TLS and digital camera. The interior orientation of the camera and internal error sources of the TLS can be determined in the laboratory to reach high accurate calibration values. However, external error sources especially atmospheric and object related errors could not be considered in the laboratory and need on-site calibration to be removed as well.

Unnikrishnan and Hebert (2005) described the algorithm to estimate the extrinsic calibration parameters of a camera with respect to a laser rangefinder using checkerboard calibration targets. In this method, a plane fits to the manually selected targets on the checkerboard pattern and aligned it with the detected pattern of the image using optimization algorithm. Moussa et al. (2012) proposed an automatic procedure to combine TLS and digital camera based upon free registration using ASIFT and RANSAC algorithm to match reflectance image and RGB image. Absolute camera orientations are obtained on the basis of space resection method. Lichti et al. (2010) presented the self-calibration of the range camera with respect to the rangefinder in a free-network bundle adjustment using signalized targets. Variance component estimation is applied to optimally re-weight observations iteratively. Pandey et al. (2012) proposed the automatic targetless extrinsic calibration of a Velodyne 3D laser scanner and Ladybug3 omnidirectional camera on the basis of the mutual information (MI) algorithm to estimate extrinsic calibration parameters by maximizing the mutual information between the reflectivity values of the laser scanner and intensity values of the camera image. Taylor and Nieto (2012) proposed a method to compute extrinsic and intrinsic calibration parameters between camera and LIDAR scanner. This approach utilizes normalised mutual information to compare images with the laser scans projections. Particle swarm optimization algorithm is applied to optimally determine the parameters.

In this research, we focus on highly accurate estimation of the extrinsic parameters between TLS and digital camera which is necessary and preliminary step for deformation monitoring. Two different strategies of target- and MI based are applied. In target-based approach, focal length of the camera, exterior orientation parameters between TLS and camera (position and orientation; 6 DOF), exterior orientation parameters between TLS and laser tracker (scale, position and orientation; 7 DOF) and target coordinates are estimated with high accuracy. Laser tracker (LT (figure 1, left)) measurements are carried out with superior accuracy and independently. In addition, its measurements are considered as a reference coordinate frame. In MI-based approach, extrinsic calibration parameters between TLS and digital camera are obtained with adaptation and modification of the Pandey's work to our research purpose by considering horizontal angle measurement of the TLS as additional parameter into the transformation matrix.

2. DATA ACQUISITION, INTERFACING AND PRE-PROCESSING

In the target-based approach, data acquisition step comprises image measurements, TLS measurement and LT measurements. In the MI-based approach, it consists of the RGB image from the digital camera and generated reflectance image from the TLS reflectivity values.

Images are captured with Nikon D-750 24.3 megapixel digital camera and centre of each target is computed based upon detection of the four centriods of the circles within each target and performing averaging. For instance, an exemplary target with detected centriods is illustrated in figure 2. Afterwards, extracted image measurements are rectified based on well known Brown's equations to eliminate the effects of radial and decentring distortions.

Figure 2. Calibration room (left) depiction of detected centroid of circles using sub-pixel target mode of the PhotoModeler software (right)

TLS point targets accurately acquired using "Fit target" mode of the Z+F LaserControl software.

The horizontal angle measurement of the TLS (Az) is defined as a 3x3 rotation matrix to rotate TLS coordinate frame around its Z-axis to the digital camera coordinate frame (Al-Manasir and Fraser (2006)). It is written for each captured image and can be considered as additional observation in the adjustment procedure. As can be seen in figure 1 (right), TLS coordinate frame, digital camera coordinate frame and Az are depicted. LT measurements are utilized as an additional observation and they are obtained by pointing to the mounted corner cubes which are located at the centre of each target.

Reflectance images can be generated based on the scanning matrix and central perspective representation. In the first approach, each 3D data is assigned to one pixel based on the scan resolution. It is quite simple and fast. However, as a drawback, straight lines appear as curved lines (Meierhold et al., 2010). In the second approach, TLS data is projected to a virtual image plane on the basis of the collinearity equation (Moussa et al., 2012).

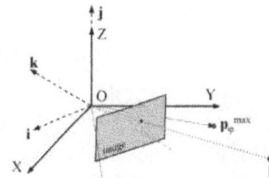

Figure 3. Definition of coordinate systems (Meierhold et al., 2010)

As can be seen in figure 3, the maximum and minimum horizontal angle of the TLS is determined to reduce the size of the entire scan of TLS data to project into the image space. In the figure 4, rectified image and generated reflectance image from TLS data are depicted.

Figure 4. Rectified image (left), Reflectance image from TLS (right)

3. METHODOLOGY

3.1 Target-based calibration

In the target based approach, focal length of the camera, extrinsic parameters between TLS and camera, extrinsic parameters between TLS and LT and target coordinates are estimated as unknown parameters. In order to obtain exterior orientation parameters between TLS and digital camera, space resection bundle adjustment is employed based upon collinearity equations to determine the condition that a perspective centre, a point in the image space and its corresponding coordinate in the object space are on a straight line (equations 1 & 2). Since aforementioned equations are non-linear with respect to the parameters, it needs to be linearized to compute parameters iteratively. Therefore, initial starting values are estimated using direct linear transform (DLT) in combination of RANSAC algorithm to robustly estimate the parameters. In addition, the mathematical model to estimate the exterior orientation parameters between TLS and LT is solved based on similarity transformation (equation 3). Furthermore, additional constraint is defined to compute target point coordinates in the object space (equation 4). Thus, least square solutions are computed by means of the Gauss-Markov model (GM) and Gauss-Helmert model (GH). Therefore, four target functions (equations 1-4) are determined as follow:

$$F_x = x - f\frac{r}{q} \tag{1}$$

$$F_y = y - f\frac{s}{q} \tag{2}$$

$$F = \lambda R_{\kappa'\varphi'\omega'}\begin{bmatrix}X_{TLS}\\Y_{TLS}\\Z_{TLS}\end{bmatrix} + \begin{bmatrix}X_c'\\Y_c'\\Z_c'\end{bmatrix} - \begin{bmatrix}X_L\\Y_L\\Z_L\end{bmatrix} \tag{3}$$

$$F = \begin{bmatrix}X\\Y\\Z\end{bmatrix} - \begin{bmatrix}X_{TLS}\\Y_{TLS}\\Z_{TLS}\end{bmatrix} \tag{4}$$

where

$$\begin{bmatrix}r\\s\\q\end{bmatrix} = R_{\kappa\varphi\omega} * \begin{bmatrix}R_3(Az)\begin{bmatrix}X_{TLS}\\Y_{TLS}\\Z_{TLS}\end{bmatrix} - \begin{bmatrix}X_c\\Y_c\\Z_c\end{bmatrix}\end{bmatrix} \tag{5}$$

$$R_{\kappa\varphi\omega} = R_\kappa . R_\varphi . R_\omega \tag{6}$$

In equations (1-6), (x, y) are the target coordinates in the image space, (X, Y, Z) are the target coordinates in the object space, $(X_{TLS}, Y_{TLS}, Z_{TLS})$ are TLS point coordinates of the targets, (X_C, Y_C, Z_C) are the translations between TLS and digital camera, $(\kappa, \varphi, \omega)$ are the rotation angles between TLS and digital camera, (Az) is horizontal angle measurement of the TLS, (X_C', Y_C', Z_C') are the translations between TLS and LT, (X_L, Y_L, Z_L) are LT point coordinates of the targets, $(\kappa', \varphi', \omega')$ are the rotation angles between TLS and LT and (λ) is the scale factor between TLS and LT.

GM model is a set up linear or non linear relation between observations and unknown parameters. It is merely determined by observations to estimate unknown parameters. In this type of least square adjustment, square sum of residuals are minimized for one type of observation (image measurements). GH model is more complete and sophisticated model compared to GM model and comprising all the unknown parameters and observations that can be updated as unknowns iteratively.

3.1.1 Gauss-Markov model:
In this research, GM model is solved based on equations 1 & 2. In equation 7, v is a vector of residuals, A is a matrix of the coefficients of the unknowns which is so called design matrix, ΔX is the reduced vector of parameters (unknown extrinsic parameters) and L^0 is the reduced vector of observations. In equation 8, F_x^0 and F_y^0 are the target functions of the equations 1 and 2 which are substituted for the initial values.

$$v = A\Delta X - L^0 \tag{7}$$

$$\begin{bmatrix}v_{x1}\\v_{y1}\\\vdots\\v_{xn}\\v_{yn}\end{bmatrix} = \begin{bmatrix}\frac{\partial F_{x1}}{\partial X_C} & \frac{\partial F_{x1}}{\partial Y_C} & \frac{\partial F_{x1}}{\partial Z_C} & \frac{\partial F_{x1}}{\partial \omega} & \frac{\partial F_{x1}}{\partial \varphi} & \frac{\partial F_{x1}}{\partial \kappa}\\\frac{\partial F_{y1}}{\partial X_C} & \frac{\partial F_{y1}}{\partial Y_C} & \frac{\partial F_{y1}}{\partial Z_C} & \frac{\partial F_{y1}}{\partial \omega} & \frac{\partial F_{y1}}{\partial \varphi} & \frac{\partial F_{y1}}{\partial \kappa}\\\vdots & \vdots & \vdots & \vdots & \vdots & \vdots\\\frac{\partial F_{xn}}{\partial X_C} & \frac{\partial F_{xn}}{\partial Y_C} & \frac{\partial F_{xn}}{\partial Z_C} & \frac{\partial F_{xn}}{\partial \omega} & \frac{\partial F_{xn}}{\partial \varphi} & \frac{\partial F_{xn}}{\partial \kappa}\\\frac{\partial F_{yn}}{\partial X_C} & \frac{\partial F_{yn}}{\partial Y_C} & \frac{\partial F_{yn}}{\partial Z_C} & \frac{\partial F_{yn}}{\partial \omega} & \frac{\partial F_{yn}}{\partial \varphi} & \frac{\partial F_{yn}}{\partial \kappa}\end{bmatrix}\begin{bmatrix}\Delta X_C\\\Delta Y_C\\\Delta Z_C\\\Delta \omega\\\Delta \varphi\\\Delta \kappa\end{bmatrix} - \begin{bmatrix}F_{x1}^0\\F_{y1}^0\\\vdots\\F_{xn}^0\\F_{yn}^0\end{bmatrix} \tag{8}$$

3.1.2 Gauss-Helmert model:
GH model is the combination of the observations and unknowns in the target functions and it is denoted as:

$$F(\hat{L}, \hat{X}) = w + Bv + A\hat{x} \tag{9}$$

Where $\hat{L}$ is estimated observation vector, $\hat{X}$ is estimated unknown vector, A matrix is derivative with respect to the unknown parameters, B matrix is derivative with respect to the observations and w is the vector of misclosures. Thereafter, unknown parameters are computed as follows:

$$\begin{bmatrix}k\\\hat{x}\end{bmatrix} = -\begin{bmatrix}BQ_{ll}B^T & A\\A^T & 0\end{bmatrix}^{-1} \cdot \begin{bmatrix}w\\0\end{bmatrix} \tag{10}$$

Where $\hat{x}$ is the estimated reduced unknown vector. Moreover, vector of residuals are computed by:

$$v = Q_{ll}B^T k \tag{11}$$

Where Q_{ll} is the cofactor matrix of the observations. In addition, a-posteriori variance factor $(\hat{\sigma}_0^2)$ is calculated as follows (Niemeier, 2002):

$$\hat{\sigma}_0^2 = -\frac{k^T(w + A\hat{x})}{b - u} \tag{12}$$

Where b is the number of constraints and u is the number of unknown parameters.

3.1.3 Variance Component Estimation and Statistical Test:
Variance component estimation is applied to assign optimal weights to the observations in the adjustment procedure iteratively. The statistical test is performed to investigate the adjustment results. Additionally, the uncertainty of the measurements and unknown parameters is computed. In this research, χ^2 and F test with 95% confidence level are applied to detect outliers.

3.2 Mutual Information-based Calibration

Mutual information (MI) is used to detect statistical dependencies or a measure of coupling between signals (Pompe et al., 1998). MI is defined on the basis of Shannon's definition of entropy (equation 13) and is interpreted based upon the amount of information and event that occurs, the uncertainty about the result of an event, and the dispersion of the probabilities when the event occurs (Alempijevic et al., 2006).

$$H = \sum_i p_i * \log\frac{1}{p_i} \tag{13}$$

Where p_i is the probability mass function of random variable i.

In this work, MI is determined on the basis of the entropy of reflectance image (H(A)), entropy of RGB image (H(B)) and H(A,B) as a joint entropy (equation 14) and generally it means the amount of information that A contains about B. MI is maximized by maximizing the terms H(A) and H(B) and minimizing the H(A,B).

$$MI(A,B) = H(A) + H(B) - H(A,B) \qquad (14)$$

where

$$H(A) = H(p(a)) = \sum_a p_a * \log \frac{1}{p_a} \qquad (15)$$

$$H(B) = H(p(b)) = \sum_b p_b * \log \frac{1}{p_b} \qquad (16)$$

$$H(A,B) = H(p(a,b)) = \sum_a \sum_b p(a,b) * \log \frac{1}{p(a,b)} \qquad (17)$$

In equations (15-17), a and b are the real numbers of the events that random observations A and B of a probabilistic experiment are mapped onto them (Pandey, 2014). MI based approach is an automatic procedure that is usually applied in outdoor calibration without any need of mounted targets in the field. In this approach, extrinsic calibration parameters are estimated by maximizing the mutual information between reflectance image of TLS and RGB image of the digital camera and correlation coefficients are computed. Then, different scan measurements from different horizontal angle measurements of TLS are considered in a single optimization framework and the parameters of interest are computed by means of the gradient ascent algorithm (Pandey et al., 2012).

The main goal of the author is to apply MI based approach for in situ calibration to eliminate systematic errors (e.g. clamping system) and consequently avoiding target based calibration in the field. In order to adapt pandey' algorithm to our work, Az included as additional parameter in the transformation matrix to re-project 3D point clouds to the 2D image correctly. Thus, equation 5 is utilized to perform this projection.

As a drawback of this method, it cannot be applied directly to range sensors without associated reflectivity information. Furthermore, it needs quite good initialization values of 6 DOF. Moreover, in case of speed up of the algorithm, it is significantly slower than target based approach. In addition, in order to obtain better results, reflectivity values of the TLS need to be calibrated in addition to image enhancement and filtering of the RGB image (e.g. brightness and contrast).

4. EXPERIMENTS AND RESULTS

In this research, two different approaches of target- and MI-based are investigated in two case studies. In the first case study, two different adjustment models (GM and GH models) are solved. GM model is just implemented as preliminary test to achieve primary results very fast. But, due to more complete and accurate results of the GH model compared to the GM model, merely its results from target-based approach are presented. In the second case study, MI based approach from Pandey's work is adapted to this research and its results are presented and discussed.

For the experiments, a calibration room is measured with Z+F IMAGER 5006 in the super high resolution mode with horizontal and vertical angle resolution of $0.0018°$. It has the maximum field of view of $360° \times 310°$ in horizontal and

vertical respectively and its accuracy is $0.007°$ rms. Thereafter, 84 images are captured with Nikon D750 digital camera to fully cover our calibration room. Targets are measured in the both image and object space, respectively. The number of measured targets in object space is 25 and number of measured targets in the image space is 395. LT is utilized for validation and check the accuracy of our calibration results with the super high accuracy.

4.1 Case Study I:

In this case study, least square solution is solved on the basis of GH model. Employed sensors are TLS, LT and digital camera. Observations are target coordinates in the image space, TLS coordinates of the targets, LT coordinates of the targets and horizontal angle measurements of TLS. Extrinsic parameters between TLS and digital camera (table 1), extrinsic parameters between TLS and LT (table 2), focal length and target point coordinates in object space are the unknown parameters.

6 DOF	Value	σ
ω	88.7180 (Deg.)	0.0032
φ	0.11965 (Deg.)	0.0045
κ	0.04651 (Deg.)	0.0018
X_C	-0.0021 (m)	0.0003
Y_C	0.2195 (m)	0.0002
Z_C	0.0956 (m)	0.0001

Table 1. Extrinsic parameters between TLS and digital camera (6 DOF) using GH model

7 DOF	Value	σ
ω'	0.10599 (Deg.)	0.0007
φ'	-0.0611 (Deg.)	0.0012
κ'	96.3559 (Deg.)	0.0012
X_C'	12.8276 (m)	0.0001
Y_C'	13.9031 (m)	0.0001
Z_C'	1.6952 (m)	0.0001
λ	0.9999	2.03e-05

Table 2. Extrinsic parameters between TLS and LT (7 DOF) using GH model

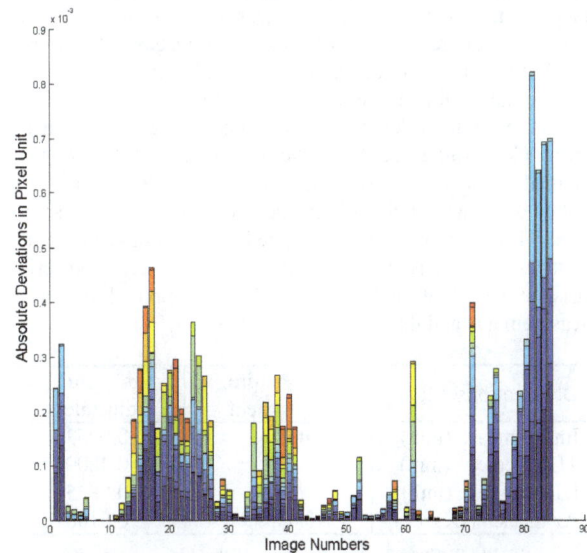

Figure 5. Absolute deviations of the re-projected estimated TLS data and estimated image targets in pixel unit

Figure 5 is illustrated to visualize the accuracy of the implemented space resection bundle adjustment algorithm. It shows the absolute deviations of the re-projected estimated TLS data and estimated image targets in pixel unit. As can be seen from Y-axis, the deviations are in sub-pixel range which indicates that constraints of the adjustment are fulfilled.

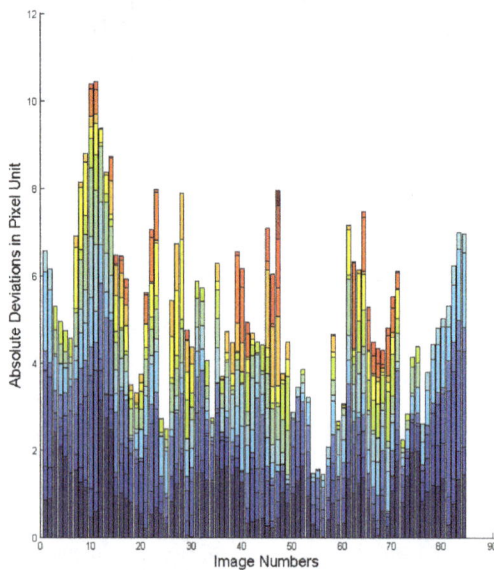

Figure 6. Absolute deviations of the estimated image targets and image targets measurements in pixel unit

In the figure 6, the absolute deviations of the estimated image targets and image targets measurements in pixel unit is depicted. X-axis corresponds to image numbers and Y-axis corresponds to absolute deviations in pixel unit. For each image, all the targets with their deviations in x and y directions are considered in one column. For instance, in the first column of the figure 6, image one contains three targets which with consideration of their deviations in x and y directions respectively, six colourful blocks are shown. In addition, the maximum magnitude belongs to image 11 since it contains six targets.

Variance component estimation leads to obtain accurate standard deviations of the observation. As we can see in table 3, standard deviation of the image measurements is in sub-pixel range since the resolution of the captured images is 0.006 mm. In addition, standard deviation of the TLS measurements is close to half millimeter since we were close to the targets (less than 6 meter) in our laboratory and experiencing less systematic errors. Concerning the Az of the TLS is a bit worse than its nominal value in the user manual that is $0.007°$ since it is written down with 0.001 decimal degree from the display screen of the TLS. Moreover, as we expected, standard deviation of the LT measurements is close to 0.1 mm. Furthermore, A-posteriori variance factor of unit weight $(\hat{\sigma}_0^2)$ is computed for entire measurements and that is equal to 0.8975.

Observations	σ – before adjustment	σ – after adjustment
Image meas. (mm)	0.0243	0.0053
TLS meas. (mm)	1.0	0.4809
LT meas. (mm)	0.1	0.0835
Az meas. (Degree)	0.03599	0.0100

Table 3. Standard deviations of the observations

As can be seen in figure 7, residuals for all type of the observations (image measurements, TLS measurements, LT measurements and horizontal angle measurements of the TLS) are illustrated. Furthermore, some of the LT measurements

residuals are too large which we will investigate them in the future.

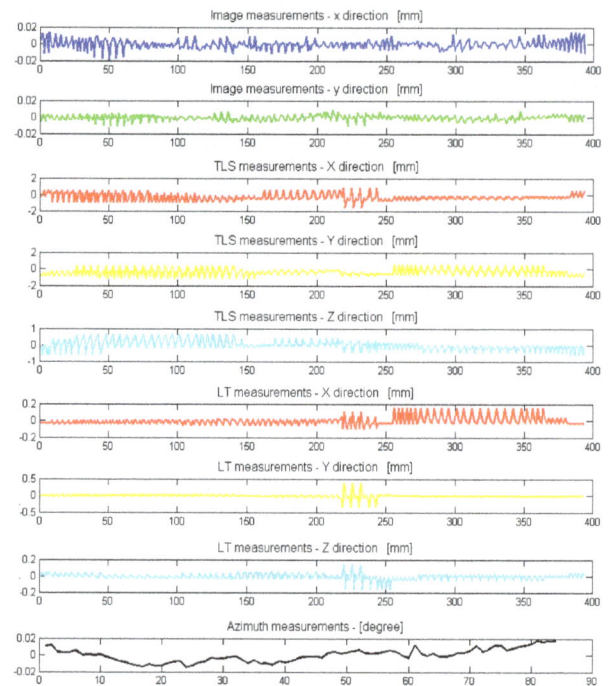

Figure 7. Residuals of the observations

In figure 8, re-projection of the downsampled point clouds into the rectified image by the usage of calibration parameters is illustrated.

Figure 8. Re-projection of downsampled point clouds into the rectified image using estimated extrinsic parameters of the calibration

4.2 Case Study II:

In the second case study, MI based approach as an alternative approach is applied to compare it with high accurate target based approach. In addition, author is investigating about applicability of MI based approach for in field calibration. In this work, we examine the MI based approach merely for one image. Extrinsic parameters between TLS and digital camera are indicated in table 4.

6 DOF	Value
ω	88.7189 (Deg.)
φ	0.12316 (Deg.)
κ	0.04646 (Deg.)
X_C	-0.0023 (m)
Y_C	0.2207 (m)
Z_C	0.0964 (m)

Table 4. Extrinsic parameters between TLS and digital camera (6 DOF) – MI based approach

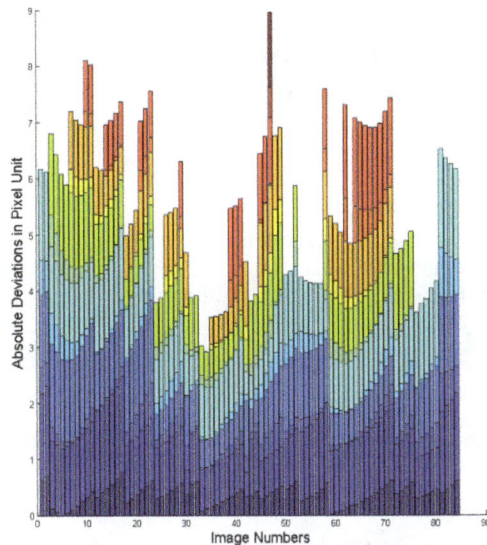

Figure 9. Absolute deviations of the re-projected estimated TLS data and estimated image targets in pixel unit - MI based approach

In the figure 9, absolute deviations of the re-projected estimated TLS data and estimated image targets in pixel unit is indicated. As can be seen, estimated extrinsic parameters in MI based approach is not as accurate as target based approach and it leads to increasing the deviations.

ΔX_C (m)	ΔY_C (m)	ΔZ_C (m)	$\Delta \omega$ (Deg.)	$\Delta \varphi$ (Deg.)	$\Delta \kappa$ (Deg.)
0.0002	-0.0012	-0.0007	-0.0009	-0.0035	5.0e-05

Table 5. Deviations of 6 DOF in the two aforementioned approaches

In the table 5, deviations of extrinsic calibration parameters (6 DOF) between two approaches are indicated. These deviations can be related to the sensors noises or uncertainties of the measurements and unknown parameters. Furthermore, it can be due to remaining outliers and also considering just one image in the MI-based approach.

CONCLUSION

The main objective of this research is to obtain extrinsic parameters between fused sensors (TLS, digital camera and LT). Two different methodology of target based and MI based are applied. As a result of the first methodology, GH model is more accurate comparing to the GM model since we have this possibility to use different types of the observations with different weights in the non linear relations to the parameters. In addition, variance component estimation assists us to automatically assign adequate weights to the observations iteratively and consequently arising high accurate adjustment results. Moreover, statistical tests are beneficial due to rejecting outliers and large residuals which are above the pre-determined test value. MI based approach is come up with the lower accuracy results compared to the target based approach and it did not fully satisfy our purpose for in situ calibration and needs more efforts and investigations.

In the future work, MI based approach for numerous scans and images should be investigated. In addition, high accurate extrinsic parameters from target based approach are utilized in deformation monitoring and analysis to exploit the possibility of images and TLS data simultaneously.

REFERENCES

Alempijevic, A., Kodagoda, S., Underwood, J., Kumar, S., & Dissanayake, G. (2006). Mutual information based sensor registration and calibration. In *Intelligent Robots and Systems, 2006 IEEE/RSJ International Conference on IEEE*, pp. 25-30.

Al-Manasir, K., & Fraser, C. S. (2006). Registration of terrestrial laser scanner data using imagery. *The Photogrammetric Record, 21*(115), 255-268.

Lichti, D. D., Kim, C., & Jamtsho, S. (2010). An integrated bundle adjustment approach to range camera geometric self-calibration. *ISPRS Journal of Photogrammetry and Remote Sensing*, 65(4), 360-368.

Meierhold, N., Spehr, M., Schilling, A., Gumhold, S., & Maas, H. G. (2010). Automatic feature matching between digital images and 2D representations of a 3D laser scanner point cloud. *Int. Arch. Photogramm. Remote Sens. Spat. Inf. Sci, 38*, 446-451.

Moussa, W., Abdel-Wahab, M., & Fritsch, D. (2012). An Automatic Procedure for Combining Digital Images and Laser Scanner Data. *International Archives of the Photogrammetry, Remote Sensing and Spatial Information Sciences, 39*, B5.

Niemeier, W. (2002). Ausgleichungsrechnung: eine Einführung für Studierende und Praktiker des Vermessungs-und Geoinformationswesens. Walter de Gruyter, Berlin.

Pandey, G., McBride, J. R., Savarese, S., & Eustice, R. (2012). Automatic Targetless Extrinsic Calibration of a 3D Lidar and Camera by Maximizing Mutual Information. *Proceedings of the AAAI National Conference on Artificial Intelligence*, pp. 2054-2056. C++ source code is available from: http://robots.engin.umich.edu/SoftwareData/ExtrinsicCalib

Pandey, G. (2014). An Information Theoretic Framework for Camera and Lidar Sensor Data Fusion and its Applications in Autonomous Navigation of Vehicles, *Doctoral dissertation, The University of Michigan*. Available from: http://deepblue.lib.umich.edu/handle/2027.42/107286

Pompe, B., Blidh, P., Hoyer, D., & Eiselt, M. (1998). Using mutual information to measure coupling in the cardiorespiratory system. *Engineering in Medicine and Biology Magazine, IEEE, 17*(6), 32-39.

Schneider, D., & Maas, H. G. (2007). Integrated bundle adjustment of terrestrial laser scanner data and image data with variance component estimation. *The Photogrammetric Journal of Finland, 20*, 5-15.

Taylor, Z., & Nieto, J. (2012). A mutual information approach to automatic calibration of camera and lidar in natural environments. In *Australian Conference on Robotics and Automation*, pp. 3-5.

Unnikrishnan, R., & Hebert, M. (2005). Fast extrinsic calibration of a laser rangefinder to a camera. *Technical report CMU-RI-TR-05-09, Robotics Institute, Carnegie Mellon University*

16

THE ASSESSMENT OF ORTHOPHOTO QUALITY WITH RESPECT TO THE STRUCTURE OF DIGITAL ELEVATION MODEL

M.Modiri [a], H.Enayati [b], M. Ebrahimikia [c], *

[a] Professor at Malek Ashtar University of Technology, Esfahan, Iran- mmodiri@ut.ac.ir
[b] M.s degree of Photogrammtry, Khaje Nasir university Tehran, Iran
[c] M.s degree of Photogrammtry, The university of Tehran, Iran - moj_ebrahimikia@yahoo.com

KEY WORDS: orthophoto, height displacement, orthophoto elongation, Orthophoto precision, Orthophoto quality, DEM structure

ABSTRACT

Orthophoto is an image which is being corrected geometrically so each object has to be situated on the corrected place consequently. Choosing the best DEM structure with respect to the area topographic is the most challenge which has more important role when dealing with rough surfaces displacements in duration of orthophoto procedures. The Lower DEM resolution makes points density lower and makes the procedure faster but cause to decreasing the product precision in compare to choosing the other one. However if a fine resolution DEM cause to very delicate displacement corrections aside of the other benefits but it makes to appear some undesired visualized errors like as elongation error especially in an areas which are hidden with some obstacles and there are lacks of data in an imaging. For preventing of such error in DEM structure calculation and earning the most benefits, we found and execute some solutions. In other word we answered to this question that what DEM resolution is the best for orthophoto production. In the following we have done some tests. First a dense DEM of a topographic area calculated and edited accurately then its density was reduced in some steps gradually. At each stage the root mean square error (RMSE) of interpolated heights of points which were laid in the distance between the corresponding DEMs pixels has been calculated respectively. Two interpolation methods (Nearest neighbour and Bilinear interpolation) have been used in this test. Decreasing the DEMs density or increasing the pixel size made the amounts of errors high and the rate of this changing dependent on the kind of topography directly. So we divided the area into some reasonable topographic classes then calculated our results for each class separately. The result of each strategy compared with each other and presented in both numerical tables and some illustrated images.

Because of the relation between horizontal precision of orthophotos which are existed in the standard producing instruction and the accuracy of the DEM which are mostly related to its density, the suitable resolution for producing different scale orthophotos at each kind of topographic class have been calculated from mentioned methods consequences and shown as a final result.

---

* Corresponding author

1. INTRODUCTION

Orthophotos are one of the most applicable georefrenced products that usually produced based on aerial or satellite images Having. Geographical and radiometrical data together made the high number of Orthophoto usages in prior stages of most engineering plans which have caused to have more attention to both of the geometric and quality aspects. They are produced in different resolution or scales. Its geometrical resolution highly affected by digital elevation model (DEM) and the quality of geometrical corrections like as aerial triangulation (AT) algorithm so they are important factors in orthophoto generations. The error of AT due to some obvious effects on orthophotos like as displacements. In this paper after passing a review of generation of orthophotos and its necessary steps and factors like as georefrencing procedure and digital elevation model (DEM), we have focused on DEM structure influences on the quality and precision of orthophotos. The precision of DEM is related to images scale and resolution, radiometric quality Of images in matching procedure, DEM grid spacing and the way of its producing, Simard(1997). The best amount of DEMs density is one of the largest challenges in orthophoto producing which makes the time of producing low or high. If decreasing the density makes the procedure faster but cause to reach unreal DEM and incorrect rectification on orthophotos either incorrect position of features and relative displacements between same features on coverage images. At the next part of this paper we have a review on how the orthophoto is made based on corresponded DEM and images orientation factors as well as the effect of DEMs density on horizontal displacements of orthophotos. For explaining more about the effects of applying a suitable DEM structure on removing of the most height differences of captured images, we have tried to apply some unreasonable resolution of digital elevation model (DEM) in orthophoto producing and results of them have shown respectively. In the third part, the permissible amount of orthophoto error is expressed then with respect to the importance of DEM precision, the best amount of DEMs density estimated for each kind of topography and scales of orthophotos.In the result part, outputs of previous sessions summarized and reviewed totally then a table about the suitable amount of DEMs density for each scale of orthophoto at each kind of topographic classes is presented.

2. ORTHOPHOTO PRODUCTION AND DEM IMPORTANCE

In this session, principles of orthophoto with respect to images orientation and DEM are presented. The importance of using a suitable DEM and effects of making any deformation on it are explained theoretically and practically.

2.1 Orthophoto

During of orthophoto production, related images with their orientation and DEM are used. At these process effects of height displacements and tilt angle effects on captured images are removed, (Chapter14,2014). In fact at this way DEM cells whom size are dependent on the orthophoto size are imaged on the vertical datum. DEM pixels are got values from one of interpolation algorithm on gray values of corresponded DEM cells on orthophoto(fig1),(Leica photogrammetry,2008). With the hypothesis that using a correct georefrencing parameters and a correct DEM cause to place each feature on its true position correctly (maximum error of DEM based on the scale of orthophoto would be defined).

2.2 Shape and precision of DEM at orthophoto production

At this part, the effect of DEM and imaging configure on orthophoto are studied. For assessing how much they are related to each other,effect of any changes on DEMis illustrated on orthophoto and calculated consequently.

2.2.1 DEM height changes on image rectification
In fig (2), point1 on a captured image is an image of a place which is higher than its surrounded area. This cause to make some height displacements at orthophoto production. The true place of that point relied on 1'. During of rectifying, using a true DEM, all matched points are respective to each other like as point1 on the captured image and point1 on DEM. The image of this point on Height Datum is point 1' which its gray value gets back on captured image base on an interpolation algorithm. Next, this DEM was changed and smoothed, fig(2).

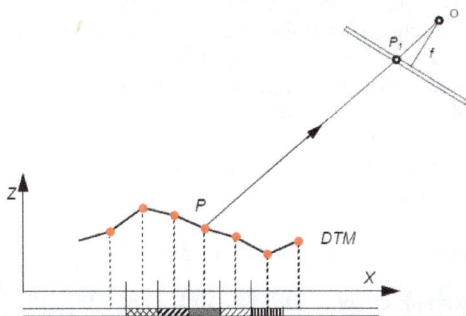

Figure1, getting gray value to rectified image(orthophoto)

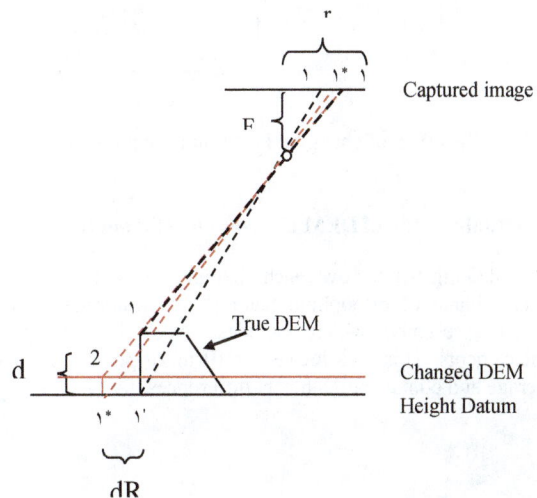

Figure2, The effect of changing DEM on orthophoto production

Now point2 on this model is replaced to point1 and imaged on a different place point1* instead of the true point 1' on orthophoto With supposing that the distance between 1' and 1* as dR, DEM height difference as dh, focal length as F and the distance between a point on a captured image to image center as r, fig(2); Eq2 shows the relation between DEM points height changes and coordinate displacements.

$$\frac{dh}{dR} = \frac{F}{r} \tag{1}$$

$$dR = \frac{r \, dh}{F} \tag{2}$$

With the hypothesis that the focal length is fixed any changes at DEM features height in the corner of image lead to more displacements in compare to those which are near to image center.

2.2.2 the effect of DEM height changes on orthophoto mosaic

At this session the effect of DEM height changes on orthophoto mosaic continuously is studied. In fig(3),points1 and 1' are two corresponded points on photos left and right. They are related to a height place(point1 on DEM) which point2 is its base. During of orthophto production with using a true DEM, both of points have to move to point2 (image of the base of the feature). if DEM is changed like what was shown in fig(3), the place of intersection of those points (1,1') with changed DEM moved to two different places(1 and 1') on orthophoto which cause to make a difference dR between two same features on orthophoto mosaic.

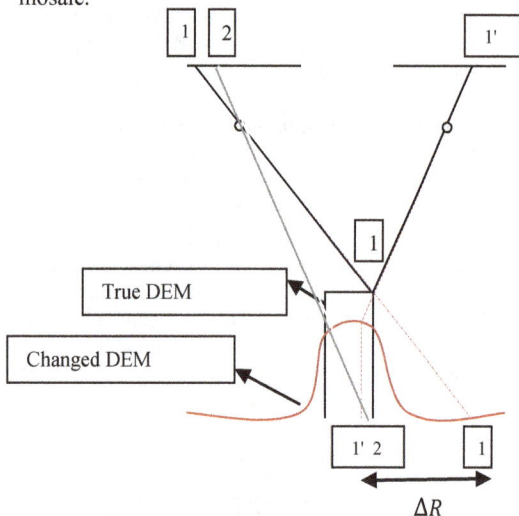

Figure3, The effect of changing DEM on orthophoto mosaic

2.3 visually study of DEM changes on orthophoto

For explaining better how much changing DEM could change the coordinate of orthophoto features, an orthophoto mosaic sheet is presented which includes some clear elongation displacements. This area located in 10 to 20 percent of image coverage and contains high height differences, fig (4).

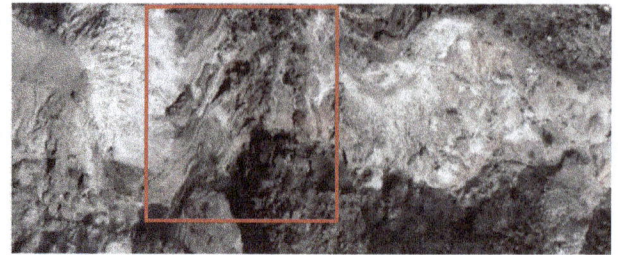

Figure4, Part of first orthophoto

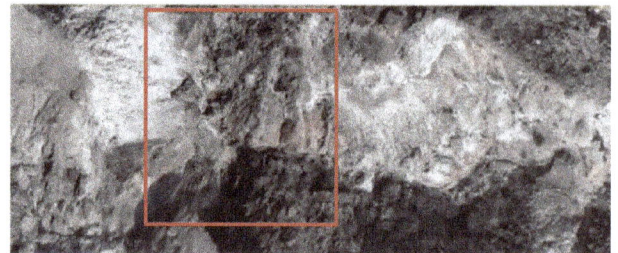

Figure5, orthophoto with sparser DEM(15 m)

Differences of highest to lowest place are about 100 meters in 1.4 meter of distance, so using a suitable DEM is an important factor. The existed elongation in this image is because of the lack of data in the captured image which was used in this orthophoto. Fig(5) shows another orthophoto of this area which is produced with a smoother DEM. The density of used DEMwas decreased in fig (5) so many places in fig(4) is remain originally without any rectifying process on it. With comparison of two figures (4,5), any differences on DEM because of its density cause to some clear changes on orthophoto. The coordinate of the shown point on the first orthophoto(fig(4)) is 671881.57,3020168.72, DEM value at this point is 68.15meter and on the other orthophoto(fig(5)) with sparser DEM density is 671883.16,3020171.37 and its DEM value is 51.57 meter. The distance of this point from the center of the image is 18.87 millimeter and with considering eq2, the difference of displacement on two orthophotos is calculated as 3.085 meter.

$$\frac{18.87 * (68.15 - 51.57)}{101.4} = 3.085$$

3. THE SUITABLE DEM DENSITY

Digital elevation model is an important factor in producing orthophoto which its precision has a direct influence on orthophoto quality. Amount of this depends on the rate of the area topographic and cause to more error on orthophoto at height places in compare to flats (eq(2)). So attention to DEM density with respect to the type of area is very important. At this session, the suitable DEM density is accessed. First a precise DEM of the area is chosen then the density of it is decreased continuously. At each step height values of all first points are calculated based on two current interpolation algorithms and compared with the first ones. With respect to the orthophto scale, relation between possible horizontal error in orthophoto and height DEM error, amount of possible DEM error and its density would be reached

3.1 DEM accuracy on orthophoto with respect to

For defining maximum effect of DEM on orthophoto, Eq(2) is studied again. Eq(2) can transform to Eq(3) If entire of image has been used in orthophoto producing and want to estimate the maximum DEM effect on orthophoto producing with supposing of using ultracam camera, f=101.4 , D=61.78 mm   (D:

maximum Radial distance from image center) With assuming the correct matching, incorrect amount of DEM resolution without respect to topographic type is related to dh directly. So choosing any incorrect DEM pixel size cause to make high differences between calculated and real height values and dR error on orthophoto.

$$dR = 0.61\, dh \qquad (3)$$

Eq(4) shows the total amount of error in orthophoto in different stage of production.
(Total Orthophoto RMSE)2 = (Triangulation RMSE)2+(RMSE of DEM)2

$$(4)$$

For calculating dR, the amount of average triangulation calculation error is subtracted from the total error of each orthophoto ,eq(5).
dR=

$$\sqrt{(\cdot.3 * \cdot.\cdot\cdot\backslash * Scalemap)^2 - (1/3 * \cdot.3 * \cdot.\cdot\cdot\backslash * Scalemap)^2}$$

$$(5)$$

The error of DEM on orthophoto is calculated base on Eq(4) and supposing that the average differences of features on orthophoto mustn't be more than 0.3mm in map scale as well as the triangulation error is about 1/3 of the total error.

RMSE of ultracam D dR=0.61dh	RMSE of DEM(m)	Total RMSE(m)	Scale
0.46	٠,٢٨	0.3	1:1000
0.92	0.56	0.6	1:2000
2.3	1.4	1.5	1:5000

Table1. Amount of permissible DEM height error in orthophoto production based on eq(3,5)

3.2 Case study and data preparation

The study area comprises 44940 pixels (DEM values) of flat to too high topographic area, for doing the interpolation calculation in different densities, this area is divided to some same size patterns then the height values of points inside of each of them are calculated based on two interpolation algorithm(Nearest Neighbor and Bilinear). The size of patterns grows after each step with a pixel and the error values as well as averaging of errors (RMSE) are calculated separately. Fig (6) shows area DEM and fig (7) shows area slope map.

Figure6, Case study DEM(m)

Figure7, Slope map (percent)

Because of the interpolation error is highly dependent to area topographic type, the area is divided to some topographic classes and the average error is calculated for each group separately. Topographic classes are grouped as:
1- More than 70 degree of slope
2- 50<Slope degree<70
3- 35<Slope degree<50
4- Slope degree<35
These classes are grouped for high slope cliffs, high slope mountains, hills and flat areas. Because of the high topographic area includes a few pixels and are important in true orthophoto production, puts away of calculations. Fig (8) shows classified study area based on defined classes.

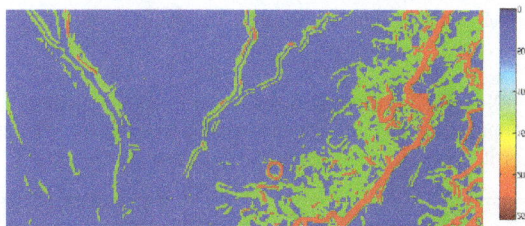

Figure8, Classified slope map based on mentioned classes

At this classifing, pixels in groups 2,3 and 4 are colored with red, green and blue separately.
Total area pixels are 44940, pixels of second class are 2554, pixels of third class are 8621 and there are 33765 pixels in fourth class.

3.3 Interpolation

At this session the result of each interpolation is presented.

3.3.1 Bilinear interpolation results

The result of precision for Bilinear interpolation on classes 2 to 4 is presented in table 2.

Bilinear Interpolation				
Interpolation resolution (m)	RMSE(m) General 44940pix	RMSE(m) Class2 2554pix	RMSE(m) Class3 8621pix	RMSE(m) Class4 33765pix
1	0.12828	0.43572	0.08612	0.04477
1.5	0.22408	0.70805	0.15729	0.08451
2	0.31824	0.96429	0.23404	0.12912
2.5	0.38878	1.12415	0.31575	0.17490
3	0.46914	1.30385	0.42683	0.20867
3.5	0.50801	1.33846	0.48095	0.25616
4	0.56093	1.42665	0.56372	0.28704
4.5	0.60273	1.48178	0.62108	0.32145
5	0.69539	1.67113	0.70094	0.39907

Table2. Bilinear interpolation precision at each of classes with increasing step of a pixel

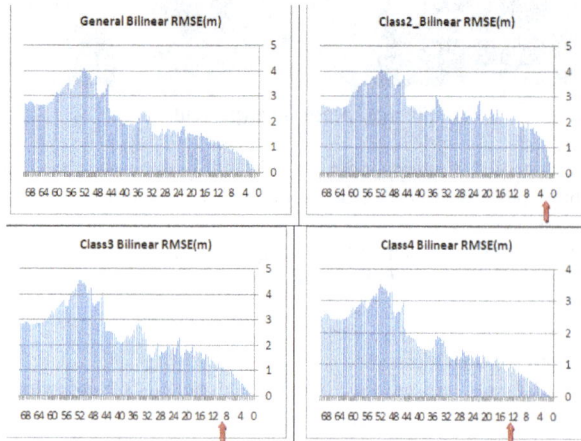

Figure9, Plots of precision changes in Bilinear interpolation in each increasing step. Arrows show the maximum bound of reasonable relation between grid spacing and errors (Horizontal axis: DEM grid spacing, vertical axis: amount of errors)

3.3.2 Nearest Neighbor Results

Like as previous method the result of precision for Nearest neighbor interpolation on classes 2 to 4 is presented in table 3.

Nearest Neighbor Interpolation				
Interpolati on resolution (m)	RMSE(m) General 44940pi x	RMSE(m) Class2 2554pix	RMSE(m) Class3 8621pix	RMSE(m) Class4 33765pi x
1	0.12829	0.43572	0.08612	0.04477
1.5	0.36895	1.23047	0.23071	0.09606
2	0.38768	1.23650	0.25972	0.12274
2.5	0.60911	1.91764	0.44941	0.20107
3	0.61604	1.88487	0.48431	0.21380
3.5	0.77334	2.23574	0.71578	0.30074
4	0.76831	2.19252	0.70815	0.32078
4.5	0.90864	2.44190	0.89602	0.40934
5	0.95253	2.54670	0.92748	0.44767

Table3. Nearest neighbor interpolation precision at each of classes with increasing step of a pixel

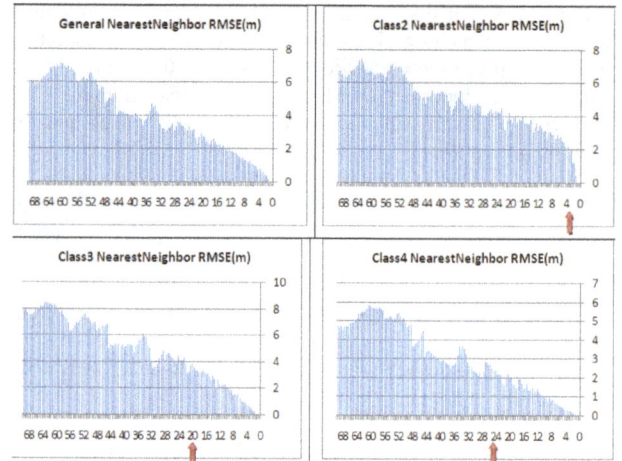

Figure10, Plots of precision changes in Nearest neighbor interpolation in each increasing step. Arrows show the maximum bound of reasonable relation between grid spacing and errors(Horizontal axis: DEM grid spacing, vertical axis: amount of errors)

4. RESULTS

The precision of rothophotos like as the other geometrical product are dependent on the used data in duration of production.

DEM and the procedure quality are two important factors in orthophoto production. DEM cause to correct height displacements and the quality and the precision of it affect orthophotos directly. The height of each pixel must calculate based on an interpolation algorithm on some related DEM height values. At this paper, orthophoto production based on captured images, images orientation and DEM was studied and geometric equations and mathematical models precisely explained. The role of DEM was carefully took attention with choosing a sheet of orthophoto which included some elongation errors and the effect of its changes with DEM was studied. With respect to DEM effects on orthophoto feature coordinate displacements and eq(1,2), amount of decreasing of density must be done with respect to map scale carefully. In the other test, the relation between the precision of interpolated height values, the density, and interpolation algorithm and slope rate changes has been done. At this test a precise dense DEM of an almost high topographic area prepared and at some steps its density has been decreased. The amount of differences between calculated and real height values have been assessed at each topographic classes and algorithms. Taking a review of bilinear algorithm results in table2 shows that increasing DEM pixel size lead to reducing amount of interpolation precision. This manner continues in all plots of fig(9) until DEM density is more sparse and there isn't any reasonable relation between them. In fig(9), the plot of area without attention to topographic classes shows error values (vertical axis) to interpolation distance (horizontal axis) in a curve line but its increasing manner until 26 m of resolution is clear. However because of this plot considers all of topographic classes could be much important in decision. Other plots at this figure show reasonable manners for classes of 2 to 3. In plot of class2, the linear manner was removed and took a cure shape so faster than other classes. This form shows high sensivity of this class to decreasing the DEM density. At plot of the other classes (3, 4), this manner can be seen but the rate of changes is slower and the linear manner has been kept until larger DEM pixel size. All plots show that the manner of changes hasn't been reasonable at the end tail of figures and we can't consider these values in our results. Both interpolation algorithm results (Nearest neighbor and bilinear) have the same manner and error values in nearest neighbor are larger and stricter than the other. However choosing DEM large cells cause to decrease the height meaning relation between the grid points isn't possible and proposed to fixing this bound value for larger orthophoto scales. In fig(11), error values at each interpolation algorithms at each DEM grid space are shown. Nearest neighbor algorithm shows higher error values in steeper areas and the discontinuously is clear between its output values whereas Bilinear interpolation outputs because of interpolating in two dimensions are smoother and the shape of area in estimating the errors with increasing the grid space is kept.

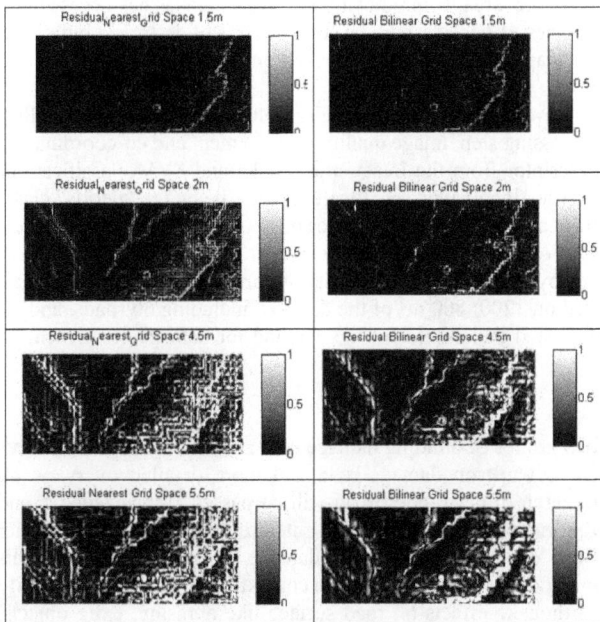

Figure11, Error values on DEM pixels at each interpolation algorithm(Left plot: Nearest neighbor interpolation, Right plot: Bilinear interpolation)

With comparison of both of two interpolation algorithms and error plots, maximum possible errors at each scale of orthophotos in case of using Ultracam images has been extracted and presented in table(4) consequently.

DEM Resolution(m)with respect to each topographic class			Maximum Possible Error(meter)	Orthophoto Scale
Topographic Class1	Topographic ۲Class	Topographic Class3		
١	٣	۵٫۵	٠٫۴۶	١:١٠٠٠
١٫۵	۵٫۵	٩٫۵	٠٫٩٢	١:٢٠٠٠
۴	١١٫۵	٢۴٫۵	٢٫٣	١:۵٠٠٠
۴	١١٫۵	٢۴٫۵	۴٫۶	١:١٠٠٠٠

Table2. Calculated suitable DEM resolution for each scale of Orthophotos

REFERENCES

BC Ortho-Image specification;Base mapping and cadastre,GeoBC; March2011.

Chapter14: Orthophotography-Gis-Lab.gis-Lab.info/docs/books/aerial-mapping/cr1557_14.pdf.Access 10/2014.

Degaard Nielsen.M, "True orthophoto generation", IMM-Thesis;Technical university of Denmark.

[8] Report to the U.S. Geological Survey on Digital Orthoimagery, ASPRS, 2005, http://nationalmap.gov/report/ASPRS_Report_on_Digital_Orthoimagery.pdf

"Guidelines for Best Practise and Quality Checking of Ortho Imagery", European Commission, Directorate General, Joint Research Centre, Institute for the Protection and Security of the Citizen, Monitoring Agriculture with Remote Sensing Unit.Issue 2.3,(2004). http://marsunit.jrc.it/Mapping/.

Honkavaara.E, Markelin.L, Marttinen.J and Vilander.M, "External Quality Control of Medium-scale Orthophoto Production",case Finnish Land Parcel Identification System' Nordic Journal of Surveying and Real Estate Research , 2004,VOL 1, p131:P143.

Honkavaara.E, Kaartinen.H, Kuittinen.R, , Huttunen.A, and Jaakkola.J "Quality of FLPIS Orthophotos", Reports of the Finnish Geodetic Institute, (1999). 99:1.

Leica photogrammetry suiteproject manager;users guide February 2008.

Pala.V, Arbiol.R "True Orthoimage Generation in urban areas, Institute cartographic of catalunya;P309:P314.

Simard.G "Accuracy of digital orthophotos", A report submitted for the degrss of masters in engineering; the university of New Brunswick; February 1997.

17

AUTOMATIC BLOCKED ROADS ASSESSMENT AFTER EARTHQUAKE USING HIGH RESOLUTION SATELLITE IMAGERY

H. Rastiveis [a], *, E. Hosseini-zirdoo [a], F.Eslamizade[a]

[a] Dept. of Geomatics Engineering, School of Eng., University of Tehran, hrasti@ut.ac.ir

KEY WORDS: High resolution Satellite Image, Earthquake, Classification, Texture Analysis, Roads damage map

ABSTRACT

In 2010, an earthquake in the city of Port-au-Prince, Haiti, happened quite by chance an accident and killed over 300000 people. According to historical data such an earthquake has not occurred in the area. Unpredictability of earthquakes has necessitated the need for comprehensive mitigation efforts to minimize deaths and injuries. Blocked roads, caused by debris of destroyed buildings, may increase the difficulty of rescue activities. In this case, a damage map, which specifies blocked and unblocked roads, can be definitely helpful for a rescue team.

In this paper, a novel method for providing destruction map based on pre-event vector map and high resolution world view II satellite images after earthquake, is presented. For this purpose, firstly in pre-processing step, image quality improvement and co-coordination of image and map are performed. Then, after extraction of texture descriptor from the image after quake and SVM classification, different terrains are detected in the image. Finally, considering the classification results, specifically objects belong to "debris" class, damage analysis are performed to estimate the damage percentage. In this case, in addition to the area objects in the "debris" class their shape should also be counted. The aforementioned process are performed on all the roads in the road layer.In this research, pre-event digital vector map and post-event high resolution satellite image, acquired by Worldview-2, of the city of Port-au-Prince, Haiti's capital, were used to evaluate the proposed method. The algorithm was executed on 1200×800 m² of the data set, including 60 roads, and all the roads were labelled correctly. The visual examination have authenticated the abilities of this method for damage assessment of urban roads network after an earthquake.

1. INTRODUCTION

Natural disaster always threat people life, property, etc. Among all, earthquake is important than the others and make irreparable damagebecause it's unpredictable and suddenly happened and population grows in city that has not suitable safety. Lifeline of urban infrastructure is the most sensitive because of the important role in rescue and assistance to the victims of an earthquake of great importance. Evaluating and identifying intact ways of damage and determining the degree of destruction is the most parts after the earthquake need to the rescue. Fast and accurate knowledge of the location of roads and the destruction of roads is valuable information for rescue and reconstruction work after the earthquake. A damage map contains information such as destruction of intact and damaged buildings or roads and, also may show the degree of destruction. For determining and identifying the destruction of roads and ways correctly can used object data and image. Depending on the type and quality of the object data, and image or a combination of both of them can be used. Image data contains aerial and satellite images and object data include vector map. The reason of time consuming ground working is not suitable in disaster management.Comprehensive information, complete and accurate geomantic knowledge after earthquake by data such as arial and satellite image, ground map data and LiDAR data. The goal of this research is offer algorithm in order to identify the intact and damaged roads, automatically, and determine damage degree, and at the end, generate the damage map. This research used pre-event vector data and post-event satellite image.

2. LITERATURE REVIEW

There are many researches for making damage map after earthquake in last three decades. These studies can be divide into two groups of building damage map and road damage map. This paper work on damage road and more detailed overview on research of destruction road will be passed. Samadzadegan and Zarrinpanjeh used QuickBird satellite images before and after Bam earthquake for roads damage map generation. In this method, they used spectral and contextual information to identify earthquake effects on road surface like abrasion, extra objects, vegetation, obstruction by using classified algorithms. According to the existence of these effects of road surface they finally concludes that which road is intact and which one is blocked(Samadzadegan and Zarrinpanjeh 2008).

In one of the last researches, Bahr used WorldView satellite images and LiDAR coverage after the earthquake in Port-au-Prince, the city of Haiti, to reach Four Goals:1-Identifying the condition of homeless people who needs help, 2-locating the region. 3-Buildings existence and recognizing the buildings that has been ruined. 4-locating the roads through which the ground groups can transport. The role of analyses is an important issue in this study (Bahr 2014)

Another method that used both pre and post-event high resolution satellite image of QuickBird pre-event vector map in order to analyze the road image .. The aim of HaghighatTalab study is to introduce new method of automatic recognizing damage roads in urban area which uses genetic algorithm to find features of optimal texture. For the next step they attempted to determine the obstruction by one of supervise classification algorithms. Finally they analyzed the roads which were blocked by using Fuzzy Inference System and they also report 0.67 total accuracy and 0.67 Kappa(Haghighattalab, Mohammadzadeh et al. 2012).

3.   BLOCKED ROADS DETECTION

Basically determining an undamaged road from damaged road is done by using satellite images in two automatic ways :using (processing image algorithms)and visual interpretation (observation is done by human operators).Everyday engineering science geomatic developments ,algorithm discovery ,new formula ,presenting new idea by the experts of discipline an one hand ,and problems like time -consuming and intolerable process and necessity of experienced faculty on the other hand make the visual interpretation be less attracted in contrast to other ways. Another problem of visual interpretation is the time-consuming and intolerable process and also human mistakes like tiredness and their inaccuracy. So they tried to do the process automatically by producing new techniques and idea and new seawares like digital image processing.

Among the information sources which are used for making damage map satellite images is more attracted by experts and damage manegment than other methods because of having constant region coverage and having no security limitation and quick access and having spatial resolution .Using this images in order to determine safe road from damaged roads ,the changes can be done through three ways: shadow analysis ,texture analysis and analysis of change detection which will be described one by one in the next paragraph.

3.1 Shadow Analysis

Shadow has an essential value in this method .In satellite image shadow is depend of the time of camerawork and the location of sun in the sky .shadow can be used as one data to determine the existence of debris in the road .So when the uniform shape and size of shadow has been changed it can be the sign debris of the building in that road .But shadow only can be the sign of building safety of one side of the road and the existence of shadow is not exactly shows the safety of the entire road.

In this study, the existence of shadow in one road is considered as the safety of only that part of the road.

3.2 Texture Analysis

Information of images of one object is divided into three groups:1-spectral data 2_textural data 3_structural data.
Textural data is the most essential and powerful data among this three division .Textural data helps us to determine debris analysis .There is no actual definition for texture. But according to its different usages there is different meaning for texture. Texture is information more one pixel that shows the relation between the data of one pixel with its surrounding pixel .In other words pixel is not a texture per se, but it has grey degree. A texture can have features like grading, size, direction, density, equality, uniformity, harshness and non-roughness. So we can identify different damaging parts of the road like vehicle, raid surface safety, trees and also debris by texture analysis. In this study, textural data are used to determined earthquake effects in the surface of roads.

3.3 Changes Disclosure

Changes disclosure methods can be employed to determine destructed paths. Through these methods, two images from before and after the earthquake are compared against each other leading to determination of the changed terrains. Various algorithms including image difference and image classification difference are presented for changes disclosure using satellite images. In order to distinguish blocked paths from un-blocked ones, these algorithms can be utilized as well.

Generally, calculation included in these methods can be conducted on pixel or object levels. Disclosure conducted on pixel level provides advantages such as more simple and less time consuming processing, while when object level is used, processing would be more complex due to segmentation step and pixel to image object conversion. The disadvantage of using the pixel level is the impact of co-coordination accuracy on the results which imposes the requirement of coordination with accuracy level pertinent to the pixel size.

4.   PROPOSED METHOD

In this paper, a novel method for providing damage map based on surveying before the earthquake and high resolution WorldView II satellite images after it, is presented. As it is observable in flowchart 1, firstly in pre-processing step, image quality improvement and co-coordination of image and map are performed. In the second step, after extraction of texture descriptor from the image after earthquake and SVM classification, different terrains are detected in the image. Then based on available terrains on the surface of each path and extraction of candidate paths from the output of classification, the paths conditions in regard of getting blocked of remaining un-blocked is evaluated leading to providence of paths damage map. The following includes details on every mentioned step.

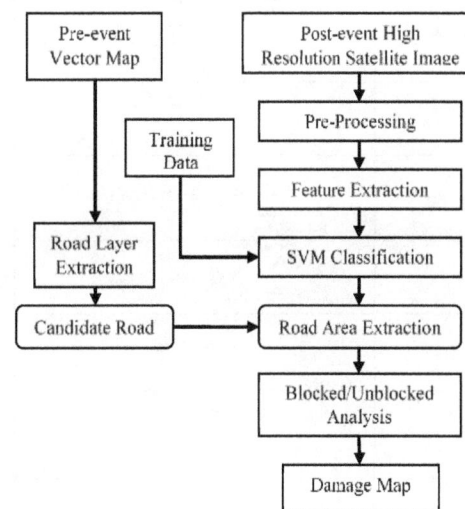

Fig.1: The flowchart of the suggested method for detection of un-blocked paths using satellite images with high separation capability

4.1 Pre-Processing

In order to improve the image quality, histogram balancing is conducted on the image from after earthquake situation, also in order to be able to relate the map and the image they are co-coordinated.

4.2 Image Classification

After pre-processing, in order to achieve terrain detection in the path area, image classification is conducted. Through this step, a set of educational data are selected for each class on the image under study. Then via extraction of pertinent texture descriptor and utilization of a proper classification algorithm, the image from after the earthquake is classified into six classes including path, building, shade, tree, debris, and vehicle. In the suggested method, texture and spectral descriptors are utilized in the classification procedure. Texture descriptors include Haralick descriptors while spectral descriptors include three color bands of R, G, and B of the high resolution satellite image after the earthquake. The SVM approach is deployed for classification which is deemed to be a consummate tool for this problem. The following sections include the details on descriptor extraction and classification methods.

4.2.1 Descriptors Extraction

Various methods for texture analysis on images have been developed among which structural and statistical approaches can be referred to. In previous studies, satellite images are employed in destruction detection more than statistical approaches, however the capabilities of these methods is well established on a considerable level. Multiple texture descriptors can be extracted through statistical methods such as Co-occurrence matrices, Statistical moments, and Markov random fields.

In this study, the parameters extracted from Co-occurrence matrices are utilized to demonstrate the texture behavior. Multiple texture parameters are extracted from this matrix among which six descriptors including contrast dissimilarity, variance, entropy, homogeneity and mean which provide more desirable capability level in separation between classes are employed. Table 1 presents the equations used for calculation of these parameters based on the Co-occurrence matrix. (1)

Descriptors	Formulation
homojeneity	$\sum_{i,j=0}^{N-1} \dfrac{P_{i,j}}{1+(i-j)^2}$
mean	$\mu = \dfrac{\sum_{i,j=0}^{N-1} P_{i,j}}{N^2}$
entropy	$\sum_{i,j=0}^{N-1} P_{i,j}\,(-\ln P{i,j})$
variance	$\sum_{i=0}^{N-1}\sum_{j=0}^{N-1} (J-\mu j)^2 * P_{i,j}$
dissimilarity	$\sum_{i,j=0}^{N-1} P_{i,j}\,|i-j|$
contrast	$\sum_{i,j=0}^{N-1} P_{i,j}\,(i-j)^2$

Table 1. Extracted descriptors using the Co-occurrence matrix

In addition mentioned features in table 1, some of the texture descriptors from other methods can be utilized as well, also three color bands of R, G, and B of satellite images can be employed as a characteristic for classification.

4.2.2 SVM Classification

The SVM algorithm was developed by Vladimir Vapnik in 1963 and was extended to fit the nonlinear state by Corinna, Cortez, and Vapnik.(2,3) SVM employs a meta separator surface thorough maximization of the separator surface distance from each side on each class, a surface which is called the optimum separator surface. Fig.2 shows the separator line between two hypothetical classes.

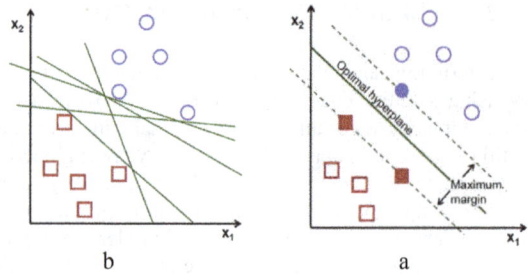

Fig 2. Meta Separator surface between two classes. A: Optimum separator surface. B: Meta Separator surfaces.

Postulate that n educational samples and two classes are available, in a manner that the following relation can be written for each sample:

$$D = (x_i, y_i) \mid x_i \in R_p,\, y_i \in \{-1, 1\}\}_{i=1}^{n}$$

In this relation x_i is a real vector and y_i stands for each class output for the educational sample I, the objective is to find an optimum Meta separator surface which separates the two classes. The equation of this separator surface can be written as w.x-b=0 , and for the two classes under consideration the relations are presented in following form.

w.x-b=1   (1)
w.x-b=0   (2)
w.x-b=-1   (3)

In the following relation if Y is equal to one, the sample belongs to the first class, while if it is equal to minus one; the sample will be belonging to the second class. In order to prevent the points from getting to the peripheries relations 1, 2, and 3 are written as follows.

w.x-b$\geq$1   (4)
w.x-b$\leq$-1   (5)

Finally, relations 4 and 5 can be written as follows:

$y_i\,(w.x_i\text{-}b)\text{-}1 \geq 0$ (6)
$y_i\,(w.x_i\text{--}b) \geq 1\text{-}\xi\_i$ (7)

With the conversion of equations 6 and 7 the problem of surface detection is conclusively reduced to minimization of Min[

$$\frac{|w|^2}{2} + c\sum_{i=1}^{r}\xi - i\,]$$

, in which C is the regulation parameter in optimization and ξ_i determines the level of imperfection in classification, which is defined through |w| minimization and Lagrange coefficient utilization(Gunn 1998)

4.3 Path Extraction

Considering the fact that in the pre-processing step, the maps are co-coordinated with the images from the condition after the earthquake, it is possible to transfer each coordinates from the map to image or vice versa. After classification, the points representing beginnings and ends of each path are detected in the image, and then through considering the buffer space equal to the width of the path, the path region is extracted from the classified image.

4.4 Path Blockage Analysis

In this step, after extraction of path region for each candidate path based on the number of pixels for each class, it is possible to evaluate the path blockage. In this study, the ratio between the amounts of debris on the path surface to the entirety of the path surface is employed as the detector parameter. If the ratio stands above a pre-defined amount the path is considered to be blocked while if this condition is not met the path is deemed to remain un-blocked. Eventually, the blocked paths can be shown with different colors in comparison with un-blocked paths and the path destruction path is provided.

5. TEST AND RESULTS ANALYSIS

The suggested algorithm is tested on the high resolution satellite image after Haiti quake in 2010 and the original maps of the region. In this section, firstly the characteristics of the under study region are presented and afterwards the obtained results are discussed.

5.1 The Case Study Region

In this study, high resolution satellite image after Haiti earthquake in 2010 and the vector data sets before the earthquake are utilized to evaluate the suggested algorithm. Within these data sets, the case study region with the area equal to 1200x800 square meters is selected and the paths contained in the region with the number equal to sixty are evaluated.

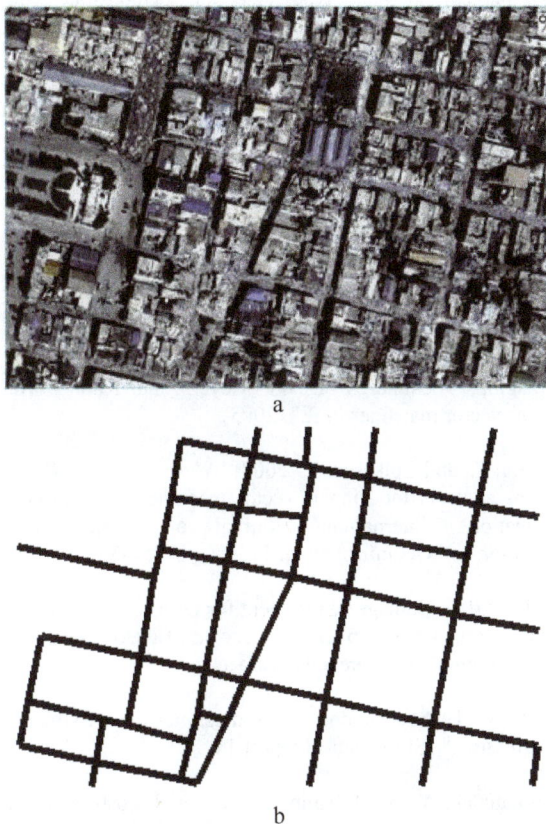

a

b

Fig 3. Study area. a. Post-event high resolution satellite imagery. b.Pre-event vector map

5.2 Obtained Results

After pre-processing, educational data sets are gathered from the under study region and pertinent descriptors are extracted to obtain desirable classification. In this study, 18 texture descriptors along with 3 spectral descriptors are used to form the description space in SVM classification method. In Fig.4 some of the extracted descriptors are depicted.

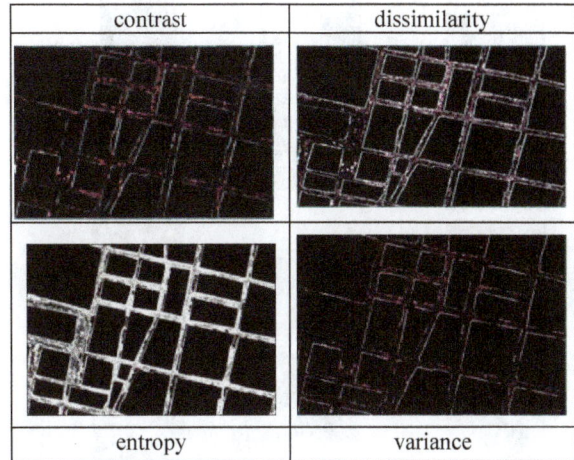

| contrast | dissimilarity |
| entropy | variance |

Fig 4. Some of the texture descriptors extracted from satellite images for classification

After extraction of pertinent descriptors, the image is classified into the mentioned six classes. The results of the classification are delineated in Fig.5.

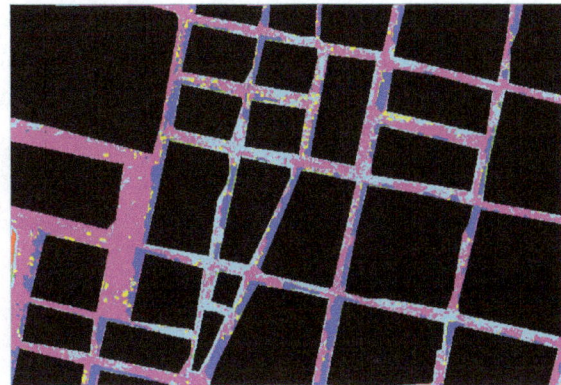

Fig 5. Image classification after quake

After classification and extraction of path regions based on the aforementioned approach, the potential blockage of paths is detected. Fig.6 shows some of the paths which are not considered blocked, while Fig.7 depicts the paths which are considered as blocked ones by the algorithm.

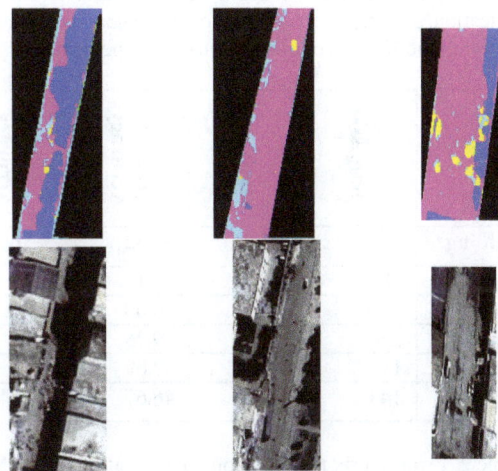

Fig 6. Some paths deemed un-blocked by the algorithm

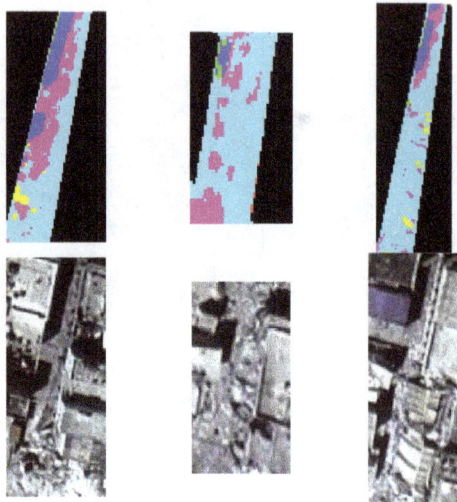

Fig 7. Some paths deemed blocked by the algorithm

After running the algorithm for the entirety of the 60 selected paths, 23 were reported to remain un-blocked while the other 37 were detected as blocked paths; Fig.8 shows the obtained damage map from the algorithm.

Fig 8. The obtained destruction map from the suggested algorithm

5.2 Accuracy Evaluation

Two accuracy evaluation steps are presented; in the first step the validity of SVM classification is included. Conveniently, classification methods accuracy is evaluated via error matrices. Table 2, includes the resultant error matrix from the classification. The reported accordance percentage is around 83 percent with the Kappa coefficient equal to 75 percent.

Confusion Matrix	Commission	Omission	Producer Accuracy	User Accuracy
building	26.64	0.58	99.42	73.36
tree	18.42	96.49	3.51	81.58
road	5.31	4.00	96.00	94.69
shadow	.29	0.75	99.25	99.71
car	15.73	42.31	57.69	84.27
debrise	14.02	53.33	46.67	85.98

Table 2. Error matrix resulted from classification via SVM method

		Algorithm	
		Blocked road	Unblocked road
Reference	Blocked road	37	1
	Unblocked road	1	23

Table 3. Comparison between the destruction maps from visual interpretation and the suggested method.

Eventually, with a comparison between the destruction maps from actual surveying and the suggested method, an accordance percentage about 97 percent is achieved which is the result of only 2 paths detected not correctly. The evaluations further reveal that building regions lead to the majority of the occurred errors and the previously assessed classification error does not impose a considerable impact on the final results.

6.   CONCLUSION AND PROPOSITIONS

In this paper, an algorithm is presented to distinguish between blocked and un-blocked paths using WorldView II satellite images with high separation capability after earthquake occurrence and original maps of the region. A validity percentage of 97 percent in path type detection demonstrates high capabilities of the developed algorithm, moreover, the capacity of the algorithm can be enhanced through increment in image separation level; thus, it is recommended that the algorithm be checked for other data sets as well. Furthermore, considering the importance of utilized descriptors in detection of different terrains, it is suggested that other type of descriptors be utilized for a comparison on final results.

REFERENCES

Shapiro, Linda, ed. Computer vision and image processing. Academic Press, 1992.

Karatzoglou, Alexandros, David Meyer, and Kurt Hornik. "Support vector machines in R." (2005).

Kavzoglu, T., and Colkesen, I., (2009) 'A   kernel functions analysis for support vector machines for land cover classification', International Journal of Applied Earth Observation and Geoinformation, 11, (5), pp. 352-35

Bahr, T. (2014). Damage Assessment for Disaster Relief Efforts in Urban Areas Using Optical Imagery and LiDAR Data. EGU General Assembly Conference Abstracts.

Gunn, S. R. (1998). "Support vector machines for classification and regression." ISIS technical report 14.

Haghighattalab, A., A. Mohammadzadeh, et al. Post-earthquake road damage assessment using region-based algorithms from high-resolution satellite images. Remote Sensing, International Society for Optics and Photonics.

Samadzadegan, F. and N. Zarrinpanjeh (2008). "Earthquake destruction assessment of urban roads network using satellite imagery and fuzzy inference systems." The international archives of the photogrammetry, remote sensing and spatial information sciences 37(B8): 409-414.

18

A PHOTOGRAMMETRIC APPRAOCH FOR AUTOMATIC TRAFFIC ASSESSMENT USING CONVENTIONAL CCTV CAMERA

N. Zarrinpanjeh [a], F. Dadrassjavan [b], H. Fattahi [c*]

[a] Islamic Azad University of Qazvin - nzarrin@qiau.ac.ir
[b] University of Tehran - f.javan@nikarayan.com
[c] Tehran Traffic Control Company - saran_hf @yahoo.com

KEY WORDS: Photogrammetry, Automatic Traffic Assessment, CCTV Camera

ABSTRACT

One of the most practical tools for urban traffic monitoring is CCTV imaging which is widely used for traffic map generation and updating through human surveillance. But due to the expansion of urban road network and the use of huge number of CCTV cameras, visual inspection and updating of traffic sometimes seems to be ineffective and time consuming and therefore not providing real-time robust update. In this paper a method for vehicle detection accounting and speed estimation is proposed to give a more automated solution for traffic assessment. Through removing violating objects and detection of vehicles via morphological filtering and also classification of moving objects at the scene vehicles are counted and traffic speed is estimated. The proposed method is developed and tested using two datasets and evaluation values are computed. The results show that the successfulness of the algorithm decreases by about 12 % due to decrease in illumination quality of imagery.

1. INTRODUCTION

Acquiring reliable information about traffic condition in urban road network is crucial for traffic monitoring and management. One of the most practical tools for such task is the use of conventional closed circuit cameras (Song 2006). As transportation flow is monitored by visual inspection the current situation of traffic is assessed and then updated. But due to the expansion of urban road network and the use of huge number of CCTV cameras, visual inspection and updating of traffic sometimes seems to be ineffective and time consuming and therefore not providing real-time robust update. On the other hand traffic monitoring tools such as speed detectors, mechanical vehicle counter and radar based tools are found successful for such purposes but they are so expensive compared to conventional CCTV cameras to be installed in each and every corner of urban area. In this paper an automated approach suitable for conventional CCTV camera is proposed to evaluate road traffic to decrease the influence on human surveillance. The proposed method is focused on counting the number of moving vehicle and also estimating the instant speed of the traffic flow in specific periods of time (Ferrier etal., 1994).

In this paper, reviewing methods of road traffic assessment a camera based approach for road traffic assessment is presented. The proposed method is tested and evaluated using real CCTV imagery.

2. A REVIEW ON ROAD TRAFFIC MONITORING

There are many operational systems to facilitate the process of road traffic monitoring. These methods are designed and developed to measure the road traffic condition in terms of vehicle counting and speed estimation. In general all traffic monitoring systems could be divided into two categories of active direct measurement systems and passive vision based optical measurement systems. Active Direct sensor consider mechanical, electrical or beam emitting tools to detect vehicle on the road. Solutions in this category require installation of specific tools at the place of surveillance and as a result are considered as expensive solutions. On the other hand camera based system use indirect approaches

2.1 Active Direct Systems

One of mostly used tools for traffic control for past few decades are Magnetic loops (Dhar, 2008). These devices are installed inside each traffic lane and counting vehicle passing over them. Some variants of the magnetic loop have been used to classify vehicles as well. Another efficient tool for traffic monitoring is microwave radars. Microwave radars detect vehicles through direct measurement procedures. Most microwave radar use the Doppler principle to detect vehicles. According to the Doppler principle the difference in frequency between the transmitted and received signals is proportional to the speed of the vehicle.

Laser based systems are for reliable counting, classifying and measuring speed of vehicles. On the other hand, Ultrasonic sensors consider sound waves to determine the presence or distance of an object. By measuring the time taken for the sound echo to return the distance of an object could be computed.

---

* Corresponding author

2.2 Passive Optical

Camera based systems are able to detect, count and classify vehicles. These systems use video image processors to identify vehicles and their traffic flow parameters by analysing imagery through image processing algorithms. Moreover, passive infrared detectors do not transmit energy but use an energy sensitive photon detector located at the optical focal plane to measure the infrared energy emitted by objects in the detector's field of view. Thus, when a vehicle enters the detection region of the device, it produces a change in energy which is sensed by the photon detector. This system can only detect the presence of vehicles and is not able provide any information regarding the speed of the vehicle. Change in weather conditions such as fog, rain or snow results in performance degradation of these systems.

3. PROPOSED PHOTOGRAMMETRIC APPROACH FOR TRAFFIC MONITORING

In this study a vision based system based for traffic monitoring is proposed. A general overview of the algorithmic flow is present in figure 1. The algorithm starts with video stream reception and verification. At the first step the camera is initialized in terms of definition of the scene, object coordinated system and area of interest. The information is provided by user to the algorithm to initially construct image object transformation. The algorithm tries to keep the initialization configuration and restore it in case of violation. At the next step stability of the camera is verified. Any abrupt change in camera position and rotation angles is not tolerated and speed detection is stopped until camera stability is stablished. Then the illumination status is analysed which defines the acquisition exposure condition in order to tune vehicle detection parameters. The next step is dedicated to scene coordinate restitution. The algorithm constantly verifies if the xy coordinate systems is still valid so the speed detection precession is guaranteed. At the final step vehicle detection is applied on the verified image frame.

The flowchart of vehicle detection is presented in figure 2. Vehicle detection starts with Barescene generation from video stream. Barescene is defined as the image of road without any moving objects. Barescene is generated through median filtering of frames in different time epochs. Next, camera stability is checked during Barescene generation. Subtracting image frame from Barescene a trace of moving objects is generated. Through morphological filtering, trace of objects turn into detectable and countable entities. These entities are verified and classified with respect to size and area of the detected object. The position of each entity considered as the centre of detected object and is stored along with frame epoch and a view of detected entity. Processing the extracted entities position compared to frame rate the instant speed of vehicles are computed. It is recommended to compute speed from all available frames and consider the average speed as result speed value.

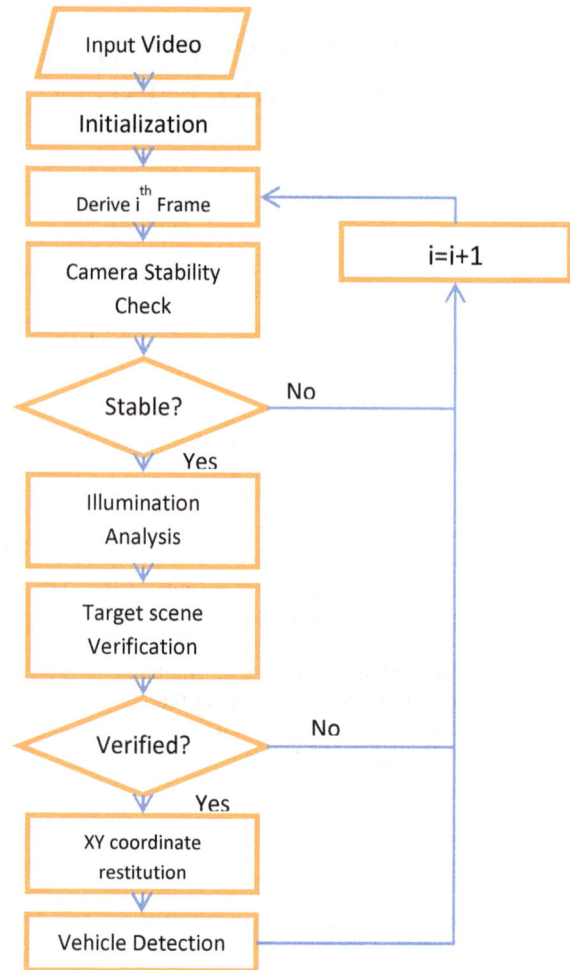

Figure 1: The flowchart of general processing of the algorithm.

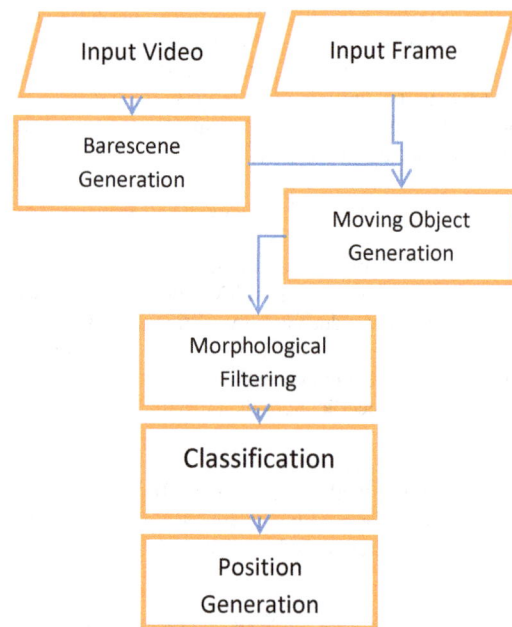

Figure 2: Flowchart of vehicle detection using image frames.

3.1 Challenges of Vehicle Detection

The most challenging issue in the traffic monitoring is that the process of information generation should be applied fast enough to be considered real-time or near real-time. If the process of implication of algorithms is time consuming the results whether robust and reliable are not found suitable for real world applications. Moreover CCTV images usually suffer from low spatial resolution and the effects of compression applied to facilitate image transport and storage which reduces the information content of data and restricts the solutions to be logged. On the other hand CCTV cameras are built for surveillance and monitoring and therefore they are mostly equipped to rotate and zoom. Although these necessary capabilities provide widen monitoring perspectives but they extremely violate photogrammetric geometry in terms of ruining exterior and interior orientation of camera.

In normal situations the process of vehicle detection and traffic monitoring is not too hard to implicate but the problems occur when camera and environment conditions are far from what it is called ideal. Camera angle and more specifically the tilt angle of image acquisition is very important. If camera is set to be vertical with zero tilt angle the traffic assessment and vehicle counting is restricted due to small field of view of the camera. In reverse horizontal acquisition would intensify the scale variation due to tilt angle which is not suitable for geometric computations. On the other hand direct solar radiation at sunrise and sunset which results blooming effect reduces the information content of imagery and violates vehicle detection. This also happens in case of front view of vehicles with light on during the night. Additionally, as illumination intensity decrease during the night, vehicle detection and speed estimation get more sensitive especially in case of camera movement where relation between primary and secondary scene is hard to generate due to low information content of the imagery. In these cases it is recommended to use cameras pre-set option to skip restitution procedure.

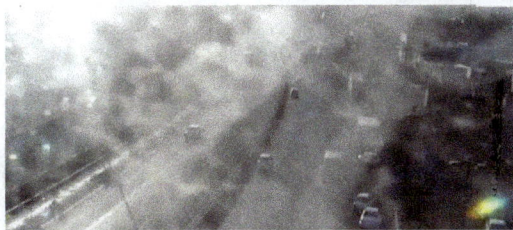

Figure 3: Information content reduction due to illumination problems.

Figure 4: Blooming effect of car lights.

4. TESTS AND RESULTS

To test the capability of the proposed method tests are defined and applied using real data from Tehran Traffic Control Company and National Road Traffic Control Organization.

At first image frames are processed to generate Barescene. 11 frames within 10 seconds are chosen and pixels with median values in all 11 frames are assigned to generate Barescene. To avoid mistakes from comment and direction indicators on the screen 25 percent from top and bottom of the image is removed. It should be considered that in case of heavy traffic jams road Barescene generation should be applied on higher numbers of frames. Figure 5 shows a sample frame of the CCTV images. Figure 6 shows the results of Barescene generation on the selected frame.

Figure 5: A frame of the CCTV footage.

Figure 6: Barescene generation using median filtering.

One of the most critical processing of vehicle detection is the establishment of a robust transformation between image and object spaces. For the purpose four GCPs are measured on the screen. The object coordinate of GCPs are measured through field surveying. Assuming that the road surface is smooth enough to be considered planar, 2D projective model is chosen to transform 2D image coordinates into 2D ground coordinate system (equation 1). This enables the conversation of car movement in image space detected by the algorithm in different frame to speed of the vehicle.

$$X = \frac{a_1 i + a_2 j + a_3}{a_7 i + a_8 j + 1}$$
$$Y = \frac{a_4 i + a_5 j + a_6}{a_7 i + a_8 j + 1} \tag{1}$$

Where X and Y represent object coordinate system and i and j are refer to as image coordinates. Coefficients are computed by the use of GCPs. Figure 7 shows the position of GCPS measured on the bed of the road connected to each other in shape of a foursquare.

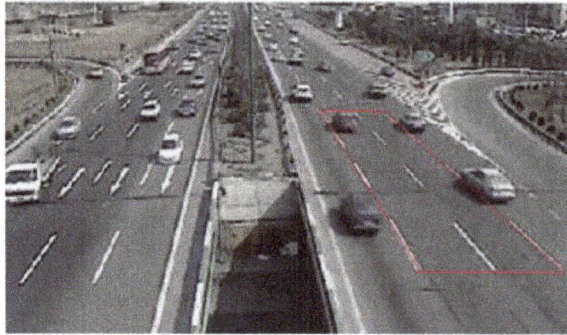

Figure 7: Definition of ground coordinate system.

To minimize the computational cost of vehicle detection on the huge data sets especially in case of high definition images, the area of interest is defined and any necessary processing is restricted to the area of interest. Inserting the number of lanes and dividing the area of interest into lane numbers, individual inspection of each lane is possible. As a result over computation are avoided and real time and near real time traffic monitoring is found more feasible.

This also enables the algorithm to monitor multiple lanes even in reverse directions.

Figure 8: Definition of rea of interest and number of lanes.

Figure 9 shows the result of vehicle detection on the selected area. As illustrated 5 objects are detected on the screen and only one item is detected as a robust object for speed estimation. As a matter of fact comparing the extracted objects size and location witch coordinate system and the scale of imagery, besides counting and classifying vehicles, it is decided whether the information is robust enough to compute the speed. As shown in figure 9 only one vehicle is considered for speed estimation. The decision is made via counting the number of pixels with respect to image spatial resolution.

Figure 9: The results of vehicle detection.

To restore ground coordinate transform in case of changes in camera viewing angle SIFT algorithms are used. Figure 10a

shows the position of ground control points in form of a grid. As shown in figure 10b a change in camera viewing angle induced by human user is detectable. This mostly happens when the camera is not equipped with pre-setting systems. Therefore returning back to initial imaging angle is not executed exactly to the same point. In figure 10c the result of sift algorithm to restore grid position is shown. As tested with different viewing angle the algorithm is successful in restoring position of camera view if compared imageries enjoy more than 50 percent overlap and scene illumination is enough for object detection. In other words the algorithm is not successful images during the night.

a

b

c

Figure 10: the results of camera position restoring. a) Initial image. b) Unadjusted image. c) Restored position.

Visual inspection of results shows that the algorithm is successful to count vehicles and estimate instant speed where the mentioned visual constraints are respected. Image quality reduction by any reason reduces the success of the method. Therefore, an analytical inspection on accuracy of the estimated speed is proposed. Moreover, for factious and figurative evaluation of the efficiency of the proposed method completeness correctness and quality criteria for vehicle counting in different condition are computed and analysed. As shown in table 1 results of evaluation of vehicle counting is presented. Dataset#1 refers to imaging with adequate amount of light (daylight) and in opposite dataset#2 suffers from low illumination during sunrise and sunset.

As shown in table 1 the algorithm is 92% successful with respect to quality of results for dataset#1. In return due to illumination decrease vehicle detection quality of results has decreased to 80.2 %.

Table 1: Evaluation of results of vehicle counting

Dataset	Correctness	Completeness	Quality
Dataset#1	96.3	95.1	92.6
Dataset#2	86.9	83.4	80.2

5. CONCLUSIONS

In this paper a method for traffic monitoring of roads is proposed developed and evaluated. According to the specification of the proposed method and the results of evaluation, in conclusion the following items are mentionable.

Vehicle detection using CCTV images is a proper and inexpensive tool for traffic assessment especially in urban areas.

The results of vehicle detection and counting and also speed estimation is totally related to the quality of imaging, specifically the spatial resolution of the imagery, illumination conditions and also angle of image acquisition.

The quality of results especially in case of speed estimation is related to the stability of images, if the camera is not stable the correct state of camera should be restored. On the other hand using proper equipment especially the capability of definition of pre-sets for camera acquisition angles could be helpful to avoid stability problems.

Finally it could be concluded that the proposed method in more successful during the day since the illumination condition is ideal. By the decrease of environmental light the successfulness of this method decreases.

REFERENCES

Dhar, A., 2008. "Traffic and Road Condition Monitoring System", Technical report, Indian Institute of Technology Mumbai, India,

Ferrier , N., Rowe , S., Blake, A.,1994. "Real-time Traffic Monitoring." Proceedings of the 2nd IEEE Workshop on Applications of Computer Vision, 1994. 81- 87

Song, K., Tai, J., 2006. "Dynamic calibration of Pan–Tilt–Zoom cameras for traffic monitoring." *IEEE Transactions on Systems, Man, and Cybernetics*, Part B: Cybernetics, 36(5):1091–1103

19

CLUSTERING OF MULTI-TEMPORAL FULLY POLARIMETRIC L-BAND SAR DATA FOR AGRICULTURAL LAND COVER MAPPING

H. Tamiminia [a, *], S. Homayouni [b], A. Safari [a]

[a] School of Surveying and Geospatial Engineering, Dept. of Remote Sensing, College of Engineering, University of Tehran, Iran - (haifa.tamiminia, asafari)@ut.ac.ir
[b] Dept. of Geography, Environmental Studies and Geomatics, University of Ottawa, Ottawa, Canada - saeid.homayouni@uOttawa.ca

KEY WORDS: Kernel-Based Fuzzy C-means, Crop Classification, Polarimetric SAR Images, Multi-Temporal Data, Target Decompositions

ABSTRACT

Recently, the unique capabilities of Polarimetric Synthetic Aperture Radar (PolSAR) sensors make them an important and efficient tool for natural resources and environmental applications, such as land cover and crop classification. The aim of this paper is to classify multi-temporal full polarimetric SAR data using kernel-based fuzzy C-means clustering method, over an agricultural region. This method starts with transforming input data into the higher dimensional space using kernel functions and then clustering them in the feature space. Feature space, due to its inherent properties, has the ability to take in account the nonlinear and complex nature of polarimetric data. Several SAR polarimetric features extracted using target decomposition algorithms. Features from Cloude-Pottier, Freeman-Durden and Yamaguchi algorithms used as inputs for the clustering. This method was applied to multi-temporal UAVSAR L-band images acquired over an agricultural area near Winnipeg, Canada, during June and July in 2012. The results demonstrate the efficiency of this approach with respect to the classical methods. In addition, using multi-temporal data in the clustering process helped to investigate the phenological cycle of plants and significantly improved the performance of agricultural land cover mapping.

1. INTRODUCTION

Remote sensing offers an important source of data for studying spatial and temporal variability of the environmental parameters (Blaes et al., 2005). Crop identification from earth observing satellites is essential for monitoring food security, agricultural and economic planning. Both optical/infrared and microwave satellite data can be used for this purpose. Recently, Synthetic Aperture Radar (SAR) imagery, thanks to their potential of data collection regardless of weather and illumination conditions, has become an essential tool for crop mapping and monitoring activities (Jiao et al., 2014). Radar-based crop type classification requires earth observations with multiple polarization. In particular, Polarimetric Synthetic Aperture Radar (PolSAR) sensors can acquire data sensitive to the dielectric properties of the crop canopy and its geometric structure (i.e., the size, shape, orientation distribution of leaves, stalks and fruits) ((Soria-Ruiz et al., 2007) and (Skriver et al., 1999)). Fully polarimetric radars record the complete characterization of the scattering field. Thus, both four mutually coherent channels recorded and phase information are observed and recorded for further processing. As a result, users can synthesize any linear or nonlinear polarization and can generate other polarimetric variables (McNairn et al., 2004). Consequently, land cover classification is one of the most important applications of PolSAR data (Lee and Pottier, 2009).

In general, classification methods are divided into two categories: supervised and unsupervised or clustering. The most important issue with supervised classification approaches is providing of high quality and quantity training data, which is almost costly and time-consuming. Because of this limitation, there is a strong interest in developing of unsupervised techniques (Rignot, et al., 1992).

Clustering algorithms aim to identify the unknown structure or pattern among the data. These structures can be the natural groups or clusters within the multi-dimensional feature space by measuring similarities between different pixels' data.

The C-means clustering families are the best known and robust techniques of batch clustering models, due to using the least square models (Bezdek, et al., 2005). The most frequently used algorithms of these families are C-means and Fuzzy C-means (FCM) algorithms. Hard clustering algorithms, such as K-means, divide data into distinct classes, whereas in fuzzy clustering, every pixel has a membership of belonging to all clusters rather than belonging to one single cluster. Presence of the mixed pixels in the remote sensing data is a potential case study to use fuzzy C-means method in clustering. FCM is a technique of clustering which permits data points to cluster based on spectral similarity and mainly used in pattern recognition (Vanisri, 2004). Nonetheless, these models are relatively efficient for linearly separable patterns, the inherent nature of polarimetric data is very complex. As a result, they are not linearly separable. As a solution, the aim of this paper is to present a kernel-based fuzzy C-means clustering (KBFCM) approach to classify nonlinear data in order to map the agricultural crop lands.

Unsupervised PolSAR classification follows three major approaches. One based on the inherent statistical characteristics of PolSAR data (Bell and Hall, 1967). The second category classifies PolSAR imageries by inherent physical scattering characteristics (Zyl, 1989). This approach has the advantage of providing information for class type identification, but the classification results typically lose the details and the number of classes is normally small. In the third category, both statistical and physical scattering characteristics are considered. As a result, they can classify PolSAR data most effectively (Lee and Pottier, 2009). However many clustering methods have been proposed, they are not efficient enough to distinguish between classes, especially in agricultural crop mapping.

This paper presents a framework for classifying multi-temporal full polarimetric SAR data using kernel-based fuzzy C-means clustering method, over an agricultural region. The paper is organized as follows: In Section 2, a brief review of kernel-based clustering and the proposed method will be defined. Section 3 gives some information about used data. Section 4 presents the results and discussion of the clustering results. Finally, Section 5 concludes the paper.

2. METHODOLOGY

2.1 Kernel Principals

In the field of remote sensing, descriptive machine learning algorithms often focus on land cover classifications (Camps-Valls and Bruzzone, 2009). In machine learning, kernel methods are used for classification, clustering, regression, density estimation and visualization with heterogeneous types of data, such as time series, images, strings or objects (Schölkopf and Smola, 2002). In this subsection, a brief description of the notion of kernels is presented.

Kernel methods by using a nonlinear transform, map the dataset $x = \{x_1, x_2, ..., x_n\}$ defined over the input or attribute space X ($x \in X$) into a higher dimensional Hilbert space H, or feature space, which enable them to distinguish nonlinear data with linear methods. The mapping function is denoted as (Camps-Valls et al., 2008):

$$\varphi : X \rightarrow H, x \rightarrow \varphi(x) \qquad (1)$$

However, direct computation in the high-dimensional feature space consumes much time and sometimes even infeasible. To avoid working in the potentially high-dimensional space, the dot product can be evaluated directly using a nonlinear function in input space by means of the kernel trick (Camps-Valls et al., 2008). Every function that meets the Mercer's condition can be used as a kernel function. Usually, kernel functions are used instead of Mercer kernels or, equivalently, positive definite kernels.

The similarities between elements in feature space can be measured using inner product. For convenience we introduce the following function that does exactly that (Camps-Valls and Bruzzone, 2009):

$$K : X \times X \rightarrow \mathbb{R}, (x, x') \rightarrow K(x, x') \qquad (2)$$

which is required to satisfy for all $x, x' \in X$:

$$K(x, x') = \langle \varphi(x), \varphi(x') \rangle_H \qquad (3)$$

This function is called a kernel. The mapping φ is referred to as its feature map and the space H as its feature space.

The distances of the elements in H can be evaluated entirely in the terms of kernel evaluations (Camps-Valls and Bruzzone, 2009):

$$d^2(x, x') = \left\| \varphi(x) - \varphi(x') \right\|^2 = \langle \varphi(x), \varphi(x) \rangle +$$

$$\langle \varphi(x'), \varphi(x') \rangle - 2 \langle \varphi(x), \varphi(x') \rangle = \qquad (4)$$

$$K(x, x) + K(x', x') - 2K(x, x')$$

The most well-known kernel functions are: Radial Based Function (RBF), polynomial and linear.

The RBF kernel: $K(x, x') = exp^{-\frac{|x-x'|^2}{2\sigma^2}}$ (5)

The polynomial kernel: $K(x, x') = (< x, x' + 1 >)^p$ (6)

The linear kernel: $K(x, x') = < x, x' >$ (7)

2.2 Kernel-based fuzzy C-means algorithm

The main idea of using a kernel function in the similarity criteria is to compute the distance between pixels in feature space (Camps-Valls et al., 2008). Feature space due to its inherent properties enable us to clustering nonlinear datasets simpler and more efficient. Kernel-based algorithms, unlike to the classical clustering techniques which use the Euclidean distance, benefit from kernel distance in order to calculate the objective function. The minimized objective function formula is given by (Girolami, 2002):

$$J(X; U, C) = \sum_{j=1}^{c} \sum_{i=1}^{n} u_{ji}{}^m d^2 =$$

$$\sum_{j=1}^{c} \sum_{i=1}^{n} u_{ji}{}^m \left| \varphi(x_i) - \varphi(c_j) \right|^2 = \quad , 2 \leq c < N \qquad (8)$$

$$\sum_{j=1}^{c} \sum_{i=1}^{n} u_{ji}{}^m (K(x_i, x_i) + K(c_j, c_j) - 2K(x_i, c_j))$$

where   J = objective function
         U = fuzzy partition matrix, $u_{ji} \in [0,1]$
         C= cluster centers, C={$c_1, c_2, ..., c_c$}
         m = weighting exponent

Optimization of objective function with respect to two variables (cluster centers and partition matrix) is one of the most important issues facing the partitional clustering models (Chen and Zhang, 2004). By an alternative optimization, in each iteration, the cluster centers and fuzzy partition matrix can be calculated using the following equations:

$$u_{ji} = \frac{d_{Kji}^2(x_i, c_j)^{-\frac{1}{(m-1)}}}{\sum_{j=1}^{c} d_{Kji}^2(x_i, c_j)^{-\frac{1}{(m-1)}}} \qquad (9)$$

$$c_i = \frac{\sum_{i=1}^{m} u_{ji}^m K(x_i, c_j) x_i}{\sum_{i=1}^{m} u_{ji}^m K(x_i, c_j)} \qquad (10)$$

It is worth to note that Equation (10) is derived by using the RBF kernel function, which satisfying $K(x, x) = 1$. In order to apply other kernel functions, first the kernel function is replaced with K in Equation (8), then, this equation is optimized cluster centers and fuzzy partition matrix. For instance, the cluster centers for polynomial kernel will be obtained as:

$$c_i = \frac{\sum_{i=1}^{n} u_{ji}^m x_i (< x_i, c_j > +1)^{p-1}}{\sum_{i=1}^{n} u_{ji}^m (< c_j, c_j > +1)^{p-1}} \quad (11)$$

The kernel-based fuzzy C-means clustering procedures are shown below:

Step 1: Choose the kernel function, the maximum number of iterations, m and ε.
Step 2: Initialize the cluster centers.
Step 3: Calculate fuzzy partition matrix U according to Equation (9).
Step 4: Update cluster centers by Equation (10) or (11) depending on the kernel function and U.
Step 5: Until termination criterion satisfied (objective function (i+1) – objective function (i) < ε or maximum iterations reached stop, otherwise, go to step 3.
Step 6: Defuzzify the final U in order to obtain a hard membership decision. The highest membership was obtained by using the maximum membership decision rule.

Figure 1 shows an overview of the proposed method. It starts by preprocessing of raw L-band SAR data. This preprocessing step consist of multi-looking and speckle filtering. Then the decomposition algorithms were applied to the covariance matrix of full polarimetric SAR data, in order to extract several physically and statistically based features. These features are the linear polarization intensities in HH, HV and VV, Freeman-Durden, Yamaguchi and Cloude-Pottier decompositions. Finally, the proposed kernel-based fuzzy C-means clustering was applied on these features to map out the various crop types.

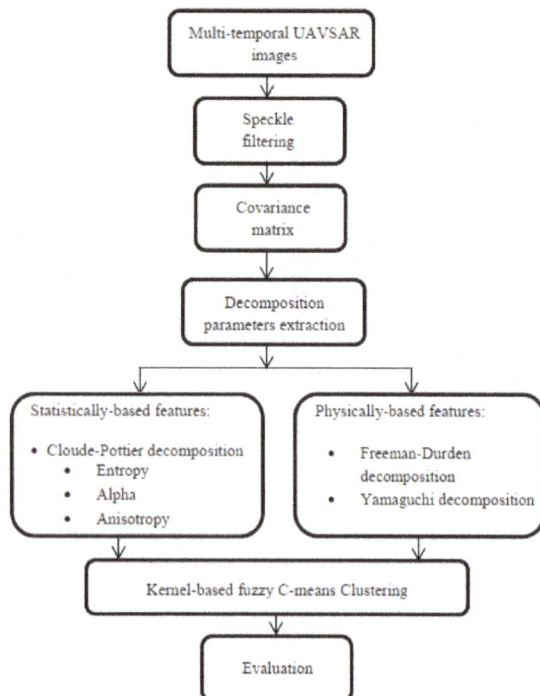

Figure 1. An overview of the proposed method

3. POLARIMETRIC SAR DATA

The proposed method has applied to four dates of polarimetric SAR data acquired by Uninhabited Aerial Vehicle Synthetic

Aperture Radar (UAVSAR), over an agricultural area near Winnipeg, in Manitoba, Canada, during June and July in 2012. The data acquisition was to support the Soil Mapping Active Passive Validation Experiment 2012 (SMAPVEX 12) mission of the JPL- NASA. The UAVSAR data were collected 14 days during which soil and moisture and vegetation conditions significantly variant. The campaign started at the period of early crop development and finished at the point where crops had reached maximum biomass (McNairn et al., 2014). The UAVSAR system is an aircraft-based fully polarimetric L-band radar system operated by the JPL Radar Science and Engineering Section. Though there are plans to fly the system aboard a UAV, such as a Global Hawk, in the future, the UAVSAR system is currently flying aboard a NASA Gulfstream III and nominally operates at 45,000 ft AGL. The system has a precision real-time GPS and sensor-controlled flight management system which allows for repeatable flight paths that remain within a 10 m diameter tube about the intended track. This flight precision allows for differential interferometric studies of dynamic phenomena. The UAVSAR radar has some unique features that are worth noting. First, it has quad-pol capability. Second, it has a range bandwidth of 80 MHz which gives the Single-Look-Complex (SLC) data a 1.66 m range and 0.6 m azimuth resolution. The third notable feature is that the antenna can be steered ±20° along the azimuth line. This allows the radar line of sight direction to be independent of the wind-induced motion of the aircraft (Rosen et al., 2006). Its high spatial resolution allows studies of backscattering from homogeneous vegetation covers. Figure 2 shows a 501×501 subsample of the color composite of Pauli decomposition.

Figure 2. Color composite of Pauli decomposition
($R=S_{hh} + S_{vv}$, $G=S_{hh} - S_{vv}$, $B= S_{hv}$)

4. RESULTS AND DISCUSSION

The proposed method has applied to the polarimetric features extracted from four datasets. To evaluate the results quantitatively, a ground truth image of the study area is used. This area consists of four dominant crop types including Wheat, Soybeans, Canola and Oats. The reference crop map of the study area is presented in Figure 3. The performance of the proposed algorithm is evaluated using actually three strategies: First, we have investigated the results of using different types of kernel functions. In second strategy we compared the best result of the kernel functions (RBF) with conventional methods. Finally, we evaluated RBF kernel with different number of dates to investigate the influence of the multi-temporal data for classification. The detailed comparisons and results are shown below.

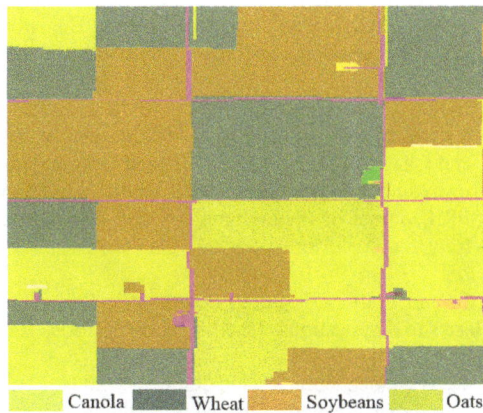

Figure 3. Reference map and its legend

Method	Overall accuracy	Kappa coefficient
KBFCM-RBF	89.95	0.86
FCM	77.22	0.69
K-means	71.85	0.6

Table 2. Comparisons between kernel-based and conventional methods

4.1 Different kernel functions evaluation

Table 1 contains accuracy parameters of using different kernel functions. The experimental results demonstrate that the accuracy of RBF kernel is more than polynomial and linear kernels. RBF kernel is the most popular kernel among the others in SAR data processing. Its efficiency can be explained by the nature of SAR data, because the statistical distribution of these data is approximately normally distributed and it is less sensitive to the noise. The next one is polynomial kernel, the experimental results showed that with increasing the kernel parameter the bigger values affect the dot products and the kernel values become unreal as a result the accuracy will decrease. In addition, the parameters of polynomial kernel are usually difficult to determine and it is more time-consuming and complicated. The linear kernel is the simplest one that takes the least time to process. It has the lowest results among other kernels.

Kernel type	Overall accuracy	Kappa coefficient
RBF	89.95	0.86
Polynomial	88.43	0.84
Linear	86.58	0.81

Table 1. Accuracy parameters of different kernel functions assessment

4.2 Comparisons with conventional methods

In this strategy we have compared the result of KBFCM with conventional methods such as K-means and FCM. Table 2 presents the overall accuracies and kappa coefficients of the results. Figure 3 shows the visual comparison of resulted classification maps. The performance of kernel-based fuzzy C-means algorithm with RBF kernel is very obvious in Table 1 and Figure 3. It can be seen that using kernel function to map input data into feature space and measuring the similarities with kernel rather than Euclidean distance can improve the clustering accuracy. Because calculating the distance with Euclidean equation is more sensitive to noise. Table 2 also represents the better accuracy for FCM algorithm with respect to K-means algorithm. This result can be explained by the inherent property of the fuzzy clustering. In fuzzy clustering, unlike to hard clustering we can consider each pixel to all classes. Existing mixed pixels especially in PolSAR data is one of the reasons for performance of the fuzzy C-means clustering.

Figure 3. Final Crop maps: KBFCM- RBF (Top), FCM (Middle), and K-means (Bottom)

4.3 Multi-temporal SAR data assessment

In this strategy we investigate the impact of using more than one date in clustering results. The results show that using multi-temporal datasets with four dates of data increase the overall accuracy up to 27% with respect to the single-date data. The overall accuracies and kappa coefficients of the clustering results are provided in Table3.

Number of dates	Overall accuracy	Kappa coefficient	Kernel type
4	89.95	0.86	RBF
3	81.73	0.75	RBF
2	71.14	0.61	RBF
1	62.14	0.47	RBF

Table 3. Accuracy parameters of multi-temporal assessment

5. CONCLUSION

The aim of this paper was to study the capability of multi-temporal UAVSAR L-band PolSAR data for crop mapping. Four full polarimetric images were acquired over an agricultural area. Several statistical and physical based decomposition features in addition to linear polarization intensities were extracted from data. A kernel-based fuzzy C-means clustering method based on several kernel functions (i.e. RBF, polynomial and linear) was applied to these features in order to classify data. The results showed that multi-temporal datasets increases the overall accuracy with respect to the single-date, two-dates and even three-dates data. Among the different kernels which used in this research, RBF kernel has much better accuracy than polynomial and linear kernels. Moreover, the proposed algorithm because of working in the higher dimensional feature space is more efficient than the other well-known classic methods such as FCM and K-means algorithms.

ACKNOWLEDGEMENTS

The authors would like to thank the SMAPVEX 2012 team for providing this data set. They also acknowledge the comments of Dr. Heather McNairn from Agriculture and Agri-food Canada, Ottawa that help to improve the paper.

REFERENCES

Bell, G. H. and Hall, D. J., 1967. A clustering technique for summarizing multi-variate data. Behavioral Science, 12(2), pp. 153–155.

Bezdek, J. C., Keller, J. M., Krishnapuram, R. and Pal, N. R., 2005. Fuzzy models and algorithms for pattern recognition and image processing. Springer, New York.

Blaes, X., Vanhalle, L. and Defourny, P., 2005. Efficiency of crop identification based on optical and SAR image time series. Remote Sens. Environ. Vol. 96, pp. 352–365.

Camps-Valls, G., Gomez-Chova, L., Munoz-Mari, J., Rojo-Alvarez, J. L. and Martinez-Ramon, M., 2008. Kernel-Based Framework for Multitemporal and Multisource Remote Sensing Data Classification and Change Detection. IEEE Transactions on Geoscience and Remote Sensing, 46(6), pp. 1822–1835.

Camps-Valls, G. and Brizzone, L., 2009. Kernel Methods for Remote Sensing Data Analysis, 1st ed. John Wiley & Sons, Chichester,U.K.

Chen, S.C. and Zhang, D. Q., 2004. Robust image segmentation using FCM with spatial constraints based on new kernel-induced distance measure. IEEE Transactions on Systems, Man, and Cybernetics _Part B: Cybernetics, 34(4), pp. 1907–1916.

Girolami, M., 2002. Mercer kernel based clustering in feature space. IEEE Transactions on Neural Networks, Vol.13, pp. 780–784.

Jiao, X., Kovacs, J. M., Shang, J., McNairn, H., Walters, D., Ma, B. and Gang, X., 2014. Object-oriented crop mapping and monitoring using multi-temporal polarimetric RADARSAT-2 data. ISPRS Journal of Photogrammetry and Remote Sensing, pp. 38–46.

Lee, J. S. and Pottier, E., 2009. Polarimetric Radar Imaging: From Basics to Applications, 1st ed. CRC Press, Boca Raton.

McNairn , H., Hochheim, K. and Rabe, N., 2004. Applying polarimetric radar imagery for mapping the productivity of wheat crops. Canadian Journal of Remote Sensing, 30(3), pp. 517–524.

McNairn , H., Jackson,T. J., Wiseman, G., Bélair, S., Berg, A. A., Bullock, P., Colliander, A., Cosh, M. H., Kim, S.-B., Magagi, R., Moghaddam, M., Adams, J. R., Homayouni, S., Ojo, E., Rowlandson, T., Shang, J., Goïta, K. and Hosseini, M., 2014. The Soil Moisture Active Passive Validation Experiment 2012 (SMAPVEX12): Pre-Launch Calibration and Validation of the SMAP Satellite. IEEE Transactions on Geoscience and Remote Sensing, Vol. 53, pp. 2784–2801.

Rignot, E., Chellappa, R. and Dubois, P., 1992. Unsupervised segmentation of polarimetric SAR data using the covariance matrix. IEEE Trans. Geosci. Remote Sensing, Vol. 30, pp. 697–705.

Rosen, P. A., Hensley, S., Wheeler, K., Sadowy, G., Miller,T., S. Shaffer, Muellerschoen, R., Jones, C., Zebker, H. and Madsen, S., 2006. UAVSAR: A new NASA airborne SAR system for science and technology research. In: IEEE Conference on Radar, pp. 24–27.

Schölkopf, B. and Smola, A. J., 2002. Learning with Kernels Support Vector Machines, Regularization, Optimization, and beyond. MIT Press, Cambridge, MA.

Skriver, H., Svendens, M. T. and Thomsen, A. G., 1999. Multitemporal C- and L-band polarimetric signatures of crops. IEEE Transactions on Geoscience and Remote Sensing, 37(5), pp. 2413–2429.

Soria-Ruiz , J., McNairn, H., Fernandez-Ordonez, Y. and Bugden-Storie, J., 2007 . Corn monitoring and crop yield using optical and RADARSAT-2 images. IEEE International Geoscience and Remote Sensing Symposium, pp. 3655–3658.

Vanisri, D., 2014. A novel kernel based fuzzy C-means clustering with cluster validity measures. International journal of computer science and network solutions, 2(12), pp. 18–26.

Yu, J., Yan, Q., Zhang, Z., Ke, H., Zhao, Z. and Wang, W., 2012. Unsupervised classification of polarimetric synthetic aperture radar images using kernel fuzzy C-means clustering. International journal of image and data fusion, 3(4), pp. 319–332.

Zyl, J.J. Van, 1989. Unsupervised classification of scattering mechanisms using radar polarimetry data. IEEE Transactions on Geoscience and Remote Sensing, Vol. 27, pp. 36–45.

20

EVALUATION OF EFFECTING PARAMETERS ON OPTIMUM ARRANGEMENT OF URBAN LAND USES AND ASSESSMENT OF THEIR COMPATIBILITY USING ADJACENCY MATRIX

S. Vaezi [a,], M.S. Mesgari [a], F.Kaviary [b]

[a] Dept. of Geomatic Engineering, K. N. Toosi University of Technology, Valy-Asr Street, Mirdamad Cross, Tehran, Iran
s.vaezi@mail.kntu.ac.ir
[a] Dept. of Geomatic Engineering, K. N. Toosi University of Technology, Valy-Asr Street, Mirdamad Cross, Tehran, Iran
mesgari@kntu.ac.ir
[b] Dept. of Geomatic Engineering, K. N. Toosi University of Technology, Valy-Asr Street, Mirdamad Cross, Tehran, Iran
farnaz_kaviari@mail.kntu.ac.ir

KEYWORDS: Land-use, Sustainable Urban Development, GIS, Density Levels, Descriptive Analytical Method, Adjacency Matrix

ABSTRACT

Todays, stability of human life is threatened by a set of parameters. So sustainable urban development theory is introduced after the stability theory to protect the urban environment. In recent years, sustainable urban development gains a lot of attraction by different sciences and totally becomes a final target for urban development planners and managers to use resources properly and to establish a balanced relationship among human, community, and nature. Proper distribution of services for decreasing spatial inequalities, promoting the quality of living environment, and approaching an urban stability requires an analytical understanding of the present situation. Understanding the present situation is the first step for making a decision and planning effectively. This paper evaluates effective parameters affecting proper arrangement of land-uses using a descriptive-analytical method, to develop a conceptual framework for understanding of the present situation of urban land-uses, based on the assessment of their compatibility. This study considers not only the local parameters, but also spatial parameters are included in this study. The results indicate that land-uses in the zone considered here are not distributed properly. Considering mentioned parameters and distributing service land-uses effectively cause the better use of these land-uses.

1. INTRODUCTION

Sustainable urban development is one of the most comprehensive concepts in recent decades.Generally speaking, it means a correct and efficient use of funds, manpower, and natural resources to make an appropriate consumption pattern . An appropriate organizational structure can be satisfactorily obtained from technical features to meet the needs of today's and future generations .So sustainable development not only considers an improvement for today's generation, but it also considers future generations .On the other hand, sustainable development is an equilibrium state among different aspects of development, which is aimed to improve the quality of human life.So sustainable urban development is a form of development which provides the continuous development of cities and urban communities for the future generations. It changes land-uses and density levels to meet the residents' needs for housing, transportation, and welfare services.This planning process makes a city environmentally inhabitable, stable economically, and socially correlated over time. Sustainable development is a solution against physical, social and economic patterns of development. Using that would prevent the destruction of natural resources, an increase of population, injustice and decreasing the quality of current and future people's life. (Isaksson, R., 2006).

In fact, it is a set of tasks which improves the human life with respect to needs of urban community. Thus, cities can be stable only if their infrastructure and land-uses are allocated

equivalently. Assessment of urban land-uses plays an important role in a local-spatial organization of a city and approaching the targets of sustainable urban development.

Two directing agents in terms of social welfare and prosperity would be assessed for characterizing the use of ground or every kind of urban activity. According to these two general factors, five criteria including per capita, utility, dependence, access and compatibility are used as criteria of urban land use planning in location of urban functions. These criteria are briefly defined below: per capita is the extent of land dedicated to a specific use of urban facilities with s specified performance level for each individual. (Omer, 2006) Utility refers to compatibility between the land use and its location. Therefore, utility is dependent on many physical, social, economic parameters and etc... (Steiner, 2008). In addition, each urban activity and land use requires access to some other activities in order to increase productivity. In fact, each land use has a set of relations with other land uses that define the affiliation of one land use to other land uses (Taleai, et al., 2007). Access refers to the access of lands to the public transportation network. (Reddy, 2014). And finally, compatibility refers to proximity of land uses that lead to improved performance of one another and separation of land uses that have negative impact on the performance of one another. Therefore, land uses should be juxtaposed in such a way that they can have the most positive interaction with their neighbouring units. (Taleai, et al., 2007) One of the goals of urban land use planning is to provide favourable proximity for land uses and to separate the incompatible uses. Therefore, assessment of compatibility between land uses, is an attempt to ensure that with the development of the city, each new use is still a good neighbour to other land uses. To achieve this end, the characteristics of each urban land use, relationships between land uses and the impact land uses on one another, must be fully investigated.

This study is an attempt to investigate the effective factors in proper arrangement of land uses and their impact on the extent of service delivery uses with a descriptive analysis approach. For this purpose, the Delphi method was used to make compatibility matrix that is presented in the next section.

2. MODELING

2.1 Compatibility Matrix

In order to model the effective parameters in land use arrangements, first a compatibility matrix is formed. In order to achieve a compatibility matrix, first, the existing land uses provided in the map are classified based on the level of performance and type of activity (according to the vision outlined in the comprehensive plan). Afterwards, The Delphi method is used as a framework for making a compatibility matrix and determine the different levels of physical compatibility between different urban land uses. Physical compatibility. Physical compatibility is in fact the compatibility compared to the other adjacent land uses without direct consideration of socio-economic conditions and their emphasis on permanent compatibility and incompatibility aspects, such as noise, pollution, shade, etc. (Taleai, et al., 2007).It should be noted that the Delphi method used in this study is an iterative process to achieve consensus of a group of experts on a particular subject matter and its assessment criteria. This method is particularly used in conditions where there are no standard criteria for assessment. (Shiftan, et al., 2003)
Different compatibility levels including high, medium, neutral, fairly incompatible and totally incompatible are used in this study and each of these levels is defined as follows.

2.1.1 High Compatible
When two fully compatible land uses come together, they greatly enhance usability and usefulness of each other.

2.1.2 Medium Compatible
When two land uses with low compatibility come together, their effectiveness is slightly increased.

2.1.3 Neutral
When two neutral land uses come together, they would have no positive or negative impact on the performance of each other.

2.1.4 Medium Incompatible
When two fairly incompatible land uses come together, they lead to decreased usability and usefulness of each other.

2.1.5 High Incompatible
Complete non-compatible applications have a very negative impact on the performance of each other and the first priority is that these two applications are not close together

AHP is one of the common methods used to quantify the compatibility matrix. This method is a powerful and flexible decision-making process for the introduction of the priorities of decision-makers and make the best decision when the qualitative and quantitative aspects of a decision are required (Masoomi, z., 2013)

Different compatibility levels for different land uses are shown in Table 1.

W	RP	I	U F	At	C M	Ad	T	E	C	R	
N	HC	MI	MI	HC	MC	MI	HC	MI	HC	HC	R
	HC	MI	MI	HC	N	MC	HC	HC	HC	HC	C
		MI	MI	HC	N	MI	MI	HI	HC	MI	EI
			MI	MI	MC	HC	HC	MI	HC	HC	T
				N	N	HC	HC	MI	MC	MI	Ad
					HI	N	MC	N	N	MC	C M
						N	MI	HC	HC	HC	At
							MI	MI	MI	MI	U F
								MI	MI	MI	I
									HC	HC	RP
										N	W

Table 1. Different compatibility levels for different land uses R as Residential, C as Commercial, E as Educational, T as Therapeutic, Ad as Administrative, CM as Cultural and Mosque, A as Athletic, UF as Urban Facilities, I as Industrial, RP as Recreation and Park, W as Wasteland. (Derived from Masoomi,2014)

Although, urban planners use compatibility matrix at the level of urban zones to determine future land use in developing regions, in this study, compatibility matrix is used to assess the compatibility between the existing detailed uses in the constructed urban areas and at the level of building plaques.

2.2 Study Area

The study area is the central area of the city of Abhar, Zanjan which is located 90 km south of East Zanjan and 230 kilometers West of Tehran. Abhar is 1,540 m above sea level, with 87, 396 inhabitants in 1390, was considered the second most populated city in the province of Zanjan, after the city of Zanjan. In this study, building plaque map and subject information regarding the use of each plate, as needed baseline data, are provided and prepared for storage in GIS software.
Figure 1 shows the map of Abhar. The center of Abhar show in figure 2, which there are a lot of different land_use considered

in this study, is used for implementing and calculating the compatibility.

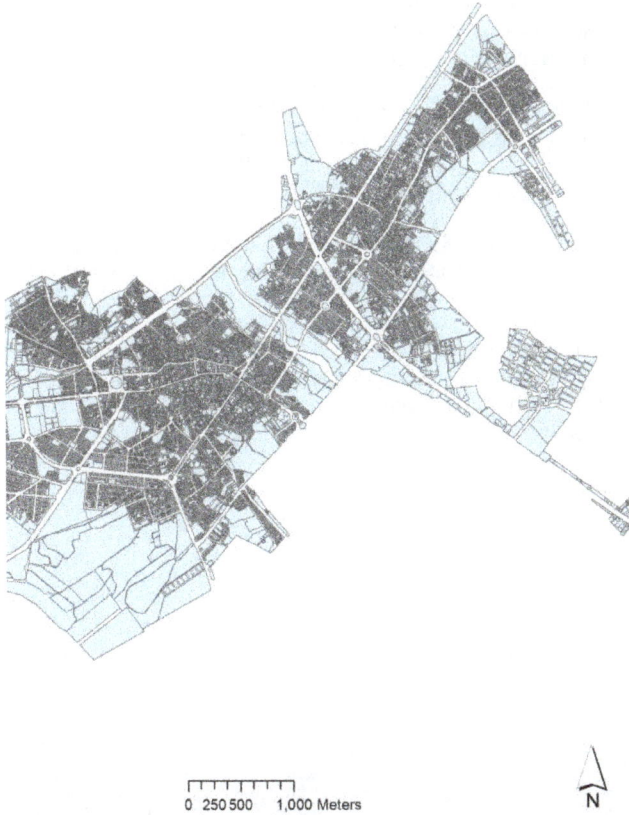

Figure 1. Abhar city map. Different applications show with different colours

Figure 2. Center of Abhar, which there are different land-use considered in this study

Given that in this study, shift from compatibility levels to incompatibility levels is more important than shift from incompatibility levels to compatibility levels, the AHP results are finally standardized as Table2

Standardized values	compatibility levels
0.43	High Compatible
0.28	Medium Compatible
0.18	Neutral
0.08	Medium Incompatible
0.04	High Incompatible

Table 2. The AHP standardized results

It should be noted that the following equation is used to calculate the compatibility number for each plaque. (Masoomi, 2014)

$$F = \left(w_1 \frac{1}{n}\sum_{i=1}^{n}\left(\frac{1}{n_i}\sum_{j=1}^{n_i}\left(Comp_{ij}\right)\right) + w_2 \min_{ij}(Comp_{ij})\right) \quad (1)$$

Where　i = Plaque　　　j = Plaque neighbours
n_i= Number of plaque neighbours
w_1= Total compatibility factor
w_2= least compatibility factor
　　　n = number of total plaque
$Comp_{ij}$ =Standarded compatibility of the plaque i with the plaque j

In order to define neighborhood, a circle with a radius of 12 meters was drawn and the plaques located within this radius were regarded as neighbor plaques. In this study, the weights are set to 1. The first part of equation 1 calculates the sum of compatibility between plaque 1 and adjacent plaques. The second part resolves the compensation of sum of the compatibilities to an acceptable level, by considering the greater effect of minimum compatibilities.

3. DISCUSSION AND CONCLUSION

Assessment of the study area's land use in terms of compatibility was conducted using Python in the ArcGIS environment. The results showed that in the study area a total of 7904 units have high incompatible and medium incompatible uses and a total of 4598 units have compatible and high compatible uses compared to the neighboring plaques uses, which could have a positive impact on the performance of neighboring land uses. Results are shown in Figure 2

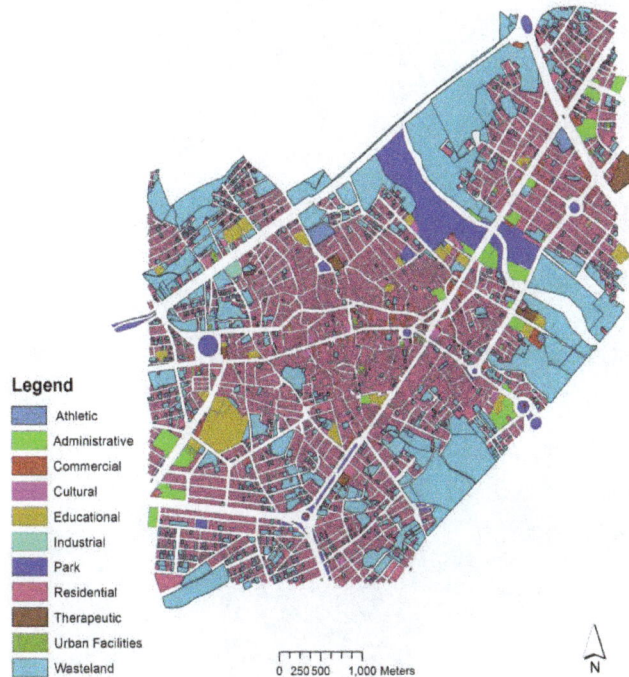

Legend
- Athletic
- Administrative
- Commercial
- Cultural
- Educational
- Industrial
- Park
- Residential
- Therapeutic
- Urban Facilities
- Wasteland

Omer, Itzhak. "Evaluating accessibility using house-level data: A spatial equity perspective." Computers, environment and urban systems 30.3 (2006): 254-274.

Reddy,P.B. (2014) planning for future urban development: land use analysis. Dehli,India: Jawaharlal Nehru Technological Univercity, Hyderabad

Shiftan, Yoram, Sigal Kaplan, and Shalom Hakkert. "Scenario building as a tool for planning a sustainable transportation system." Transportation Research Part D: Transport and Environment 8.5 (2003): 323-342.

Steiner,F. (2008). The Living Landscape: An Ecological Approach to Landscape Planning (second ed). New York: Mc Graw Hills

Taleai, Mohammad, et al. "Evaluating the compatibility of multi-functional and intensive urban land uses." International Journal of Applied Earth Observation and Geoinformation 9.4 (2007): 375-391.

Legend

- High and Medium Compatible
- High Incompatible
- Medium Incompatible

0 250 500   1,000 Meters

N

Figure 3. Results of Analysis

Considering the above-mentioned parameters and better substitution of service uses, it would be possible to take more useful advantage of them. Use of GIS system's spatial analysis can significantly help to analyze the current situation of land uses in this region. In addition, consideration of the proposed indexes in urban designing and changes in the property uses have negative impact on the performance of neighboring plaques and in fact are placed in incompatible and fairly incompatible category of property use. Moreover, substitution of more suitable uses, not only leads to reduction of energy consumption and development of a sustainable economy, but also leads to development of a healthier environment, more comprehensive justice, and improved quality of life. And can provide a way for the current cities to achieve sustainability in all arenas.

REFERENCES

Isaksson, R. (2006), "Total quality management for sustainable development Process based system models",Business Process Management Journal,Vol. 12 No. 5 ,pp. 632-645.

Masoomi.Z. "Modeling physical impacts of urban land-use change using optimization algorithms and spatial analyses. " PhD Thesis in Geographic Information Systems,Khaje Nasir Toosi Univercity of Technology(2014)

Masoomi, Zohreh, Mohammad Sadi Mesgari, and Majid Hamrah. "Allocation of urban land uses by Multi-Objective Particle Swarm Optimization algorithm."International Journal of Geographical Information Science 27.3 (2013): 542-566.

21

APPLICATIONS OF MEDIUM C-BAND AND HIGH RESOLUTION X-BAND MULTI-TEMPORAL INTERFEROMETRY IN LANDSLIDE INVESTIGATIONS

J. Wasowski [a] *, F. Bovenga [b], R. Nutricato [c], D. O. Nitti [c], M. T. Chiaradia [d]

[a] CNR-IRPI, National Research Council, 70126 Bari, Italy - j.wasowski@ba.irpi.cnr.it
[b] CNR-ISSIA, National Research Council, 70126 Bari, Italy - bovenga@ba.issia.cnr.it
[c] GAP srl, c/o Politecnico di Bari, 70126 Bari, Italy – (raffaele.nutricato, davide.nitti)@gapsrl.eu
[d] Dipartimento Interateneo di Fisica, Politecnico di Bari, 70126 Bari, Italy - Mariateresa.Chiaradia@ba.infn.it

KEY WORDS: COSMO-SkyMed, ENVISAT, ERS, Multi-Temporal Interferometry, Landslide

ABSTRACT

With the increasing quantity and quality of the imagery available from a growing number of SAR satellites and the improved processing algorithms, multi-temporal interferometry (MTI) is expected to be commonly applied in landslide studies. MTI can now provide long-term (years), regular (weekly-monthly), precise (mm) measurements of ground displacements over large areas (thousands of km^2), at medium (~20 m) to high (up to 1-3 m) spatial resolutions, combined with the possibility of multi-scale (regional to local) investigations, using the same series of radar images. We focus on the benefits as well as challenges of multi-sensor and multi-scale investigations by discussing MTI results regarding two landslide prone regions with distinctly different topographic, climatic and vegetation conditions (mountains in Central Albania and Southern Gansu, China), for which C-band (ERS or ENVISAT) and X-band COSMO-SkyMed (CSK) imagery was available (all in Stripmap descending mode). In both cases X-band MTI outperformed C-band MTI by providing more valuable information for the regional to local scale detection of slope deformations and landslide hazard assessment. This is related to the better spatial-temporal resolutions and more suitable incidence angles (40°-30° versus 23°) of CSK data While the use of medium resolution imagery may be appropriate and more cost-effective in reconnaissance or regional scale investigations, high resolution data could be preferentially exploited when focusing on urbanized landslides or potentially unstable slopes in urban/peri-urban areas, and slopes traversed by lifelines and other engineering structures.

1. INTRODUCTION

Slope hazards constitute a worldwide problem because landslides not only occur in mountains, but can also affect even modest artificial reliefs, coastal zones and river banks. It is also recognized that landslide related socio-economic losses have been increasing globally, especially in developing or recently industrialized countries (e.g., Petley, 2012).

Ground-based investigations and monitoring of terrain prone to landsliding are expensive, typically conducted only after the slope failure, and limited in terms of spatial and temporal coverage. Hence, the employment of complementary, cost-effective approaches to slope hazard detection and assessment is desirable.

Different remote sensing techniques can be used to detect and monitor ground surface displacements generated by landsliding. These include ground-based radar interferometry, air-borne and terrestrial LiDAR and air- and space-borne image matching (cf. overview by Wasowski and Bovenga, 2014a and references therein).

In this paper we focus on the application of advanced satellite MTI in landslide investigations. MTI can be considered cost-effective, especially when exploited over large areas, in as much as it can deliver great quantities of useful information for scientists and practitioners engaged in slope hazard mitigation.

MTI offers very good surveying capability of ground surface deformations, including those related to landslide activity (e.g., Colesanti et al., 2003; Hanssen 2005; Colesanti and Wasowski, 2006; Hooper et al., 2012; Motagh et al., 2013; Bally, 2013; Wasowski and Bovenga 2014a,b). The strengths of the technique include:

- Large area coverage (thousands of km^2) together with high spatial resolution (1-3 m) of the new radar sensors e.g., CSK, TerraSAR-X and multi-scale investigation option (regional to local);
- Very high precision (mm-cm) of surface displacement measurements only marginally influenced by bad weather;
- Regular, high frequency (days-weeks) of measurements over long periods (years);
- Retrospective studies using long-period (>20 years) archived radar imagery.

We discuss the performance of MTI in regional to local-scale landslide investigations by comparing the results obtained from medium resolution C-band data (ERS and ENVISAT) and from high resolution X-band data (CSK, Stripmap mode) for two landslide prone areas located in different geomorphic, climatic and vegetation settings: moderate elevation mountains of Central Albania and high mountains in Southern Gansu, north-western China. Such comparisons are still infrequent in literature and more case studies from different environments are desirable to better assess the benefits and challenges of medium and high resolution MTI applied to landslide analyses.

---

* Corresponding author

2. RADAR DATA AND MTI PROCESSING

2.1 Data

For the Chinese site we used the following SAR datasets: 32 C-band images acquired between July 2003 and August 2010 by ENVISAT ASAR; 22 X-band images acquired between November 2010 and February 2012 by CSK. For the Albanian site we used: 35 C-band images acquired between November 1992 and December 2000 by ERS; 39 X-band images acquired between May 2011 and June 2014 by CSK.

All the images were acquired in Stripmap mode along descending pass. For the CSK data we selected suitable incident angle by considering the topography of the studied areas. No such a possibility existed in case of the ERS and ENVISAT data. Specifications regarding the SAR datasets and processing results are given in Table 1.

2.2 MTI processing

We used the SPINUA (Stable Point INterferometry over Unurbanized Areas) algorithm. This PSI-like algorithm, developed for detecting and monitoring targets (PS) in non- or scarcely-urbanized areas (Bovenga et al., 2005; Bovenga et al., 2006), has been updated to increase its flexibility also in cases involving densely urbanized areas, as well as to assure proper processing of X-band data from CSK and TerraSAR-X (Bovenga et al., 2012) and from Sentinel-1A IWS mode.

SPINUA algorithm includes a patch-wise processing scheme that relies on processing small zones (usually a few km² within a radar image. The patches are selected to optimize the density and the distribution of potential coherent targets. Their small size allows using locally an approximate model for the atmospheric phase signal, which in turn ensures high processing robustness. Such *ad hoc* solutions enable obtaining quickly results on small areas using even low number of images (e.g., Bovenga et al., 2012). Moreover, in case of local scale analysis, this approach is robust against phase unwrapping errors occurring where target density is low. In case of large areas, atmospheric phase residuals are interpolated through a kriging procedure, and an *ad hoc* integration scheme is used to stitch the displacement maps retrieved on the single patches. SPINUA has been applied to investigate different studies including landslides, subsidence processes and post-seismic deformations (Bovenga et al., 2006; Nitti et al., 2009; Reale et al., 2011).

In this work, similar, relatively high coherence thresholds were adopted during processing to assure the quality and comparability of C- and X-band results (Table 1). Indeed, in all cases low standard deviations of movement velocity estimates (on total PS samples) were obtained.

Study area	Sensor	Radar band	Incidence angle (°)	Coherence threshold	Mean of Velocity St. Dev. (mm/yr)
China					
	ASAR	C	23	>0.75	0.26
	CSK	X	40	>0.85	0.63
Albania					
	ERS	C	23	>0.70	0.20
	CSK	X	30	>0.80	0.25

Table 1. Specifications for SAR datasets and processing results

3. ASSESSMENT OF MTI RESULTS FROM C-BAND AND X-BAND DATA

3.1 Mountains of Southern Gansu, China

The study area, located in southern-most part of Gansu Province, China, is dominated by steep mountains with elevation ranging from ~1000 to 4000 m (Figure 1) and semi-arid climate. Annual average precipitation is 434 mm, most of which occurs as rainfall in June-September period with occasionally intense storm episodes (Tang et al., 2011).

Vegetation cover is limited. Grass is common in the highest elevation areas which also include some forest and shrub. The barren ground is also common on the high steep slopes and it generally corresponds to rocky outcrops. The cultivated land is present on the middle-lower elevations. The area is scarcely populated, but contains many small towns, villages and infrastructure concentrated within the major river valleys.

3.1.1   Regional scale assessment of MTI results: The first CSK-based results concerning the north-western part of the study area (~40 km² around the town of Zhouqu) were presented by Wasowski and Bovenga (2014b) and Wasowski et al. (2014). High density of radar targets (>1000/km²) and detection of active slope movements were reported. Radar visibility and landslide maps were also presented and discussed.

Here we consider the larger area corresponding to the full CSK frame (~1700 km²), and along with X-band results we present the outcomes of C-band data processing. The comparison of the results (Figure 1) shows that in both cases quite different densities of radar targets are obtained (~13 PS/km² and >300 PS/km², respectively for C- and X-band data). Most PS are motionless and concentrate along the middle-lower slopes of major river valleys. A number of large clusters of moving PS are also present. Many of these have the characteristically elongated shape and correspond to very large slow landslides.

Furthermore, the evaluation of the results shows that, in terms of landslide detection capability, X-band MTI clearly outperforms C-band MTI. In particular, at the small scale of Figure 1, the CSK and ENVISAT results allow the recognition of, respectively, 18 and 6 large-sized active landslides.

3.1.2   Local scale assessment of MTI results: Such assessment is useful, because landslides are local scale features. Here we consider the north-western portion of the study area amounting to ~53 km², for which we had an inventory of larger landslides mapped from GE™ (Figure 2). In this case the PS densities obtained from C- and X-band data are, respectively, ~108 PS/km² and ~1450 PS/km². Furthermore, evident differences exist in the distribution of C-band and X-band PS, with the former notably lacking on E and SE-facing slopes.

The scarcity of PS and/or their inhomogeneous distribution have direct influence on landslide detection capability. Using C-band MTI, the activity of only one out of 25 mapped landslides was detected; two other large landslides contained few moving PS, but their number and distribution were unsuitable for a slope hazard assessment. Two smaller slides with several non-moving PS were classified as inactive. The X-band MTI allowed the detection of 6 moving landslides; 10 additional slides contained an adequate number of suitably distributed non-moving PS, so that their inactivity was indicated.

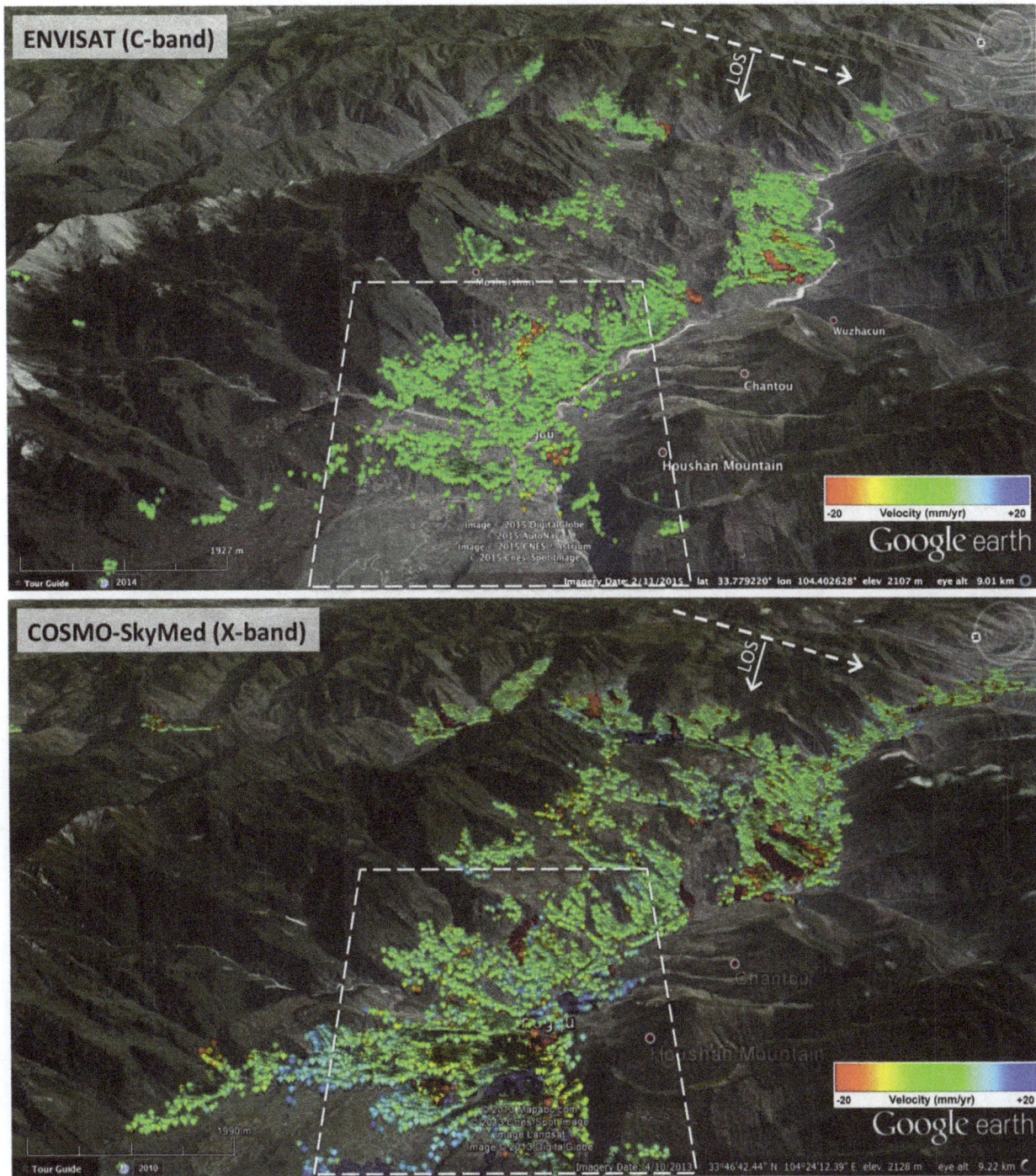

Figure 1. Regional scale (~1700 km²) overviews of the distribution and average line of sight (LOS) velocity of radar targets (PS) in Southern Gansu Mountains, China. Velocity values are saturated to ±20 mm/year. Yellowish-reddish and bluish dots represent radar targets moving, respectively, away from and toward the satellite sensor. White dashed line polygons indicate the areas shown in Figure 2. Background images are from Google Earth™. (Upper) MTI results obtained from ENVISAT ASAR data. (Lower) MTI results obtained from CSK data.

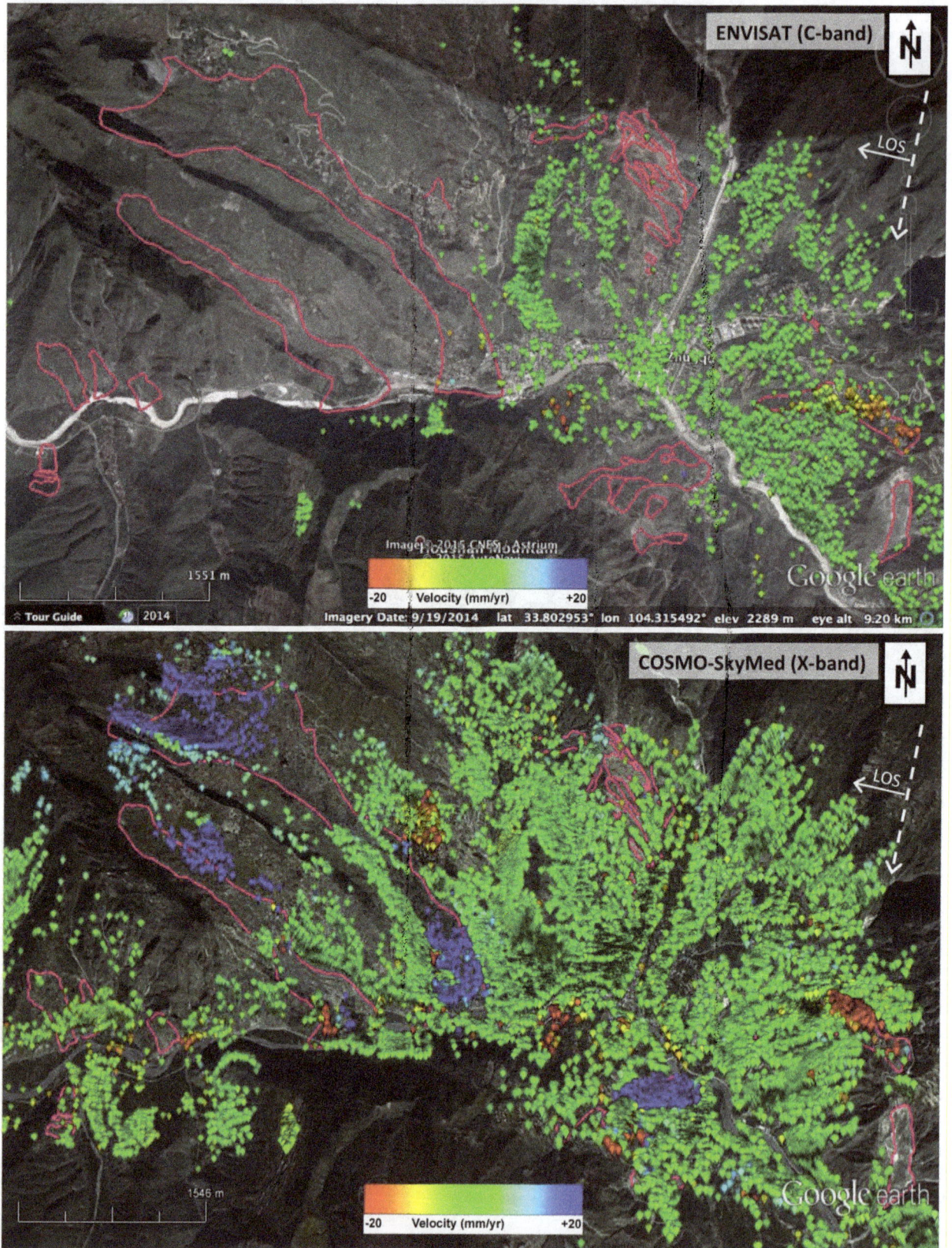

Figure 2. Local scale (~53 km²) overviews of the distribution and average line of sight (LOS) velocity of radar targets in the Zhouqu area, Southern Gansu Mountains, China. Velocity values are saturated to ±20 mm/year. Yellowish-reddish and bluish dots represent radar targets moving, respectively, away from and toward the satellite sensor. The outlines of 25 landslides (in pink), mapped using Google Earth™, are from Wasowski and Bovenga (2014b). Background images are from Google Earth™. (Upper) MTI results obtained from ENVISAT data. (Lower) MTI results obtained from CSK data

3.2 Mountains of Central Albania

The mountains of interest are located just to the east of Tirana, the country's capital (Figure 3). In comparison with the Chinese site, the topography is lower (typically below 1500 m) and slopes less steep. However the average annual precipitation in the Albanian mountains is much higher and can reach over 2000 mm (Meco and Aliaj, 2000). Winter represents the most rainy season with snow precipitation being common.

In relation to the relatively wet climate, the vegetation cover is more abundant (mainly trees and shrub) than in Southern Gansu. The GE™ imagery also reveals that barren ground can locally be common, especially in the areas most prone to landsliding and erosion (Figures 3, 4). Grass is less abundant, while the amount of cultivated land appears insignificant, because the mountains are very scarcely populated.

3.2.1 (Sub)regional scale assessment of MTI results: We build upon the recent study of Wasowski et al. (2015, in press) who presented the first MTI results for Central Albania based on Stripmap CSK data. In particular, for the sake of comparison with X-band, MTI we present and discuss the results derived from processing of C-band ERS data.

Figure 3a,b provides a wide-area overview of C-band and X-band MTI results for the Central Albania region. In both cases a large cluster of PS stands out in the western-most part of the region (lowland), which corresponds to the city of Tirana. However, PS appear very scarce in the mountainous area to the east. In particular, only few small clusters of PS are derived from C-band data. The number of PS clusters obtained from X-band processing is about three times greater. Finally, some of the PS groups show elongated shape, which, especially in case of moving PS, can be associated with slope movements. This possibility is further explored by examining MTI results at a local scale.

3.2.2 Local (slope) scale assessment of MTI results: Here we focus on a single mountain for which a considerable number of PS was derived from both C-band and X-band data (Figure 3c,d). In both cases movements were detected on north and west facing slopes. The number of C-band PS was much lower than that of X-band PS, but when considered within the geomorphic context of instability indicated from GE™ optical imagery (Figure 3), the information obtained from ERS data was sufficient for a meaningful interpretation. Only CSK data provided information (PS) for the middle-lower part of north facing slope, where slowly moving PS indicated the presence of instability conditions.

4. DISCUSSION AND CONCLUSIONS

The outcomes of this study, and in particular the comparisons of the results derived from C- and X-band data, provide useful insight on the applicability of MTI in regional to local scale slope hazard investigations, including the capability of detecting active landslides. Although the C- and X-band datasets were not coeval, such comparisons are still of interest, because slope movements are very common and persistent in both Central Albania and Southern Gansu mountains in relation to the high seismic activity, presence of weak geological materials and recurrence of hydrological triggers (Meco and Aliaj, 2000; Dijkstra et al, 2012; Wasowski et al., 2014).

Despite the significant environmental differences between the studied areas, in both cases X-band MTI proved to be more effective in terms of providing better coverage of features of interest (landslides). This was observed on all scales, from regional, sub-regional to local, and can be ascribed to much better spatial (and temporal) resolution, as well as to the higher incidence angles (40° and 30°) of X-band CSK data.

The above findings are consistent with what has been reported in the literature on X-band MTI applied to landslide investigations in other mountain ranges such as e.g., the Pyrenees (Notti et al., 2010; Herrera et al., 2013), Alps (Bovenga et al., 2012), Apennines (Wasowski and Bovenga, 2014a,b). However, with shorter revisit times, improved applicability of medium resolution C-band MTI in landslide investigations is expected from new sensors such as the ESA's Sentinel-1A. In particular, the full Sentinel-1 constellation will provide a revisit time of six days, and this should guarantee coherent interferometric phase measurements over a wider range of ground surfaces (e.g., Morishita and Hanssen, 2015).

Furthermore, the presented examples show that, while simple slope visibility maps can suffice to define suitable sensor acquisition geometry and estimate *a priori* the MTI potential to provide useful results in high mountain, semi-arid environments with limited vegetation cover, in densely vegetated mountains like those in Albania, more comprehensive MTI feasibility assessments based also on land cover / land-use information (cf. Cigna et al., 2014; Wasowski and Bovenga, 2014b) could be needed. A detailed MTI feasibility assessment can be recommended especially for very local, slope scale investigations.

MTI studies that rely on the use of multi-sensor and multi-band data are attractive, because such approach can deliver complementary information. However, in applied MTI the choice of the investigative approach will ought to be tailored to the specific study region conditions, landslide types and their likely dynamics, keeping in mind the objectives of the work and budget constraints. Considering that landslides are local scale features, one can argue that, in general, higher spatial resolution imagery will likely lead to improved MTI applicability (more detailed slope instability assessment). Temporal resolutions are also of practical importance e.g., for timely assessment of slope hazard via high frequency measurements, and for revealing detailed patterns of landslide activity and their relationships to causative and triggering factors.

ACKNOWLEDGEMENTS

Work carried out using CSK® Products© of the Italian Space Agency (ASI), delivered under a license to use by ASI. CSK imagery for the Chinese and Albanian sites were obtained within the projects, respectively, COSMO-SkyMed AO ID 1820 and "Studio su instabilità del terreno sull'area di Tirana (Albania)". European Space Agency (ESA) provided ERS and ENVISAT data (CAT-1 C1P2653 Advanced SAR interferometry techniques for landslide warning management). We thank Tom Dijkstra (BGS) and Xingmin Meng (University of Lanzhou) for valuable discussions on geology and landslide processes in Gansu Province, as well as Spartak Kucaj (Polytechnic University of Tirana) for his help within the Albanian project.

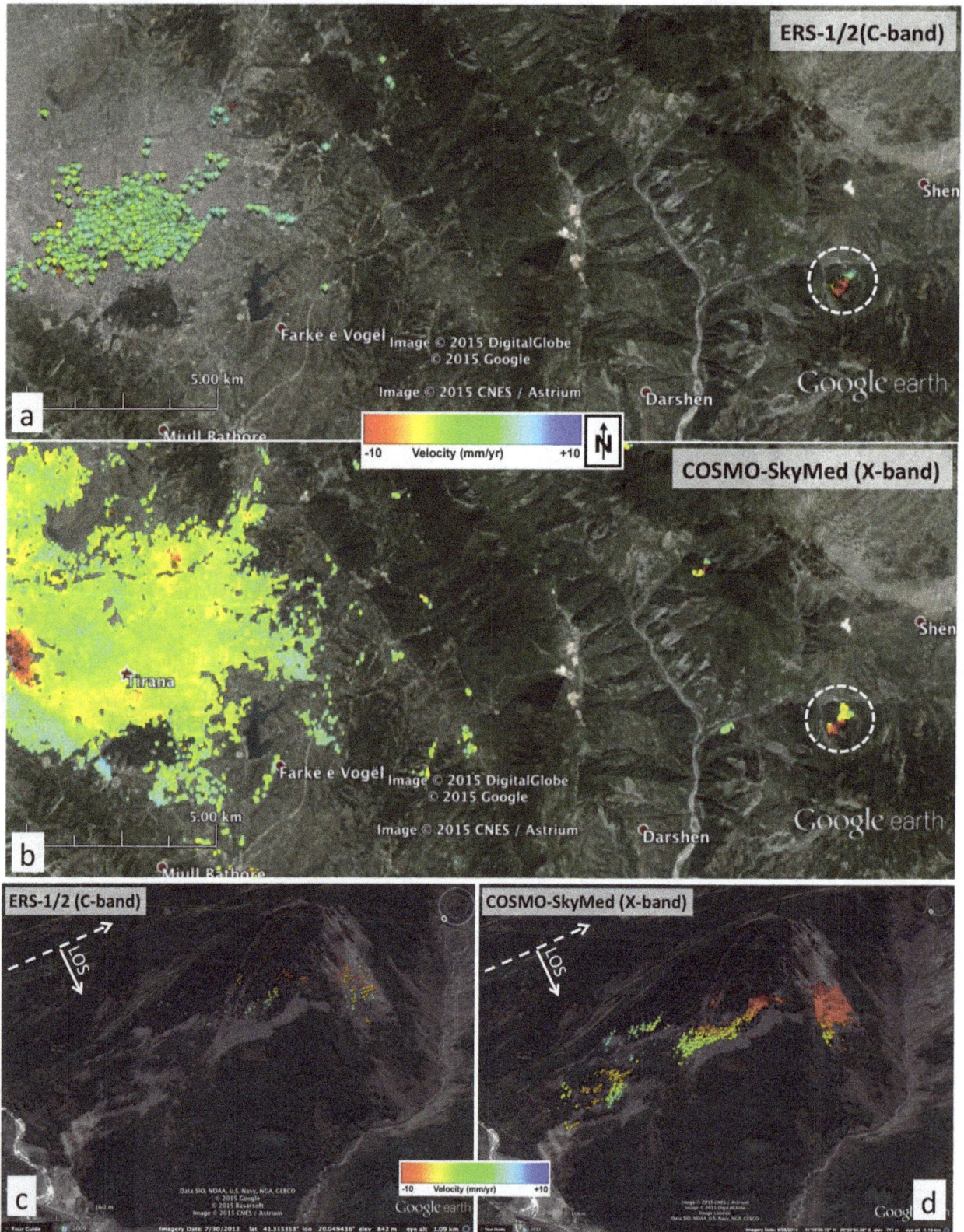

Figure 3. MTI results obtained from ERS and CSK data for the Central Albanian mountains. The distribution and average line of sight (LOS) velocity of radar targets are shown. Velocity values are saturated to ±10 mm/year. Yellowish-reddish and bluish dots represent radar targets moving, respectively, away from and toward the satellite sensor. (a) and (b) sub-regional scale overviews; (c) and (d) MTI results for a single hillslope (location marked by dashed circles in upper figures). Background from Google Earth™

REFERENCES

Bally, P. (Ed), 2013. Satellite Earth Observation for Geohazard Risk Management - The Santorini Conference - Santorini, Greece, 21-23 May 2012. ESA Publication STM-282 doi:10.5270/esa-geo-hzrd-2012

Bovenga, F., Nutricato, R., Refice, A., Wasowski, J., 2006. Application of Multi-temporal Differential Interferometry to Slope Instability Detection in Urban/Peri-urban Areas. *Engineering Geology*, 88(3-4), pp. 218-239.

Bovenga, F., Refice, A., Nutricato, R., Guerriero, L., Chiaradia, M.T., 2005. SPINUA: a flexible processing chain for ERS / ENVISAT long term Interferometry. Proceedings of ESA-ENVISAT Symposium, September 6-10, 2004, Salzburg, Austria. ESA Sp. Publ. SP-572, April 2005, CD.

Bovenga, F., Wasowski, J., Nitti, D.O., Nutricato, R., Chiaradia, M.T., 2012. Using Cosmo/SkyMed X-band and ENVISAT C-band SAR Interferometry for landslide analysis. *Remote Sensing of Environment*, 119, pp. 272-285.

Cigna, F., Bateson, L., Jordan, C., Dashwood, C., 2014. Simulating SAR geometric distortions and predicting Persistent Scatterer densities for ERS-1/2 and ENVISAT C-band SAR and InSAR applications: Nationwide feasibility assessment to monitor the landmass of Great Britain with SAR imagery. *Remote Sensing of Environment*, 152, pp. 441-466.

Colesanti, C., Ferretti, A., Prati, C., Rocca, F., 2003. Monitoring landslides and tectonic motion with the permanent scatterers technique. *Engineering Geology*, 68, pp. 3-14.

Colesanti, C., Wasowski, J., 2006. Investigating landslides with space-borne Synthetic Aperture Radar (SAR) interferometry. *Engineering Geology*, 88(3-4), pp. 173–199.

Dijkstra T. A., Chandler J. H., Wackrow R., Meng X. M., Ma D. T., Gibson A., Whitworth M., Foster C., Lee K., Hobbs P.R.N., Reeves H. J., Wasowski J., (2012) Geomorphic controls and debris flows - the 2010 Zhouqu disaster, China. Proceedings of the 11th ISL & 2nd North American Symposium on Landslides, 3-8 June, 2012, Banff, Canada.

Hanssen, R., 2005. Satellite radar interferometry for deformation monitoring: a priori assessment of feasibility and accuracy. *Int. Journal of Applied Earth Observation and Geoinformation*, 6, pp. 253-260.

Herrera G., Gutiérrez F., García-Davalillo J. C., Guerrero J., Notti D., Galve J.P., Fernández-Merodo J.A., Cooksley, G., 2013. Multi-sensor advanced DInSAR monitoring of very slow landslides: The Tena Valley case study (Central Spanish Pyrenees). *Remote Sensing of Environment*, 128, pp. 31–43.

Hooper, A., Bekaert, D., Spaans, K., Arıkan M.T., 2012. Recent advances in SAR interferometry time series analysis for measuring crustal deformation. *Tectonophysics*, 514–517, pp. 1-13.

Meço S., Aliaj S. (2000): Geology of Albania. Gebruder Borntraeger, Berlin (ISBN 978-3-443-11028-4). 246p.

Morishita Y., Hanssen R., 2015. Temporal decorrelation in L-, C- and X-band satellite radar interferometry for pasture on drained peat soils. *IEEE Trans. Geosci. Remote Sens.*, 53(2), pp. 1096–1104.

Motagh, M., Wetzel, H.-U., Roessner, S., Kaufmann, H., 2013. A TerraSAR-X InSAR study of landslides in southern Kyrgyzstan, Central Asia. *Remote Sensing Letters*, 4(7), pp. 657-666.

Nitti, D.O., Bovenga, F., Refice, A., Wasowski, J., Conte, D., Nutricato, R., 2009b. L- and C-band SAR Interferometry analysis of the Wieliczka salt mine area (UNESCO heritage site, Poland). Proceedings ALOS PI 2008 Symposium, 3 – 7 Nov., 2008, Rhodes, Greece. ESA SP-664, January 2009.

Notti, D., Davalillo, J. C., Herrera, G., Mora, O., 2010. Assessment of the performance of X-band satellite radar data for landslide mapping and monitoring: Upper Tena valley case study. *Natural Hazards and Earth System Sciences*, 10, pp. 1865–1875.

Petley, D., 2012. Global patterns of loss of life from landslides. *Geology*, 40(10), pp. 927–930.

Reale, D., Nitti, D. O., Peduto, D., Nutricato, R., Bovenga, F., Fornaro, G., 2011. Post-seismic Deformation Monitoring With The COSMO/SKYMED Constellation. *IEEE Geoscience and Remote Sensing Letters*, 8(4), pp. 696-00.

Tang, C., Rengers, N., van Asch, T.W.J., Yang, Y.H., Wand G.F., 2011. Triggering conditions and depositional characteristic of a disastrous debris flow event in Zhouqu city, Gansu Province, northwestern China. *Natural Hazards and Earth Systems Sciences*, doi:10.5194/nhess-11-2903-2011.

Wasowski J., Bovenga F., 2014a. Remote Sensing of Landslide Motion with Emphasis on Satellite Multitemporal Interferometry Applications: An Overview. In T. Davies Ed. *Landslide Hazards, Risks and Disasters*. pp. 345-403. http://dx.doi.org/10.1016/B978-0-12-396452-6.00011-2 Elsevier Inc.

Wasowski J., Bovenga F., 2014b. Investigating landslides and unstable slopes with satellite Multi Temporal Interferometry: Current issues and future perspectives. *Engineering Geology*, 174, pp. 103–138.

Wasowski J., Bovenga F., Dijkstra T., Meng X., Nutricato R., Chiaradia M.T., 2014. Persistent Scatterers Interferometry Provides Insight on Slope Deformations and Landslide Activity in the Mountains of Zhouqu, Gansu, China. In K. Sassa et al. (eds.), *Landslide Science for a Safer Geoenvironment*, Vol. 2: Methods of Landslide Studies, pp. 359-364. Springer International Publishing.

Wasowski J., Bovenga F., Nutricato R., Nitti D.O., Chiaradia M.T., Kucaj, S., Strati B., 2015 (in press). High resolution satellite multi-temporal interferometry for detecting and monitoring landslide and subsidence hazards. Proceedings 10th Asian Regional Conference of IAEG, Kyoto, 26-27 Sept., 2015.

22

MODELLING TEMPORAL SCHEDULE OF URBAN TRAINS USING AGENT-BASED SIMULATION AND NSGA2-BASED MULTIOBJECTIVE OPTIMIZATION APPROACHES

Mohammadreza Sahelgozin [a] *, Abbas Alimohammadi [b]

GIS Dept., Geoinformation Technology Center of Excellence, Faculty of Geodesy&Geomatics Engineering
K.N.Toosi University of Technology, Tehran, Iran
[a] sahelgozin@mail.kntu.ac.ir   [b] alimoh_abb@kntu.ac.ir

KEY WORDS: Subway Systems, Temporal Schedule, Modelling, Agent-based Simulation, Multi-objective Optimization

ABSTRACT

Increasing distances between locations of residence and services leads to a large number of daily commutes in urban areas. Developing subway systems has been taken into consideration of transportation managers as a response to this huge amount of travel demands. In developments of subway infrastructures, representing a temporal schedule for trains is an important task; because an appropriately designed timetable decreases *Total passenger travel times*, *Total Operation Costs* and *Energy Consumption of trains*. Since these variables are not positively correlated, subway scheduling is considered as a multi-criteria optimization problem. Therefore, proposing a proper solution for subway scheduling has been always a controversial issue. On the other hand, research on a phenomenon requires a summarized representation of the real world that is known as *Model*. In this study, it is attempted to model temporal schedule of urban trains that can be applied in Multi-Criteria Subway Schedule Optimization (MCSSO) problems. At first, a conceptual framework is represented for MCSSO. Then, an agent-based simulation environment is implemented to perform Sensitivity Analysis (SA) that is used to extract the interrelations between the framework components. These interrelations is then taken into account in order to construct the proposed model. In order to evaluate performance of the model in MCSSO problems, Tehran subway line no. 1 is considered as the case study. Results of the study show that the model was able to generate an acceptable distribution of Pareto-optimal solutions which are applicable in the real situations while solving a MCSSO is the goal. Also, the accuracy of the model in representing the operation of subway systems was significant.

1. INTORDUCTION

Urban growth results in a significant increase in the distances between locations of people's residence and locations of services such as education, occupation, shopping, health and recreation. This leads to a large number of daily commutes in urban areas; usually from home to work or school in early morning hours and in the opposite direction in the evenings. A suitably developed transportation infrastructure is a requirement to satisfy the huge amount of inter-cities travel demands. Nowadays, urban rail transportation systems (Subway Systems) are under consideration of urban transportation managers. Consumption of non-fossil fuels, reduction of the general operation costs as well as reliability and predictability of the travel times are significant advantages of the subway systems. Therefore, developments of the urban subway infrastructures are vital to improve satisfaction of both the transportation managers and passengers.

In addition to the optimization of the rail route and locations of stations, design of a temporal schedule for train operations is an important task for effective use of the system. An appropriately designed timetable would certainly improves the quality of service in subway systems (Yang et al., 2014). Three common criteria including *Total passenger travel times*, *Total Operation Costs* and *Energy Consumption of trains* are in the main concentration of the policy makers as the service quality indicators. There are two-sided interrelations that exist between each of these criteria with the temporal schedule of subway systems. For example, total passenger travel times is considered as an effective factor in the timetable design process; on the other hand, total passenger travel times is affected by the timetable too.

Similarly, total operation costs and energy consumption of trains have also the same interrelations with the temporal schedule of trains.

Furthermore, these criteria are not positively correlated with each other. Therefore, train scheduling is considered as a multi-objective optimization problem (Chang et al., 2000). Therefore, proposing a proper solution for subway scheduling that simultaneously optimizes all of the mentioned criteria has been always a controversial issue. However, finding a suitable solution for a phenomenon requires a perception of the phenomenon that can be gotten by a generalized representation of the real world. This representation is achieved with the help of a *Modell*.

In this study, it is attempted to model temporal schedule of urban trains operation in a way that it could be applied in Multi-Criteria Subway Schedule Optimization (MCSSO) problems. In order to accomplish this, a conceptual framework is designed in Section 3. The parameters that are related to subway scheduling are represented in the framework. These parameters are extracted by reviewing the literature. Then, an agent-based simulation environment is implemented in Section 4 that is a tool for performing further Sensitivity Analysis (SA). In this study, SA is used for understanding interrelations between the proposed framework components as well as their influences on the service quality indicators. In section 5, interrelations between the framework components in addition to a number of basic laws of physics are summarized in a set of mathematical functions which together represent a model for temporal schedule of subway systems. In order to evaluate performance of the developed model and how it can help to solve MCSSO problems, Tehran

---

* Corresponding author

subway line no. 1 is considered as the case study. By feeding the proposed model with the real data, the evaluation results are obtained that are provided in Section 6. Finally, conclusion of the study is provided in Section 7.

2. RELATED WORKS

A large number of researches has been performed in scheduling of transportation systems in the recent years. Scheduling passenger trains and urban trains are a major part of these studies. The large number of studies in this scope is because of the complexities of trains scheduling problems. In this regard, Yang et al. (2014) believed that physical complexities of subway systems is the reason that subway scheduling is one of the most controversial issues.

Apart from initial solutions for trains scheduling which were usually manual, other studies which concentrated on the problems that are similar to multi-criteria train scheduling optimization are summarized in this section. In a number of studies, schedule optimization of trains was performed so that it reduces energy consumption or energy loss. They include proposing an algorithm which distributes travel times of the trains for the most efficient energy consumption (Su et al., 2013), developing a cooperative scheduling model to increase simultaneous accelerates and brakes of the consecutive trains (Nasri et al., 2010; Yang et al., 2013) and applying the genetic algorithm to decrease the simultaneous acceleration of trains in order to avoid maximum traction power of the system (Chen et al., 2005).

Furthermore, some other studies focused on the multi-objective optimization of timetables for transportation systems in order to satisfy some criteria. Decreasing travel time and operation costs in high-speed rail systems (Chang et al., 2000), travel time and energy consumption in passenger trains (Ghoseiri et al., 2004; Chevrier et al., 2013; Hu et al., 2013), operation costs and energy consumption in a freight transportation system (Lau et al., 2013) and decreasing travel time, operation costs and energy

consumption in a sustainable road network design problem (Kim et al., 2012) are some examples in this scope.

In addition, there were some attempts in developing models and approches for trains scheduling. Shaoquan et al. (2009) represented an optimization model for initial schedule of passenger trains that is based on an improved genetic algorithm. Also, Sels et al. (2013) introduced an objective function for the total passenger travel times in subway systems in order to overcome the lack of mathematical modelling in a sustainable scheduling. Furthermore, Chang and Kwan (2005) evaluated the performance of the evolutionary algorithms including Genetic Algorithm (GA), Particle Swarm Optimization (PSO) and Differential Evolution (DE) in trains schedule optimization problems.

3. SUBWAY SCHEDULE OPTIMIZATION

In this section, a conceptual framework is proposed for MCSSO that aggregates all the parameters which are related to the operation of subway systems. Structure of the proposed framework is inspired from the concept of *Sustainable Transportation*. With regard to the definition of *Sustainable Development*, three components including *Social Equity*, *Economic Efficiency* and *Environmental Quality* of the system should be considered in sustainable transportation planning (Kim et al., 2012). For each of these components, a corresponding category is considered in the MCSSO framework: *Passengers*, *Operation Company* and *Environment*. Then, the related parameters to the subway scheduling that are extracted by literature review are grouped in these three categories (Figure 1).

Apart from categorization of the parameters, it is necessary for their interrelations to be represented in the framework. The interrelations are extracted by the sensitivity analysis and are provided in Table 1. In this table, the fact that if two parameters are correlated (positively or negatively) or not is shown.

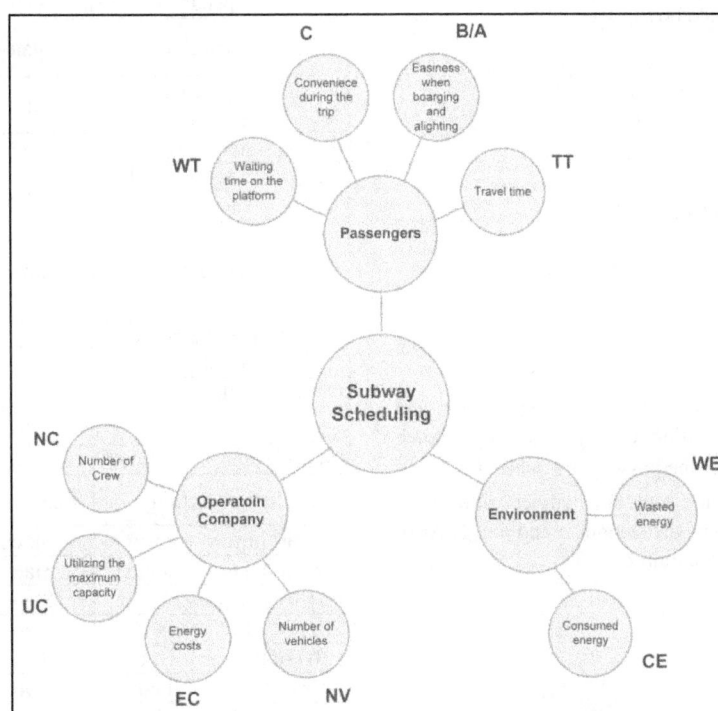

Figure 1. A conceptual framework for representing the parameters of MCSSO

Table 1. Interrelations between the proposed framework parameters

		Passengers				Operation company				Environment	
		WT	C	B/A	TT	NV	NC	UC	EC	CE	WE
Passengers	WT	----	----	----	+	-	-	+	-	-	-
	C	----	----	+	-	+	+	-	+	+	+
	B/A	----	+	----	----	+	+	-	+	+	+
	TT	+	-	----	----	-	-	+	-	-	-
Operation company	NV	-	+	+	-	----	+	----	+	+	+
	NC	-	+	+	-	+	----	----	----	----	----
	UC	+	-	-	+	----	----	----	----	----	----
	EC	-	+	+	-	+	----	----	----	+	+
Environment	CE	-	+	+	-	+	----	----	+	----	+
	WE	-	+	+	-	+	----	----	+	+	----

4. AGENT-BASED SIMULATION

In this study, the simulation environment is implemented in MATLAB 2013 in which each agent is defined as a matrix. Rows of the matrix for an agent, represent instances of that agent. The instances are identified by a unique ID number. On the other hand, each column shows a real-time state or behaviour of the instances. Table 2 represents the designed agents including their states and behaviours. Also, the permitted values of their states and behaviours are provided in the table.

In the general procedure of the simulation, there is a variable which is named as "timeStep" whose job is to count the seconds. A main loop runs iteratively that corresponds to the seconds and updates states of the agents in each second. For example, in each iteration, the values for location, velocity and acceleration of the trains are updated according to the trains' actions and some physics formulas.

In addition, some service quality indicators which are defined in the simulation environment (Eq.1) are updated in the main loop as well. This helps us to get an estimation of the service quality at the end of the simulation procedure with which the performance of the system could be assessed.

$$I_{Passenger} = \sum_{p=1}^{P} TT_p$$

$$I_{Cost} = \sum_{n=1}^{N} \sum_{i=1}^{I} ToT_i^{n,n+1} * (I * C_T - P_i^{n,n+1}) \qquad (1)$$

$$I_{Energy} = \sum_{i=1}^{I} E_i$$

Where N, I and P are the numbers of stations, trains and passengers, respectively. In addition, C_T is the capacity of each train. TT_p is the travel time for the passenger p while $ToT_i^{n,n+1}$ and $P_i^{n,n+1}$ are the travel time and the number of boarded passengers of the train i between the stations n and $n+1$. Also, E_i is the consumed energy of the train i.

Table 2. The designed agents in the simulation environment

Stations		
Specification of the Agent	**Type**	**Permitted Values**
Incoming passengers rate (east platform)	State	N^*
Outgoing passengers rate (east platform)	State	N
Incoming passengers rate (west platform)	State	N
Outgoing passengers rate (west platform)	State	N
Number of waiting passengers (east platform)	State	N
Number of waiting passengers (west platform)	State	N
Trains		
Specification of the Agent	**Type**	**Permitted Values**
Location	State	$(0, max_{Location}) \in R^{**}$
Speed	State	$(0, max_{Speed}) \in R$
Acceleration	State	$(0, max_{Acceleration}) \in R$
Number of the boarded passengers	State	N
Consumed energy	State	$\{R \geq 0\}$
ID of the last departed station	State	$\{Stations\ IDs\}$
Current action	Behaviour	$\{1 = Acceleration, 2 = Coasting, 3 = Braking, 4 = Dwelling\}$
The time remained for the current action	State	$N\ (Seconds)$
Direction of the movement	State	$\{-1, 1\}$
Passengers		
Specification of the Agent	**Type**	**Permitted Values**
Current Action	Behaviour	$\{0 = Not\ traveling, 1 = Traveling\}$
Start time of the trip	State	$N\ (Seconds)$
ID of the origin station	State	$\{Stations\ IDs\}$
ID of the last passed station	State	$\{Stations\ IDs\}$
ID of the destination station	State	$\{Stations\ IDs\}$
ID of the carrier train	State	$\{Trains\ IDs\}$
Travel time until now	State	$N\ (Seconds)$

$* N = Natural\ Numbers \qquad ** R = Real\ Numbers$

5. THE PROPOSED MODEL

5.1 Decision Variables

In addition to extracting interrelations between the parameters in the proposed framework (Section 3), the implemented agent-based simulation environment was also used to choose appropriate *Decision Variables* for MCSSO. The simulation was performed in a number of scenarios in each of which the value of a specific variable is changed significantly. Those variables that caused more clear changes in the service quality indicators were chosen as decision variables (Table 3).

Table 3. Decision Variables

Parameter	Notation
Time of acceleration for the train i between the stations n and $n + 1$	$ToA_i^{n,n+1}$
Time of braking for the train i between the stations n and $n + 1$	$ToB_i^{n,n+1}$
Time of dwell for the train i in the station n	ToD_i^n
Acceleration rate for the train i between the stations n and $n + 1$	$a_i^{n,n+1}$
Braking acceleration rate for the train i between the stations n and $n + 1$	$b_i^{n,n+1}$

5.2 Mathematical Objective Functions

The developed objective functions for Total passenger travel times (F_1), Total Operation Costs (F_2) and Energy Consumption (F_3) are represented in Eq. 2. Parameters of the model is explained in Tables 3 and 4. It should be noted that the total operation costs is considered based on the amount of the trains' capacities which is remained empty between the stations. Therefore, F_2 is the aggregated number of total boarded passengers that is going to be maximized. On the other hand, F_1 and F_3 should be minimized.

$$F_1 = \sum_{i=1}^{I} \sum_{n=1}^{N-1} \{ \frac{1}{2} * B_i^n * ToH_{i,i+1}^n + \frac{d_{n,n+1}}{a_i^{n,n+1} * ToA_i^{n,n+1}} + \frac{1}{2}$$

$$* (ToA_i^{n,n+1} + ToB_i^{n,n+1}) + ToD_i^n \}$$

$$F_2 = \sum_{i=1}^{I} \sum_{n=1}^{N-1} B_i^n * ToD_i^n$$

$$F_3 = \sum_{i=1}^{I} \sum_{n=1}^{N-1} \left(M + m * \frac{C_T}{2} \right)$$

$$* \left[\frac{1}{2} \left(a_i^{n,n+1} + f_k * g \right) * a_i^{n,n+1} \right.$$

$$* \left(ToA_i^{n,n+1} \right)^2 \right) + f_k * g * a_i^{n,n+1}$$

$$* ToA_i^{n,n+1} * ToC_i^{n,n+1}$$

$$+ \left(b_i^{n,n+1} - f_k * g \right) * ToB_i^{n,n+1} * (\frac{1}{2}$$

$$* ToB_i^{n,n+1} + a_i^{n,n+1} * ToA_i^{n,n+1}) \right]$$

(2)

5.3 Model Constraints

Some constraints are required to be taken into account in order to avoid model failures. Eq. 3 shows the considered constraints in

the model. In addition to Eq. 3, some other constraints are defined as the lower and upper boundaries of the decision variables.

- $a_i^{n,n+1} * ToA_i^{n,n+1} \leq V_{max}$

- $a_i^{n,n+1} * ToA_i^{n,n+1} = -b_i^{n,n+1} * ToB_i^{n,n+1}$

- $\frac{1}{2} a_i^{n,n+1} \left(ToA_i^{n,n+1} \right)^2 + \frac{1}{2} b_i^{n,n+1} (ToB_i^{n,n+1})^2 +$

  $\left(a_i^{n,n+1} * ToA_i^{n,n+1} \right) * ToB_i^{n,n+1} \leq d_{n,n+1}$

(3)

Where V_{max} is the maximum speed of the trains.

Table 4. Definition of the Model Parameters

Parameter	Notation
Number of stations	N
Distance between the stations n and $n + 1$	$d_{n,n+1}$
Number of the active trains	I
Average weight of each train	M
Average weight of a passenger	m
Time of the move at a constant speed for the train i between the station n and $n + 1$	$ToC_i^{n,n+1}$
Headway time between the two consecutive trains i and $i + 1$ in the station n	$ToH_{i,i+1}^n$
Boarding passengers rate for the train i in the station n	B_i^n
Capacity of each train	C_T
Coefficient of kinetic friction between the train wheels and the rail	f_k
Gravitational acceleration	g

6. EVALUATION OF THE PROPOSED MODEL

6.1 Case Study

The proposed model was tested in Tehran subway line no.1 as the case study (Figure 2). Geospatial data as well as the descriptive data were gathered in a geo-dataset with the help of official Tehran GIS data, data of Tehran Urban and Suburban Railway Operation Company, Google Earth and a field survey. In addition, data for the current scheduling of Tehran subway line no.1 were obtained by the current timetable and also by measuring average time of acceleration, dwell or braking of the trains in the field survey.

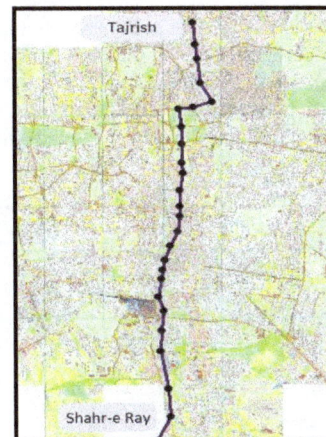

Figure 2. Study Area

6.2 Assumptions

In the optimization process, some assumptions was considered for simplifications. It is assumed that operation of all the trains are similar between each two specific stations.

6.3 Optimization Procedure: Results and Discussions

An effective approach to deal with the multi-objective problems is optimization with the multi-objective evolutionary algorithms. Also, one of the most beneficial solutions for trains schedule optimizations are the Pareto-based methods (Ghoseiri et al., 2004; Chang and Kwan, 2005). Therefore, in this study, *NSGA-II* (Deb et al., 2002) is used to solve the subway scheduling problem.

Using NSGA-II, a large number of solutions were obtained for the decision variables of the proposed model. To decrease the number of output data, *Fuzzy Subtractive Clustering* method was used to cluster the solutions. This method has two advantages: first, centre of the clusters are coincident on the Pareto front; second, the clusters are not crisply discrete. The radius of influence was set to 0.42 in the clustering process. The clustered solutions which were considered as the candidate schedules are provided in Table 5. Objective functions for the current schedule of Tehran subway line no. 1 is also represented in the table for further comparisons.

Table 5. Objective Functions for the candidate schedules in comparison to the current schedule

Solution	Objective Functions		
	Time (10^3)	Cost (10^4)	Energy (10^{10})
1	8.521	-2.563	1.087
2	7.355	-3.116	1.094
3	7.752	-2.820	1.091
4	1.671	-4.710	1.944
5	2.018	-6.056	1.838
6	6.045	-3.580	1.111
7	6.594	-3.806	1.107
8	1.889	-5.799	1.832
9	5.385	-3.716	1.126
10	1.902	-5.496	1.757
11	4.872	-4.287	1.148
Current Schedule	3.115	-4.656	1.286

To obtain a suitable vision, data in the above table are drawn in Figure 3. It is important to say that the objective functions of the candidate solutions has been normalized before they are drawn in order to avoid heterogeneity in the scales of the data. Each coloured line in the figure represents one of the cluster centers in the Table 5.

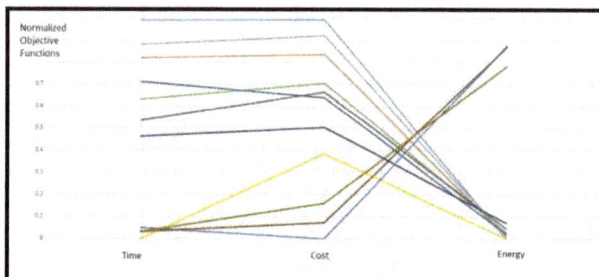

Figure 3. Normallized Objective Functions for the Candidate schedules

To choose the preferred schedule amongst the candidate schedules, it is required to consider preferences of the policy makers. Weighting the objective functions is the common way in this scope. In this study, Analytic Hierarchy Process (AHP) (Saaty, 1990) was applied as the weighting method. With regard to the obtained weights by the AHP method, the preferred schedule with the least weighted summation of the objective functions were chosen. The preferred schedule is compared with the current schedule in Table 6. The trains' speed-time graph for both schedules are represented in Figure 4.

Table 6. The Preferred Schedule

Solution	Objective Functions		
	Time (10^3)	Cost (10^4)	Energy (10^{10})
Preferred Schedule	2.922	-5.126	1.365
Current Schedule	3.115	-4.656	1.286
Percentage Change	-6.20%	-10.09%	+6.14%

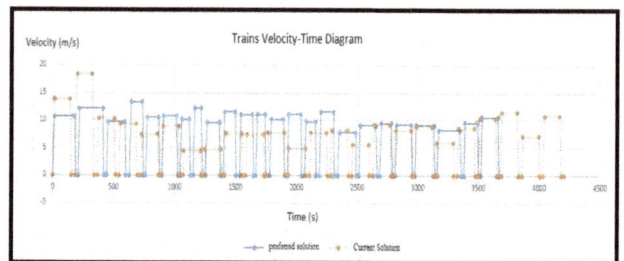

Figure 4. The Speed-Time Graph for the Preferred Schedule in comparison to the Current Schedule

As can be seen, the preferred schedule decreased the total passenger travel times and total operation costs by 6.2 and 10.1, respectively. This is the reason that the preferred schedule in Figure 4 (the blue solid line) took less time for the trains to travel from the origin station to the last station in comparison to the current schedule (the red dashed line).

7. CONCLUSION

The objective of this study was to propose a mathematical model for temporal schedule of urban trains. The motivation of developing such a model was to help the process of Multi-Criteria Subway Schedule Optimization (MCSSO). First, a conceptual framework was proposed that represents the parameters which are related to the subway schedule. Next, an agent-based simulation environment was developed to perform sensitivity analysis in order to extract the interrelations between the framework components. Then, by the help of the developed framework, a mathematical model for MCSSO was created. The decision variables and the objective functions in the model were defined in a way that they could be applied in a MCSSO process.

Tehran subway line no.1 was considered as the case study to evaluate the performance of the proposed model. Operation of the system was modelled mathematically and multi-criteria optimization of its schedule was performed. Results of the evaluation show that using the suggested model outputted an acceptable distribution of Pareto solutions. Comparing the objective functions of the preferred schedule with those of the current schedule shows reductions in the total passenger travel times and total operation costs by 6.2 and 10.1 percent, respectively. By reviewing the results, the efficiency of the proposed model can be concluded. In addition, it is provable that the developed agent-based simulation was successful in the sensitivity analysis procedure.

However, some limitations have remained with the proposed model. The model may not be compatible with the real-time data and it needs more flexibility and adaptations for instantaneous decision makings. Therefore, improving the model so that it can be compatible with the real-time data would be considerable as the future work.

REFERENCES

Chang, C. S. and Kwan, C. M., 2005. Evaluation of evolutionary algorithms for multi-objective train schedule optimization. *AI 2004: Advances in Artificial Intelligence*, Springer: 803-815.

Chang, Y.-H., Yeh, C.-H. and Shen, C.-C., 2000. A multiobjective model for passenger train services planning: application to Taiwan's high-speed rail line. *Transportation Research Part B: Methodological,* **34**(2), pp. 91-106.

Chen, J.-F., Lin, R.-L. and Liu, Y.-C., 2005. Optimization of an MRT train schedule: reducing maximum traction power by using genetic algorithms. *Power Systems, IEEE Transactions on,* **20**(3), pp. 1366-1372.

Chevrier, R., Pellegrini, P. and Rodriguez, J., 2013. Energy saving in railway timetabling: A bi-objective evolutionary approach for computing alternative running times. *Transportation Research Part C: Emerging Technologies,* **37**, pp. 20-41.

Deb, K., Pratap, A., Agarwal, S. and Meyarivan, T., 2002. A fast and elitist multiobjective genetic algorithm: NSGA-II. *Evolutionary Computation, IEEE Transactions on,* **6**(2), pp. 182-197.

Ghoseiri, K., Szidarovszky, F. and Asgharpour, M. J., 2004. A multi-objective train scheduling model and solution. *Transportation Research Part B: Methodological,* **38**(10), pp. 927-952.

Hu, H., Li, K. and Xu, X., 2013. A multi-objective train-scheduling optimization model considering locomotive assignment and segment emission constraints for energy saving. *Journal of Modern Transportation,* **21**(1), pp. 9-16.

Kim, J. H., Bae, Y. K. and Chung, J., 2012. Multi-objective Optimization for Sustainable Road Network Design Problem. In: *Int. Conference on Transport, Environment and Civil Engineering*, Kuala Lumpur (Malaysia).

Lau, H. C., Agussurja, L., Cheng, S.-F. and Tan, P. J., 2013. A multi-objective memetic algorithm for vehicle resource allocation in sustainable transportation planning. In: *Proceedings of the Twenty-Third international joint conference on Artificial Intelligence*, AAAI Press.

Nasri, A., Moghadam, M. F. and Mokhtari, H., 2010. Timetable optimization for maximum usage of regenerative energy of braking in electrical railway systems. In: *Power Electronics Electrical Drives Automation and Motion (SPEEDAM), 2010 International Symposium on,* IEEE.

Saaty, T. L., 1990. How to make a decision: the analytic hierarchy process. *European journal of operational research,* **48**(1), pp. 9-26.

Sels, P., Dewilde, T., Cattrysse, D. and Vansteenwegen, P., 2013. Expected Passenger Travel Time as Objective Function for Train Schedule Optimization. In: *Proceedings of 5th International Seminar on Railway Operations Modelling and Analysis (IAROR): RailCopenhagen2013, May 13-15, Copenhagen, Denmark.*

Shaoquan, N., Dingjun, C. and Miaomiao, L., 2009. Research on optimization model of initial schedule of passenger trains based on improved genetic algorithm. In: *Intelligent Computation Technology and Automation, 2009. ICICTA'09. Second International Conference on,* IEEE.

Su, S., Li, X., Tang, T. and Gao, Z., 2013. A subway train timetable optimization approach based on energy-efficient operation strategy. *Intelligent Transportation Systems, IEEE Transactions on,* **14**(2), pp. 883-893.

Yang, X., Li, X., Gao, Z., Wang, H. and Tang, T., 2013. A cooperative scheduling model for timetable optimization in subway systems. *Intelligent Transportation Systems, IEEE Transactions on,* **14**(1), pp. 438-447.

Yang, X., Ning, B., Li, X. and Tang, T., 2014. A two-objective timetable optimization model in subway systems. *Intelligent Transportation Systems, IEEE Transactions on,* **15**(5), pp. 1913-1921.

23

GROUNDWATER POTENTIAL ASSESSMENT USING GEOGRAPHIC INFORMATION SYSTEMS AND AHP METHOD (CASE STUDY: BAFT CITY, Kerman, IRAN)

M. Zeinolabedini[a] *, A. Esmaeily[b]

Dept. of Water Resources Engineering, Graduate University of Advanced Technology, Kerman, Mahan, 7631133131, Iran - ze.maryam@yahoo.com

Dept. of Remote Sensing Engineering, Graduate University of Advanced Technology, Kerman, Mahan, 7631133131, Iran - aliesmaeily@kgut.ac.ir

Commission VI, WG VI/4

KEY WORDS: Underground Water, AHP, GIS, Remote Sensing

ABSTRACT

The purpose of the present study is to use Geographical Information Systems (GISs) for determining the best areas having ground water potential in Baft city. To achieve this objective, parameters such as precipitation, slope, fault, vegetation, land cover and lithology were used. Regarding different weight of these parameters effect, Analytic Hierarchy Process (AHP) was used. After developing informational layers in GIS and weighing each of them, a model was developed. The final map of ground waters potential was calculated through the above-mentioned model. Through applying our developed model four areas having high, average, low potential and without required potential distinguished. Results of this research indicated that 0.74, 41.23 and 45.63 percent of the area had high, average and low potential, respectively. Moreover, 12.38% of this area had no potential. Obtained results can be useful in management plans of ground water resources and preventing excessive exploitation.

1. INTRODUCTION

Ground water is a main source for industries, communities and agricultural consumptions in the world and due to its freshness, chemical compounds, constant temperature, lower pollution coefficient and higher reliability level, considered as a basic source of supplying reliable fresh water in urban and rural areas. Nowadays, about 34% of the world's water resources belong to ground water and is an important source of drinkable water. Iran is an arid and semi arid country with very little rainfall; so that its average annual rainfall is lower than one-third of the world's average annual rainfall (Nampak et al., 2014; Rahimi & Moosavi, 2013; Magesh et al., 2012; Seif & Kargar, 2011). Therefore, recognizing these resources and optimal usage of them, means stable and permanent usage of this natural wealth. A common method for preparing ground water potential maps is based on land surveying. Recently, with the help of GIS and RS technologies, potential detection of ground water resources can be done easier, more accurate and in short-time.

GIS is a powerful tool to address a large number of spatial data and can be used in detection process of potential ground water areas. Recently, many studies conducted through indices of ground water potential models. Some of them are as follows: frequency ratio, weights of evidence and AHP. In the present study, analytic hierarchy process and GIS technique used to determine the best water resources for optimal usage of ground water resources in Baft city. AHP can be done in different ways, one of these methods is to use expert choice software which its implementation and calculation stages done automatically (Nampak et al., 2014; Magesh et al., 2012; Al-Harbi, 2001).

Moreover, different studies conducted through GIS technique. Using GIS, Abdalla, 2012, Venkateswaran et al., 2015, Dar et al., 2010, Nampak et al., 2014 and Elbeih 2015, studied underground resources of eastern and central desert of Egypt, underground resources potential in hard stones of Gadilam river basin, ground water conditions of Mamundiyar basin in Tamilandu, spatial efficiency prediction of underground resources of Langat basin in Malaysia and underground resources in Europe, respectively. Rahimi et al., 2013, Seif et al. 2011 and Yamani et al., 2014 researches were in line with the above-mentioned studies (Abdalla, 2012; Venkateswaran & Ayyandurai, 2015; Dar et al., 2010; Nampak et al., 2014; Elbeih, 2015; Rahimi & Moosavi, 2013; Seif & Kargar, 2011; Yamani & Alizadeh, 2014) . This study tries to determine the best water resources of the area (for optimal usage and preventing excessive exploitation) through analyzing effective parameters on ground water aquifers feeding, exfoliation, raster, weighting to components by means of analytic hierarchy process and combining layers in ArcGIS environment.

2. STUDY AREA

Baft is one of the oldest and most important cities of Kerman. The city is located in the southwest of the province and its center is Baft, it is 156 km away from Kerman with an area of 13162 square kilometers and located at an altitude of 2250 meters above sea level and 29 degrees 17 minutes of north latitude and 56 degrees and 36 minutes of east longitude (Figure 1). According to the last official statistics, its population is about 131567 individuals and is one of the rainiest cities of Kerman. Baft is known as the heaven of Kerman province, but in recent days, excessive usage of ground water resources resulted in water rationing. The main reason of water shortage in Baft city is plantation in surrounding plains which leads to withdrawal of considerable volume of ground water.

Figure 1. Geographical location of the study area

3. DATA AND METHODS

At first, with the help of Google Earth satellite images, the study area studied and determined. Then, in order to zone the area in terms of ground water resources potential, the following steps performed:

3.1 Parameters and layers generation

In the present study, precipitation, slope, fault, lithology, vegetation and land cover layers used to detect water resources having different potentials. Vegetation layer is a product of NDVI of Modis sensor with spatial accuracy of 250m, and Dem of Aster sensor with spatial accuracy of 30m and altitudinal accuracy of 20m. Information of 23 weather station's annual precipitation (1962-2006) presented here. Layers obtained from satellite images.

3.2 AHP processing

Analytic hierarchy process (AHP) is one of the most effective multi-criteria decision making (MCDM) techniques which helps a decision maker facing complex problems and conflict and internal multiple criteria. This method first presented by Thomas L. Saaty (1980) and like other MCDMs such as Mabeth, Electre, Smart, Promeyhee, VTA and etc. has 4 steps including 1) problem modelling and making hierarchical structure, 2) evaluating weights, 3) combining weights and 4) analyzing sensitivity. Implementing AHP method with the help of expert choice software is so easy, since accessing to it is easy and its implementation and calculation steps done automatically. In addition to the possibility of designing hierarchical diagram, decision making, designing questions, determining priorities and calculating the final weight, the above-mentioned software can analyze the sensitivity of decision making about the changes of problem parameters. In most cases, appropriate diagrams and graphs used for presenting results and performances and a user can communicate easily. In this research, after obtaining available layers through remote sensing data and other available resources, all layers analyzed in Arc Gis environment and converted to raster data. Using expert choice software, an inter-criterion and intra-criterion weight determined for layers and their classes, respectively. Finally, with the help of ArcGIS software and Raster Calculator toolbox, layers overlapped and potential map and ground water obtained (Rahimi & Moosavi, 2013; Ishizaka & Labib, 2009; Al-Harbi, 2001; Zhu & Xu, 2014).

4. RESULTS

4.1 Weighting and layers

Tables 1 to 6 show the results of weighting each parameter and according to weighting method, inter and intra-criterion weights calculated.

Inter-critrion weight	Intra-critrion weight	Precipitation level	Weighting method	layer name
0.38	0.0	Very low	More Precipitation, higher weight	precipitation
	0.1	Low		
	0.33	average		
	0.57	high		

Table 1. Weighting precipitation

Inter-critrion weight	Intra-critrion weight	Vegetation level	Weighting method	layer name
0.20	0.0	No vegetation	According to the vegetation, the area divided into 4 subareas, the more weight, the better vegetation	vegetation
	0.1	Low vegetation		
	0.33	Average vegetation		
	0.57	High vegetation		

Table 2. Weighting vegetation

Inter-critrion weight	Intra-critrion weight	Period	Lithology layer	Weighting method	Layer name
0.10	0.02	2.5 billion to 542 million years ago	Proterozoic	According to age, the area divided into 5 subareas. The more age, the more density and the lower coefficient	Lithology
	0.06	554 million to 248 million years ago	Paleozoic		
	0.19	240 million years ago	Mesozoic		
	0.31	150 million years ago	Mesozoic-Cenozoic		
	0.42	65 million years ago	Cenozoic		

Table 3. Weighting lithology

Inter-critrion weight	Intra-critrion weight	Land cover layer	weighting method	layer name
0.04	0.02	Rock	Land divided into 9 areas. The more fertility and penetration, the more coefficients	land cover
	0.03	Poor-range		
	0.04	Mod-range		
	0.05	way		
	0.07	Urban		
	0.11	Woodland		
	0.15	Garden		
	0.22	Agri		
	0.31	forest		

Table 4. Weighting land cover

Inter-critrion weight	Intra-critrion weight	Slope (degree)	weighting method	layer nmae
0.26	0.51	0-10	The area divided in to 5 subareas. The more slope, the lower coefficient.	slope
	0.27	10-25		
	0.13	25-45		
	0.06	45-65		
	0.03	65-90		

Table 5. Weighting slope

Inter-critrion weight	Intra-critrion weight	Fault density	weighting method	Layer name
0.02	0.02	0.0-0.009542	according to fault density, the area divided into 9 classes. The more fault density, the more weight	fault
	0.03	0.009542-0.27874		
	0.04	0.27874-0.0502366		
	0.05	0.0522366-0.076922		
	0.07	0.076922-0.108541		
	0.11	0.108541-0.145761		
	0.15	0.145761-0.190367		
	0.22	0.190367-0.255811		
	0.31	0.255611-0.354819		

Table 6. Weighting faults density

4.2 Classified map generation

After weighting and regarding their spatial position, each parameter classified and obtained results presented in figures 2 to 7.

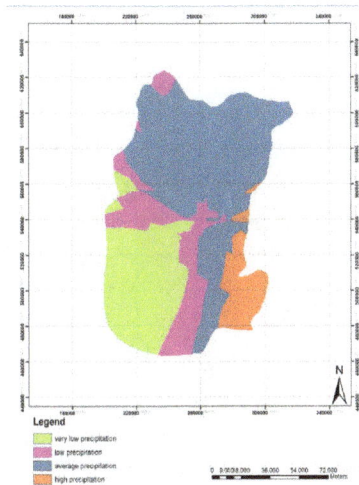

Figure 2. precipitation classification map

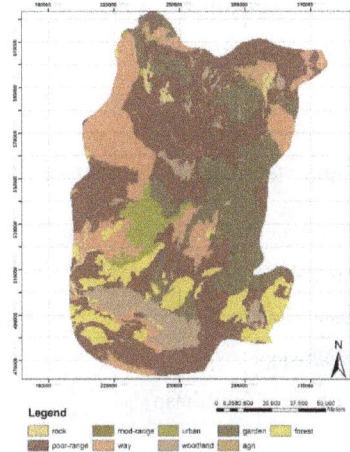

Figure 5. land cover classification map

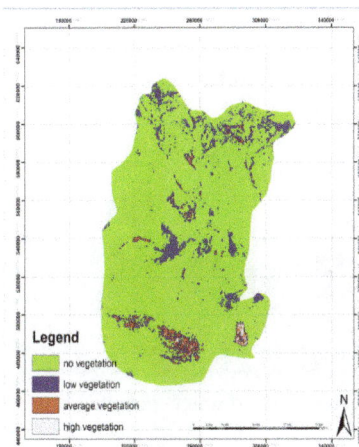

Figure 3. Vegetation classification map

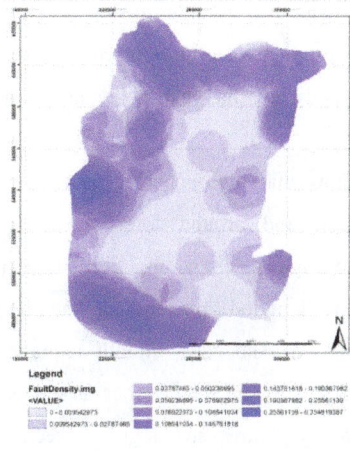

Figure 6. Fault classification map

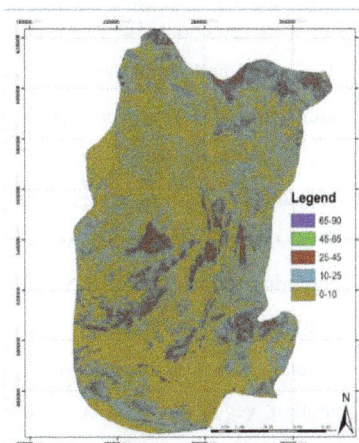

Figure 4. Slope classification map

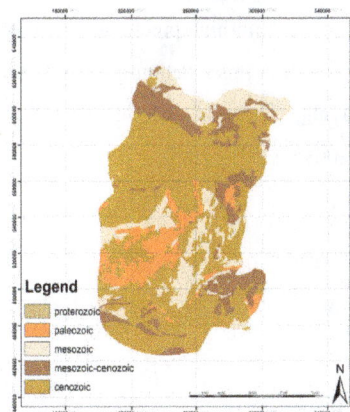

Figure 7. Lithology classification map

5. FINAL MAP

In order to determine areas having ground water resources, a weighted overlap mathematical model used. Mathematical overlap result is a suitability map which regulated in terms of the degree of suitability and included different spectra of colors.

Each layer has an inter-criterion weight. In order to prepare ground waters potential map, we can put all weights in the following formula:

GP= 0.04*RC+0.26*RS+0.02*RF+0.10*RL+0.38*RP+0.2*RV

Where   RP= raster-precipitation map

RL= raster-lithology map

RF= raster-fault map

RS= raster-slope map

RC= raster-land cover map

And

RV= raster-vegetation map

After overlapping layers, the final potential map divided into 4 areas (figure 8) and then, in order to show water potential map of the area, histogram related to it presented. It shows that only 0.74% of the study area has high potential of water resources (figure 9).

The results show that high-potential area is located in the southeast region and around forest areas (with average vegetation and high precipitation). Moreover, this zone is related to Cenozoic period (minimum age) and its slope is very low (0-10 degrees). The zone without required potential is located in central, west and southwest areas and its precipitation and slope are very low and high, respectively and our considered zone has no vegetation. Fault density in both zones is low, but since precipitation and slope layers have most coefficients, the area with high precipitation and low slope and the area with very low precipitation and high slope located in the zone with high potential and the zone without required potential, respectively.

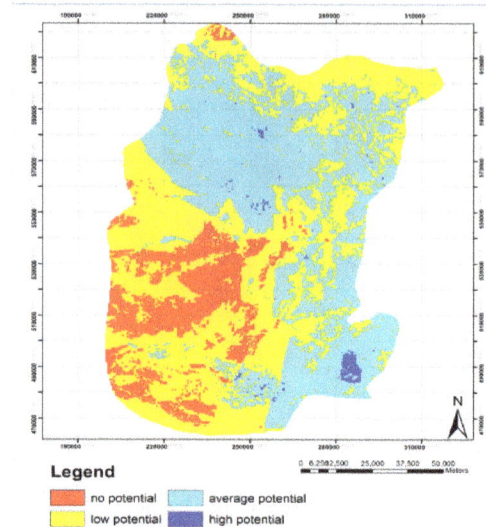

Figure 8. The final map of ground water potential

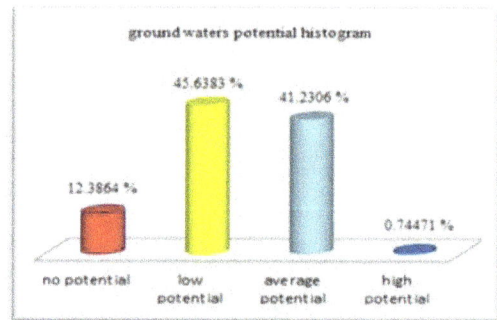

Figure 9. Ground waters potential histogram

6. CONCLUSION

The present study shows that GIS is an appropriate tool for evaluating ground water potential. Based on GIS, appropriate locations for drilling wells and ground water withdrawal can be determined. In addition, AHP used by programmers for solving complex problems of management and expert choice software makes its implementation easier (Ganapuram et al.,2009; Abdalla, 2012; Venkateswaran & Ayyandurai, 2015).

In this study, combining spatial data in GIS and expert choice software was used in order to determine areas having high potential for ground water resources. 6 different layers including precipitation, vegetation, land cover, lithology, slope and fault used. Each of them and their classes had inter and intra criterion weights, respectively. Finally, through putting 6 layers on each other, potential map prepared. Final map divided into four areas which showed 0.74, 41.23 and 45.63 percent of the area had high, average and low potential, respectively. Moreover, 12.38% of this area had no potential (figures 8 and 9).

References

Al-Harbi, K.M.A.S., 2001. Application of the AHP in project management, *International journal of project management*, 19(1), pp. 19-27.

Damavandi, A.; Rezaei F.; Panahi, M., 2011. Ground water potential zoning using RS and GIS in Sele Ben Watershed, *Second Conference in Geoscience*. Tehran, Iran (In Persian)

Dar, I.A.; Sankar, K.; Dar, M.A. 2010. Remote sensing technology and geographic information system modeling: An integrated approach towards the mapping of groundwater potential zones in Hardrock terrain, Mamundiyar basin, *Journal of Hydrology*, 394(3-4), pp. 285-295.

Elbeih, S.F. 2015. An overview of integrated remote sensing and GIS for groundwater mapping in Egypt, *Ain Shams Engineering Journal*, 6(1), pp. 1-15.

Ganapuram, S.; Kumar, G.T.V.; Krishna, I.V.M.; Kahya, E, 2009. Mapping of groundwater potential zones in the musi basin using remote sensing data and GIS, *Advanced in engineering software*, 40(7), pp. 506-518.

Ishizaka, A., Labib A. 2009. Analytic Hierarchy Process and Expert Choice: Benefits and limitations, *OR Insight*, V. 22, pp. 201-220.

Magesh, N.S.; Chandrasekar, N.; Soundranayaga J.P. 2012. Delineation of groundwater potential zones in Thenim district, Tamil Nadu, using remote sensing, GIS and MIF Techniques, *Geoscience Frontiers*, 3(2), pp. 189-196.

Nampak, H.; Pradhan, B.; Manap, M.A. 2014. Application of GIS based data driven evidential belief function model to predict groundwater potential zonation, *Journal of Hydrology*, V. 513, pp. 283-300.

Oikonomidis, D.; Dimogianni, S.; Kazakis, N.; Voudoiri, K. 2015. A GIS/remote sensing-based methodology for groundwater potentiality assessment in Timavos area Greece, *Journal of Hydrology*, V. 525, pp. 197-208.

Rahimi, D.; Moosavi, H., 2013. Potentiality of ground water resources by using AHP model and GIS technique in Shahrood-Bastam watershed, *Journal of Geography and Management*, 17(44), pp. 139-159. (In Persian)

Seif, A.; Kargar, A., 2011. Ground water potential zoning using AHP and GIS in Sirjan Watershed, *Journal of Natural Gegraphy*, 4(12), pp. 75-90. (In Persian)

Sternberg, T.; Paillou, Ph. 2015. Mapping potential shallow groundwater in the desert using remote sensing: Lake Ulaan Nurr, *Journal of Enviroments*,

Venkateswaran, S.; Ayyandurai, R., 2015. Groundwater Potential Zoning in Upper Gadilam River Basin Tamil Nadu, *Aquatic Procedia*, V. 4, pp.1275-1282.

Zhu, B.; Xu, Z. 2014. Analytic hierarchy process-hesitant group decision making, *European Journal of Operational Research*, 239(3), pp. 794-801.

Yamani, M.; Alizadeh, Sh. 2014. Ground water potential zoning using AHP in Abadeh-Fars, *journal of Geomorphology*, 1(1),pp.131-144.(In Persian)

Abdalla, F.2012. Mapping of groundwater prospective zones using remote sensing and GIS techniques: A case study from the Central Eastern Desert, Egypt, *journal of African Earth Sciences*, 70(1-2), pp.8-17

24

METHODS FOR GEOMETRIC DATA VALIDATION OF 3D CITY MODELS

D. Wagner[a,*], N. Alam[b], M. Wewetzer[c], M. Pries[c], V. Coors[b]

a University of Tehran, Tehran, Iran – dwagner4@gmx.at
b Hochschule für Technik Stuttgart – University of Applied Sciences, Stuttgart, Germany –
[nazmul.alam | volker.coors]@hft-stuttgart.de
c Beuth Hochschule für Technik – University of Applied Sciences, Berlin, Germany – [mwewetzer | pries]@beuth-hochschule.de

Commission VI, WG VI/4

KEY WORDS: Geometry, Validation, CityGML, 3D city model, Requirements, Validation Rules, Tolerances

ABSTRACT

Geometric quality of 3D city models is crucial for data analysis and simulation tasks, which are part of modern applications of the data (e.g. potential heating energy consumption of city quarters, solar potential, etc.). Geometric quality in these contexts is however a different concept as it is for 2D maps. In the latter case, aspects such as positional or temporal accuracy and correctness represent typical quality metrics of the data. They are defined in ISO 19157 and should be mentioned as part of the metadata.

3D data has a far wider range of aspects which influence their quality, plus the idea of quality itself is application dependent. Thus, concepts for definition of quality are needed, including methods to validate these definitions. Quality on this sense means internal validation and detection of inconsistent or wrong geometry according to a predefined set of rules.

A useful starting point would be to have correct geometry in accordance with ISO 19107. A valid solid should consist of planar faces which touch their neighbours exclusively in defined corner points and edges. No gaps between them are allowed, and the whole feature must be 2-manifold.

In this paper, we present methods to validate common geometric requirements for building geometry. Different checks based on several algorithms have been implemented to validate a set of rules derived from the solid definition mentioned above (e.g. water tightness of the solid or planarity of its polygons), as they were developed for the software tool CityDoctor. The method of each check is specified, with a special focus on the discussion of tolerance values where they are necessary.

The checks include polygon level checks to validate the correctness of each polygon, i.e. closeness of the bounding linear ring and planarity. On the solid level, which is only validated if the polygons have passed validation, correct polygon orientation is checked, after self-intersections outside of defined corner points and edges are detected, among additional criteria. Self-intersection might lead to different results, e.g. intersection points, lines or areas. Depending on the geometric constellation, they might represent gaps between bounding polygons of the solids, overlaps, or violations of the 2-manifoldness.

Not least due to the floating point problem in digital numbers, tolerances must be considered in some algorithms, e.g. planarity and solid self-intersection. Effects of different tolerance values and their handling is discussed; recommendations for suitable values are given.

The goal of the paper is to give a clear understanding of geometric validation in the context of 3D city models. This should also enable the data holder to get a better comprehension of the validation results and their consequences on the deployment fields of the validated data set.

1.1 INTRODUCTION

The relevance of high quality geo data is considered as a key factor for development of down-stream applications and their commercial and public usability. In the past, data quality was mostly referred as accuracy of geo data products and consistency with respect to the real world situation, e.g. (Arsanjani, Barron, Bakillah, & Helbich, 2013), (Zielstra & Zipf, 2010). Researching quality concepts for 3D data extends this definition of data quality to another field, which can be summarized as inherent or internal data quality (cf. Section 3). In this context, data quality can be defined as the grade of compliance with a predefined standard or data model, plus application and user dependent extensions.

One of the most common data models for 3D buildings on a city scale is CityGML, adopted as an OGC standard in version 2.0 in 2010. CityGML includes a semantic model in addition to the GML-based geometric model. Hence validation of consistency of semantics and geometry is a major research field. Prerequisite for investigating consistency issues is the validation of XML Schema and geometry. For schema validation, commercial tools produce reliable results, however, geometry validation of simple features such as polygons or solids is a more complicated task due to the special characteristics of the geometry model used in CityGML.

In this paper, we present algorithms and methods for geometry validation of CityGML models and discuss fundamental questions related with the task.

2. GEOMETRY MODEL OF CITYGML

CityGML 2.0 is based on GML 3.1. The geometry features available in CityGML can be regarded as a profile of the GML features (Kolbe, Gröger, & Plümer, 2005). Some additional restrictions are specified in the standard and have to be considered, such as only planar surfaces may be used, and line strings may have only straight line segments. Solids usually do not have inner shells, although it would be allowed by the standard.

The following definitions of geometric primitives are based on GML 3.1, (Coors & Gröger, 2010) and (OGC, 2006).

Points usually are *gml:Point* features and consist of coordinate triples or *gml:posList* features consisting of a list of coordinates where the number of elements can be divided by the coordinate dimension.

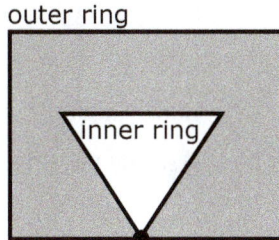

Figure 1: Inner and outer ring sharing one common point.

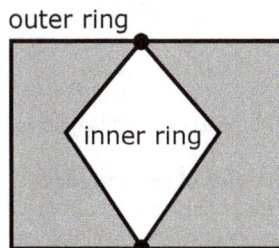

Figure 2: Inner and outer ring sharing two points, resulting in a disconnected polygon

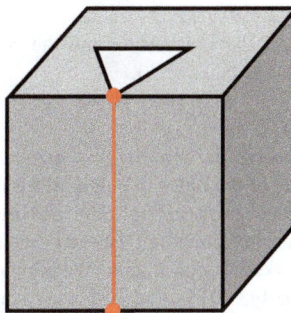

Figure 3: Extrusion of Figure 1, resulting in a non-manifold edge (red)

A *gml:LinearRing* is a finite sequence of points where the first and last point are identical (closeness) and all other points are different. Edges are implicitly defined as a straight line connection of two neighboring points. Two edges may touch each other only in their start and end points, other points of intersection or touching are not allowed (no self-intersection).

A *gml:polygon* is a surface patch with a planar *gml:LinearRing* as outer border and one or several inner rings. All rings must be coplanar, the surface of the polygon uses planar interpolation. All *LinearRing*s must not intersect with each other. The inner rings must be located completely inside the boundaries of the outer ring; they may intersect with the outer ring at one point. No inner ring may be located inside another inner ring. Polygons are orientable surfaces where the orientation is determined by the order of the points (usually the outer face is defined counter clockwise). Several connected polygons form a *composite surface* (a list of orientable surfaces).

There is still some debate about the question if outer and inner rings are allowed to touch in one single point. The problem can be considered solved for 2D polygons (Oosterom, Quak, & Tilssen, 2005). Linear rings describing a polygon may touch in one common point (Coors & Gröger, 2010). Thus, sharing a single point is normally permitted (Figure 1), whereas sharing two or more points is not allowed (Figure 2).

For 3D features, it is also makes sense to prohibit the connection of an inner ring with the outer ring in more than one point for most scenarios, because the volume of the solid would be no longer connected (Ledoux, 2013),(Kazar, Kothuri, van Oosterom, & Ravada, 2008). Consider the polygon in Figure 1 as the ground surface of a simple LoD1 building. The 3D geometry of this building would be a simple extrusion of the polygon along its normal, as shown in Figure 3. This results in a non-manifold edge (marked with the red line); the edge is shared by 4 polygons (2 from the inner and 2 from the outer ring). Therefore we consider this geometry as invalid.

A *solid* is the basis for 3D geometry. A solid is delimited by its outer shell, and may have inner shells which represent cavities inside the solid. Each shell of a solid is represented by a composite surface connected in a topological cycle (an object whose boundary is empty). Ongoing discussion (e.g. during the Quality Interoperability Experiment of OGC (Coors & Wagner, 2015)) suggest that inner shells are not used by the CityGML community, which is the reason for neglecting them in the following description.

Volumetric features such as buildings should be modeled as *Solid* features in CityGML, however, many data sets use *MultiSurface* geometry instead. In the latter case, a collection of polygons has no meaning, although we can assume in many cases that they are supposed to represent a closed volume when a building is modeled.

3. VALIDATION OF GEOMETRIC FEATURES

According to ISO 19114 data quality can be evaluated with direct or indirect methods, where the direct methods are again subdivided into internal and external (Figure 4).

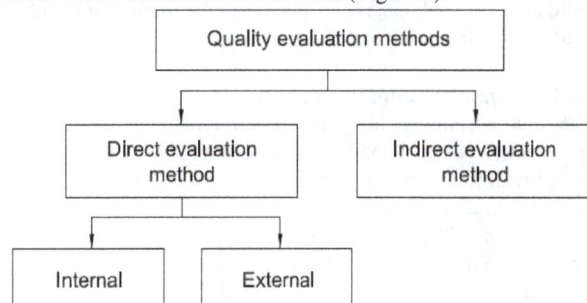

Figure 4: Classification of data quality evaluation methods (ISO 19114)

External validation is not part of this discussion. Thus the accuracy of point referring to real-world reference points is not considered. The reason for this is that most 3D models are generated from 2D data which should have this kind of quality information as meta data (e.g. cadastral data) or from 3D data with a known accuracy, usually laser point clouds or photogrammetric point clouds. The errors of the source data sets are propagated to the 3D model.

We focus on the inherent correctness and consistency of the geometry, hence validating features such as planarity of polygons or compliance of solids. The validation can be divided into two main groups of checks:

- polygon validation, which is usually applicable for *MultiSurface* elements. Each polygon is checked individually,
- and solid validation, investigating the spatial combination and topology of a group of polygons for *Solid* elements.

In both cases, a clear definition of the features concerned and a set of rules which should be adhered to, is necessary. How these can be extracted is shown in the next sections.

3.1 Point Accuracy

Independent of the element used, the accuracy of the points is determined by the number of decimal places. Depending on the generation process of the model, points might have a high number of positions after the decimal point what might result in problems when using floating point computations (Becker, 2006). It makes sense to round coordinate numbers in these cases to a useful amount, e.g. four decimal places. In a data set with Gauss-Krüger coordinates, this results in a point accuracy of 1 mm with a rounding error in the next place.

Moreover, it is important to notice that topological relationships are not stored explicitly. That means, corner points of a 3D geometric feature are stored independently for each of the bounding surfaces, e.g. three times for a cube, each representing the same point. In some cases, there might be differences which cause deficiencies in the model, e.g. it would not be watertight if it is a solid. To avoid such errors resulting from to many decimal places, the same rounding procedures as outlined above is used.

Rounding is preferred to snapping these coordinates to a common position in order to enable detection of modeling defects above a certain threshold (here: 1 mm).

3.2 Polygon Validation

CityGML is based on geometry features of GML 3.1. The definitions below are based on the detailed description in (Coors & Gröger, 2010) which is based on the GML standard, although it implies some difficulties as discussed in Section 2.

Validation of polygons according to the definitions above results in the following set of checks.

3.2.1 Minimum number of points

Although seemingly obvious, in some models degenerated *LinearRing*s are contained, e.g. consisting of only three points or less with first and last point identical, which is not sufficient to model an area. Therefore, a *LinearRing* should consist of at least 4 points. The check counts the number of entries in the sequence. The result is pass/fail including the ID of the *LinearRing* and the number of points (CP-NUMPOINTS).

3.2.2 Nullarea

The linear ring delimits an area greater than 0 (Figure 5). Collinearity is checked for all points. The result is pass/fail including the ID of the *LinearRing* (CP-NULLAREA).

3.2.3 Closeness

A *LinearRing* must be closed meaning the first and last point of the sequence defining the *LinearRing* must be identical (Figure 6). The check compares the coordinates of first and last point of the sequence. If the coordinates are not rounded, a tolerance should be defined. The result is pass/fail including the ID of the *LinearRing* (CP-CLOSE).

3.2.4 Duplicate Points

A *LinearRing* must not have duplicate points, with exception of start and end point (Figure 7). The check compares the coordinates of all points with each other. If the coordinates are not rounded, a tolerance should be defined. The result is pass/fail including the ID of the *LinearRing* and the coordinates of duplicate points (CP-DUPPOINT).

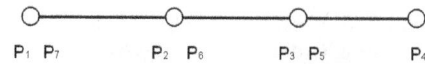
Figure 5: Degenerated LinearRing

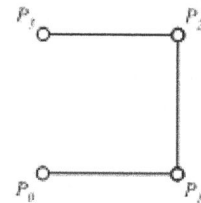
Figure 6: Closeness error of LinearRing feature

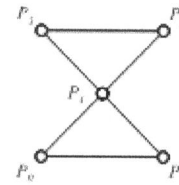
Figure 7: Self Intersection and Duplicate Point errors

3.2.5 Self-Intersection of polygon edges

Two edges can intersect only in one start-/ end-point (Figure 7). Other points of intersection or touching are not allowed (to account for rounding errors or polygons which are not perfectly planar, a small tolerance $\varepsilon \in \mathbb{R}$ is allowed). The check intersects all edges with each other. An error is detected when the result is not empty. In this case, the result is pass/fail including the ID of the LinearRing and the coordinates of the intersection point (CP-SELFINT).

3.2.6 Planarity

Checking the planarity is done in two steps. At first we fit a plane to the points of the outer ring and afterwards we calculate the distance of each point of the outer and inner rings to the plane. If the distance of one point exceeds the given tolerance ε, the polygon is marked as non-planar. We adopted the algorithm proposed by (Eberly, 2015) and use least squares where the distance is measured orthogonally to the proposed plane and not in the x-, y- or z-direction of the coordinate system, as illustrated in Figure 8 and Figure 9. Using an approach with an energy function leads to an eigenvalue problem, where the eigenvector of the smallest eigenvalue is the normal vector of the plane we are looking for. Since we are dealing with a real symmetric 3x3 eigensystem, we find the solution by applying the iterative Jacobi eigenvalue algorithm. The position vector of our plane is defined by the average of the points of the outer ring. Figure 10 shows an exaggerated warped blue quadrangle and its orange fitting plane. The dashed lines indicating the orthogonally measured distance from the corner points of the linear ring to the plane.

We use a tolerance ε of 0.01m as deviation for a point from the plane. This seems to be small for an ordinary family home. But this is mainly driven by the self-intersection algorithm for solids which intersects polygons pair wise (cf. section 3.3, Solid Self Intersection) and relies on the projection if these polygons on their fitting plane. There for the polygons should be as planar as possible, to receive reliable results. The value is based on experience and showed to be a fair trade-off between the needs of the self-intersection algorithm and existing real life models. The result is pass/fail including the ID of the *LinearRing* and the deviation in meters (CP-PLAN).

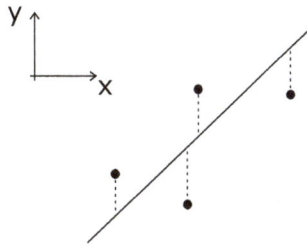

Figure 8: Line fitting done with vertical regression (in y-direction)

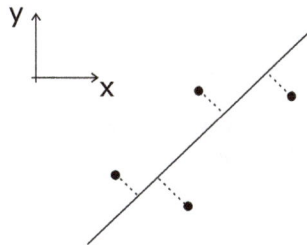

Figure 9: Line fitting done with orthogonal regression

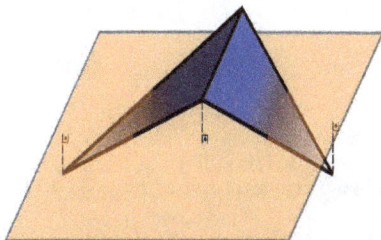

Figure 10: Overdone warpage of a quadrangle (blue) and its fitting plane (orange)

3.3 Solid Validation

The performance of the algorithms of solid checks depends on the planarity of the polygons forming the solid. Non-planar surfaces might yield incorrect results under certain conditions. To avoid these problems the tolerance should be as small as possible. However, it is possible to validate a *Solid* geometry with relatively large tolerance settings to allow only the detection of big folds.

3.3.1 Minimum Number of Polygons

The smallest solid is a tetrahedron, consisting of four triangles. Therefore, the minimum number n of polygons to define a valid solid is four, when they are situated in different planes. The result includes the ID of the erroneous geometry (CS-NUMFACES).

3.3.2 Solid Self-Intersection

The solid self-intersection check is realized by pair wise intersections of polygons of a solid. The planarity of the polygons, as described above, is mandatory, because the problem is transformed into two dimensions to avoid issues with skew warped polygons. Additionally
the shape of a surface of a non-planar outer ring is not defined in CityGML.
Let us suppose we have a triangle and a quadrangle situated as shown in Figure 11. In the first step we calculate the fitting plane of each polygon and project each polygon on its plane, as shown in Figure 12. The advantage of this procedure is that both polygons can only intersect at the intersection straight line

(dashed) of the planes, unless the planes are parallel. We intersect each polygon with the intersection straight separately and get the domain of each intersection, as shown in figure Figure 13 and Figure 14 as green line. By intersecting both domains we retrieve the intersection between both Polygons, see Figure 15.

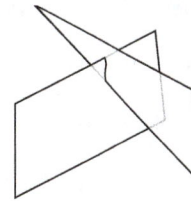

Figure 11: Initial position. A triangle (polygon 1) intersects with a quadrangle (polygon 2)

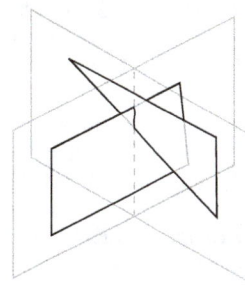

Figure 12: Intersection of the fitting planes of both polygons

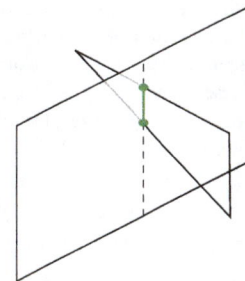

Figure 13: Intersection of polygon 1 with plane of polygon 2

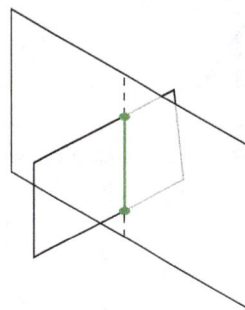

Figure 14: Intersection of polygon 2 with plane of polygon 1

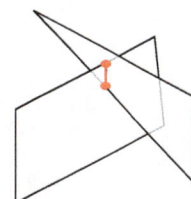

Figure 15: Combined intersection results of both polygons (intersection marked with red line)

If both polygons are located in the same plane a special treatment is necessary. In this case we intersect each edge of a polygon with the other one and determine if it is located partially or in fully inside the polygon. Merging these information lead to a 2D-domain, reflecting the area of overlapping of both polygons. This can result in a *fully embedded polygon* intersection type, where one polygon is completely contained in the other one or a simple *partially embedded polygon* intersection type as shown exemplary in Figure 16. Besides the "normal intersection" as shown in figure Figure 15 and the before mentioned we also take into account if to adjacent polygons intersect at an edge, without sharing the start or end point. We call this type of intersection "embedded edge". Like embedded polygons we distinguish between partially and fully embedded edges. Figure 17 shows different configuration types for edges and the resulting intersection types. Please note, that the black edges actually lie on top of each other and are "pulled" beside for clarity, which is also indicated by the gray dashed line. The resulting intersection is marked by the red line. Figure 17 shows a partially embedded edge, where both edges don't share a common point and no black edge is completely covered by the other one. Figure 17 shows a tricky configuration. If both edges share a common point, as indicated by the black dashed line, the intersection type will be set to *partially embedded edge*. Otherwise it will be set to *fully embedded edge*. Figure 17 shows *fully embedded edge* intersection type where the second edge is completely embedded in the first one.

Like the planarity check we are using 0.01m as tolerance to check for coincident points and 0.5° for parallel edges. This also implicates that the length of an intersection interval below 0.01m is treated as intersection point and not line. These tight error bounds result from experience and ensured reliable results especially for the intersection type embedded edge.

The result is pass/fail providing the IDs of the intersecting polygons, the type and the geometric details of the intersection (CS-SELFINT).

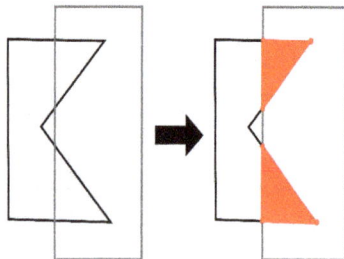

Figure 16: Partially embedded polygons intersection type (overlapping intersection in red)

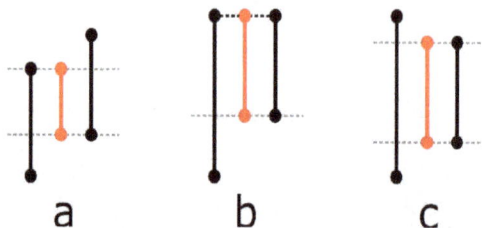

Figure 17: Partially (a, b) and fully (c) embedded intersection type (overlapping intersection in red)

3.3.3   2-Manifoldness

The shell of a solid consists of a composite surface. Therefore it must be 2-manifold. 2-mannifoldness is a complex requirement, validated by several checks.

A valid intersection of two polygons of a solid either contains a common edge, a common point of a linear ring, or is empty. Common edges and points must be elements of both polygons. Any edge of a solid must be incident to exactly two common polygons, otherwise the solid can not be 2-manifold.

Two checks compare all edges of the solid and fails if the number n of incident polygons is not equal to 2. The IDs of the solid and the edge concerned are reported. Two different error types can occur:

- $n = 1$: There is an outer edge which bounds a hole in the solid geometry, i.e. the solid is not watertight (CS-OUTEREDGE)
- $n > 2$: Topological error which violates the 2-manifoldness. In Figure 18 there is an edge shared by four polygons (CS-OVERUSEDEDGE).

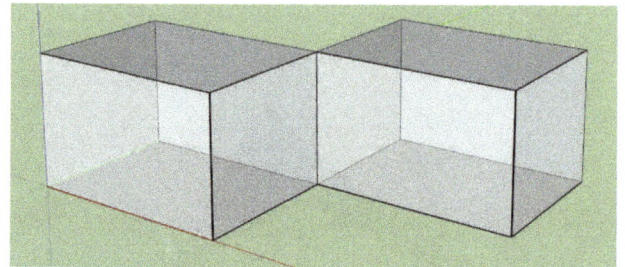

Figure 18: Topological error caused by an edge with 4 adjacent polygons

In case of a common point incident to several polygons, 2-manifoldness might not exist. This happens when the neighborhood of the point is not topologically equivalent to a disc (Figure 19). The graph $G_S = (V_P, E_P)$ of polygons and edges which are meeting in point p_i is connected for all p. Each vertex $v \in V_P$ represents exactly one polygon which contains p. Two vertices are connected with an edge $e \in E_P$ if the polygons represented by these vertices have a common edge that is bounded by p. If the graph finds more than one loop for connected polygons at a vertex then an umbrella error occurs. The result includes the ID of the solid and the coordinates of the point (CS-UMBRELLA).

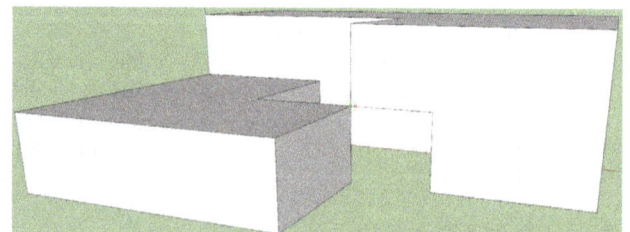

Figure 19:Topological error caused by a non-manifold point

3.3.4   Consistent orientation

The members of the composite surface forming the shell of a solid must have consistent orientation (cf. Section 2), i.e. their face normals should all be directed towards the inside of the solid or opposite. Consequently, the direction of the edges of two neighboring polygons must be opposite. In Figure 20, polygon A is anti-clockwise oriented whereas polygon B is clockwise oriented. Both polygons are incident to a common edge. The direction of the respective edges is the same which causes an error. If all or most of the edges of a polygon have wrong orientation then its orientation is wrong. The result includes the ID of the polygon (CS-FACEORIENT).

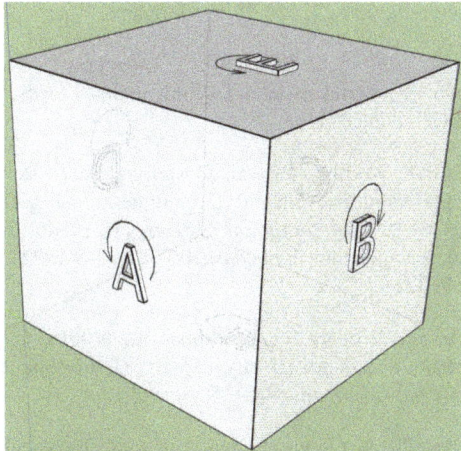

Figure 20: Inconsistent orientation of polygons

Figure 21: Orientation error of a roof polygon

By definition, normal vectors of the polygons must point towards the outside of the solid (Figure 21). If consistency of the orientation of all polygons is validated, the direction of their normal vectors must be checked. This is done by calculating a normal vector of a polygon and then intersect it with all other polygons of the solid. The number of intersections shall be odd in case it points towards the outside of the solid, even, in case it points towards the inside. The solid has valid orientation when the number of intersection points is odd. In case of error, the ID of the solid is reported (CS-FACEOUT).

3.3.5 Connected component

The shell of a solid must be connected in a topological cycle. This results in a connected geometry for each solid. Disconnected geometries can not be modeled as different part of the same solid (Figure 22). Validation of this requirement is done by generating a graph GS = (VP,EP) of polygons and edges. The result must be a connected graph which contains all polygons and edges of the solid. The check reports the ID of the solid in case of error.

Figure 22: Disconnected solid

4. HIERARCHY AND DEPENDENCY OF CHECKS

Customizing validation rules depends on user requirements. The SIG3D modeling handbook provides a guideline but the user might have own preferences or limitations depending on the application and deployment of the model. Some geometry checks are depending on others (Table 1). Checks in the first column are dependent on those marked with an 'X' in the respective row, e.g. the planarity check (F) accepts a geometry feature as input only when it has passed the checks for minimum number of points (A), closeness of the *LinearRing* (B), nullarea (C) and duplicate points (D) without errors. All polygon checks are independent of the solid checks.

Table 1. Dependencies of geometry checks

	A	B	C	D	E	F	G	H	I
A									
B									
C	X	X		X					
D	X	X							
E	X	X	X	X					
F	X	X	X	X					
G	X	X	X	X	X				
H	X	X	X	X	X	X	X		
I	X	X	X	X	X				
J	X	X	X	X	X				
K	X	X	X	X				X	
L	X	X	X	X		X	X	X	X
M	X	X	X	X				X	
N	X	X	X	X				X	

Legend:
A   Minimum number of points
B   Closeness
C   Nullarea
D   Duplicate points
E   Self-Intersection of polygon edges
F   Planarity
G   Minimum Number of Polygons
H   Solid Self-Intersection
I   Edge adjacent to less than two polygons
J   Edge adjacent to more than two polygons
K   Inconsistent Orientation
L   Orientation towards inside of solid
M   Unconnected components
N   Non-2-manifold point (umbrella)
Checks A-F are polygon checks, H-N are solid checks

Solid checks are generally performed only for *Solid* geometries and are only executed when all polygons have a minimum number of four points, are bounded by closed LinearRings without duplicate points, and have an area greater than zero. Planarity is only required for solid self-intersection and determination of correct orientation of all polygons.
MultiSurface geometries can also be checked with the solid checks, if required. This might be helpful in situations where real-world solids have been modeled as *MultiSurface*

geometries but their solid characteristics is requested for further analysis.

5. CONCLUSION AND OUTLOOK

Geometry definitions are given by standards such as ISO 19107 and GML. Their interpretation for 3D geometry is not unambiguous. Consequently, rules for valid 3D geometries may differ along with user requirements.

Validation of these rules should be done by a modularized approach, where for each restriction one or more check routines are applied. As the definition of 3D geometries is not unambiguous, a suggestion for solid geometry in CityGML models is discussed, where we recommend to prohibit common points of inner and outer rings of a polygon.

Based on this, we describe a set of basics checks. These checks can be combined to satisfy different users' needs and enable testing of complex requirements such as water-tightness of solids. Besides the basic methods used, we point out the importance of tolerances during the validation process as well as the dependency of some checks on other lower level checks.

The set of checks presented above for geometry validation of CityGML models is implemented in the software package CityDoctor (Wewetzer et al., 2013) in JAVA and C++ as proof of concept. Tests on real-world data sets and on synthetic models have been done extensively, latest during the CityGML Quality Interoperability Experiment of OGC (Coors & Wagner, 2015), and showed that general requirements for 3D city models can be validated with this approach.

The success of the strategy was confirmed in comparison with other approaches for geometric validation. Future development should focus on validation of semantic features and the coherency of semantics and geometry.

ACKNOWLEDGEMENTS

Special thanks to Shelly and David Eberly of Geometric Tools, for sharing and offering their knowledge about least squares fitting.

The project CityDoctor was funded by German Ministry for Education and Research (BMBF) and performed as a joint research project of University of Applied Sciences Stuttgart (HFT) and Beuth Hochschule für Technik Berlin.

REFERENCES

Arsanjani, J. J., Barron, C., Bakillah, M., & Helbich, M. (2013). Assessing the Quality of OpenStreetMap Contributors together with their Contributions. *Proceedings of the AGILE.* http://www.agile-online.org/Conference_Paper/CDs/agile_2013/Short_Papers/SP_S4.2_Arsanjani.pdf

Becker, P. (2006). *Errors in Floating-point Calculations.* http://petebecker.com/js/js200007.html (27. Sep 2015)

Coors, V., & Gröger, G. (2010). Handbuch für die Modellierung von 3D Objekten - Teil 1: Grundlagen (Regeln für valide GML Geometrie-Elemente in CityGML). *SIG3D Quality Wiki.* http://wiki.quality.sig3d.org/index.php/Handbuch_f%C3%BCr_die_Modellierung_von_3D_Objekten_-_Teil_1:_Grundlagen_(Regeln_f%C3%BCr_valide_GML_Geometrie-Elemente_in_CityGML) (27. Sep 2015)

Coors, V., & Wagner, D. (2015). CityGML Quality Interoperability Experiment des OGC. *Publikationen der Deutschen Gesellschaft für Photogrammetrie, Fernerkundung und Geoinformation e.V.,* 24, Köln, Germany.

Eberly, D. (2015). *Geometric Tools.* http://www.geometrictools.com/Documentation/LeastSquaresFitting.pdf (27. Sep 2015)

Kazar, B. M., Kothuri, R., van Oosterom, P., & Ravada, S. (2008). On valid and invalid three-dimensional geometries. *Advances in 3D geoinformation systems*, pp. 19–46. Springer. http://link.springer.com/chapter/10.1007/978-3-540-72135-2_2 (27. Sep 2015)

Kolbe, T. H., Gröger, G., & Plümer, L. (2005). CityGML: Interoperable access to 3D city models. *Geo-information for disaster management,* pp. 883–899, Springer.

Ledoux, H. (2013). On the validation of solids represented with the international standards for geographic information. *Computer-Aided Civil and Infrastructure Engineering*, 28(9), 693–706.

OGC (2006). *OpenGIS implementation specification for geographic information — simple feature access.* Open Geospatial Consortium.

Oosterom, P., Quak, W., & Tilssen, T. (2005). About Invalid, Valid and Clean Polygons. P. F. Fisher (Ed.) *Developments in Spatial Data Handling* (Vol. Part 1, pp. 1–16). Leicester, UK: Springer Berlin Heidelberg.

Wewetzer, M., Falkenhausen, J., Wagner, D., Alam, M. N., Pries, M., Coors, V., & Fischer, J. (2013). Verbundprojekt CityDoctor - Entwicklung von Methoden und Metriken zum Qualitätsmanagement virtueller Stadtmodelle. *Forschungsbericht 2012 - Angewandte Forschung zur Stadt der Zukunft,* pp. 15–21. Berlin, Logos Verlag.

Zielstra, D., & Zipf, A. (2010). A comparative study of proprietary geodata and volunteered geographic information for Germany. *13th AGILE international conference on geographic information science,* http://agile2010.dsi.uminho.pt/pen/shortpapers_pdf/142_doc.pdf (27. Sep 2015)

25

INVESTIGATION ON CANOPY HEIGHT AND DENSITY DIFFERENTIATIONS IN THE MANAGED AND UNMANAGED FOREST STANDS USING LIDAR DATA (CASE STUDY: SHASTKALATEH FOREST, GORGAN)

Sh. Shataee [a]*, J. Mohammadi [b]

[a] Associate Professor, Forestry Department, Forest Sciences Faculty, Gorgan University of Agricultural Sciences and Natural Resources, 386, Gorgan, Iran - Shataee@gau.ac.ir
Assistance Professor, Forestry Department, Forest Sciences Faculty, Gorgan University of Agricultural Sciences and Natural Resources, 386, Gorgan, Iran - Mohammadi@gau.ac.ir

KEY WORDS: Canopy height, Canopy density, Lidar, Managed forest, Unmanaged forest, Dr. Bahramnia forest plan of Gorgan

ABSTRACT

Forest management plans are interesting to keep the forest stand natural composite and structure after silvicultural and management treatments. In order to investigate on stand differences made by management treatments, comparing of these stands with unmanaged stands as natural forests is necessary. Aerial laser scanners are providing suitable 3D information to map the horizontal and vertical characteristics of forest structures. In this study, different of canopy height and canopy cover variances between managed and unmanaged forest stands as well as in two dominant forest types were investigated using Lidar data in Dr. Bahramnia forest, Northern Iran. The in-situ information was gathered from 308 circular plots by a random systematic sampling designs. The low lidar cloud point data were used to generate accurate DEM and DSM models and plot-based height statistics metrics and canopy cover characteristics. The significant analyses were done by independent T-test between two stands in same dominant forest types. Results showed that there are no significant differences between canopy cover mean in two stands as well as forest types. Result of statistically analysis on height characteristics showed that there are a decreasing the forest height and its variance in the managed forest compared to unmanaged stands. In addition, there is a significant difference between maximum, range, and mean heights of two stands in 99 percent confidence level. However, there is no significant difference between standard deviation and canopy height variance of managed and unmanged stands. These results showd that accomplished management treatments and cuttings could lead to reducing of height variances and converting multi-layers stands to two or single layers. Results are also showed that the canopy cover densities in the managed forest stands are changing from high dense cover to dense cover.

---

* Corresponding author

1. Introduction

The Hyrcanian original and natural forests in the northern Iran are almost high dense canopy cover and multi-layers vertical structure with high variance. Forest management plans are certificated approaches for applying forest managing and silviculture treatments programs to optimal forest management and keeping the forest stand natural composite and structure. The forest managers are interesting to aware of forest composite and structure situation before and after silvicultural and management treatments. In order to investigate on stand differences made by management plan, comparing of these stands with unmanaged stands as a nature of forests is necessary.

Traditional way to measure forest stucture and composits variables is field surveying by different methods and tools. These measurements are generally expensive, time-consuming and labour intensive, as well as difficult to perform, especially in mountainous and dense forests. Use of remote sensing tools and sources with different cappabilites and abilities is an alternative way particularly in large areas. Improving satellite data sources and classification methods offer new opportunities for obtaining more accurate forest biophysical maps. Rapid improvements in remote sensing technology have led to various types of sensors, such as multispectral, hyper spectral, ultraviolet, thermal sensors, light detection and ranging (Lidar), radio detection and ranging (radar), and other sensors. Each type of sensor has been designed for specialized purposes, tasks, and different applications. These new potential sources have been shown to be appropriate tools to assess and monitor forest attributes with reasonable accuracy levels.

The optical sensors in visible and infrared wavelengths are producing information based on registering reflectances of objects in diferent radiometric, spectral, and spatial resolutions. The optical aerial/satellite imagery usually presents two-dimensional spectral information and reflectance responses of a canopy cover's surface. Launching of many satellites providing data of submetric ground resolution can be usefull for accurate detection forest variables. However, LiDAR data from an airborne laser scanner (ALS) provides semi three-dimensional data set relating to canopy cover and canopy height. Nasset (1998 and 2002) demonstrated that lidar data for plot-based estimation error of maximum and mean canopy height with full canopy closure is less than 0.5 meters. Capacity of LiDAR data to estimate vertical and horizontal canopy structure of stands was studied in the some research such as Coops et al (2007).

In this study, different of canopy cover density and crown height variances between managed and unmanaged forest stands were investigated in Dr. Bahramnia forest management plan in managed forest stands and unmanged forest stands as well as in two dominant forest types using Lidar data.

2. Material and Methods:

2.1. Study area

The study area comprises 1100 hectares in the Southeast part of the Golestan province, Eastnorth of Iran, with elevation ranging from 270 to 740 m above mean sea level (Figure 1). The research was done on managed (parcls of 4 to 24 seri1) and unmanged (parcls of 2 to 7 of seri 2 Dr. Bahramnia forestry plan) stands. The previous forest silviculture treatment method in managed stands was shelterwood cutting (1972-1992), however, they are   currently treating by single tree selection and cutting (close to nature) method.

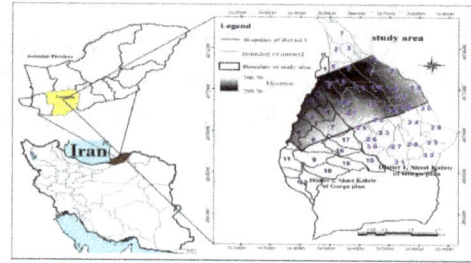

Fig.1: Location of study area in the Golestan province of Iran

2.2. Field data

We applied a systematic random sampling method to collect field data with 150×200 meter network (3.33% sampling intensity) and plots were circular with 0.1 ha. Totally, the 308 plots (219 plots in managed and 87 plots in unmanged stands) were measured in study area. In each plot, hieht of all trees with a diameter greater than 12.5 cm at breast height was measured by Vertex VL 402 device. The geographic center of each plot was recorded using a differential GPS (Trimble R3) device.

2.3. Lidar data:

Rayan Naqshe Company acquired the laser scanner data for this study under leaf on canopy condition on the 12th of October 2011using Riegl LMS Q5600 laser scanning system. The laser wavelength and mean point density are 1064 nm and about 4 m², respectively. In addition, the flight elevation was approximately 1000 m above the ground. For more information please look at the table 1.

Table 1: LMS Q560 Riegl charestrictistics

30	Minimum Range
20	Accuracy
10	Precision
NIR (1069)	Laser Wavelength
±22.5 , ±30	Scan Angle Range
≤ 0.5 mrad	Laser Beam Divergence
240≥	Laser puls speed (KRZ)
4	average measurement density
<50	Vertical accuracy
<30	Accuracy horizontal
IMU-lle	IMU
400≥	IMU sampling rate

2.4. Procesing Lidar data:

Rayan Naqshe Company using RayAnalyze software analyzed the full-wave form data. The delivery consist of point data coordinate together with additional parameters width, amplitude, number of target within laser beam and total number of target within laser beam. In addition, the points cloud Lidar data using RiProcess software classified in first, last and other pulses. Rayan Naqshe Company accomplished initial processing of laser data. After processing, it observed that there are many noises in the last, first and other pulses. Therefore, all outliers with exceeder than 50 meters in last, first and other pulses were removed using Terrascan software.

The last pulses data were used to create to the bare earth surface using Kraus and Pfeiffer and fusion software. A triangulated irregular network (TIN) was used to create a digital terrain model (DTM) using X, Y and Z (height) values of the individual terrain ground points acquired from Kraus and Pfeiffer algorithm (1998) with spatial resolution of 1m. The

DTM accuracy was assessed using 90 ground elevation points recorded by DGPS. The results showed that mean elevation differences between DTM and DGPS points were less than 40cm. In addition, a one-meter resolution digital surface model (DSM) was created using first and last pulses data by TIN algorithm in Fusion software. Then, canopy height model (CHM) was created using DTM and DSM models and applying 2 meters height break in Fusion software. The accuracy of CHM was assessed by height of 90 trees measured by Vertex VL 402 device. The results showed that the mean height difference between CHM and height of ground trees was less than 90cm. Then, according to previous study (Naesset, 2002, 2004) and Hureich and Thoma (2008), height metrics and canopy cover metrics were extracted using Fusion software with considering pixel size of 31.623×31.623 m corresponding to plot size (1000 m^2). Canopy height structure metrics in plot level were extracted by Fusion software (Table 2).

Table 2: Plot based canopy height structure metrics

Height metrics	
$h_{p0}, ... h_{p90}$	Height Percentiles
Maximum Height of points in plots	Hmax
Minimum Height of points in plots	Hmin
Mean Height of points in plots	Hmean
Median Height of points in plots	Hmedian
Coeficient variance of height points in plots	Hcv
Standard deviation of height points in plots	hSD
skewnes of height points in plots	Hskeewness
kurtosis of height points in plots	Hkurtosis
variance of height points in plots	Hvariance

Statistical significant analysis:

In order to comparing the hieght change and percent canopy cover status in the managed and unmanaged stands, General statistical measures and significant analyses were done by independent T test between two stands as well as between two same dominant forest types (i.e. Carpinus-Parrotia and Parrotia-Carpinus).

Results:

Table 3 shows the statistics measures of canopy cover metrics extracted on Lidar data in managed stand (seri 1) and unmanged stands (seri2). The statistics measures of canopy cover metric of forest types extracted from Lidar data in unmanaged and managed stands are showed in tables 4 and 5.

Table 3: Statistics measures of canopy cover metric extracted from Lidar data in managed and unmanged stands

Attribute	Seri	Plots	Mean (%)	Min (%)	Max (%)	Range (%)	SD (%)
Canopy cover (%)	I	219	98.992	52.811	100	47.189	3.389
	II	87	99.491	97.883	100	2.117	0.444

Table 4: Statistics measures of canopy cover metric of Carpinus-parrotia forest type extracted from Lidar data in managed and unmanged stands

Attribute	Seri	Stand type	Mean (%)	Min (%)	Max (%)	Range (%)	SD (%)
Canopy cover (%)	I	Carpiuns-parrotia	99.03	85.53	100	44.47	2.13
	II		99.49	98.37	99.98	1.61	0.37

Table 5: Statistics measures of canopy cover metric of Parrotia-carpinus forest type extracted from Lidar data in managed and

Attribute	Seri	Stand type	Mean (%)	Min (%)	Max (%)	Range (%)	SD (%)
Canopy cover (%)	I	Parrotia - Carpiuns	99.29	97.06	100	2.94	0.70
	II		99.36	97.88	99.97	2.08	0.54

unmanged stands

Results of significant different of percent canopy cover (extracted by Lidar data) between managed and unmanged stands using independent T-test showed that there is not significant different between two stands in 95 percent confidence level (table 6) as well as in same forest type (table 7 and 8).

Table 6: Results of significant analysis of percent canopy cover in managed and unmanged stands

Attribute	T statistics	Significant level
Percent canopy cover	-1.67	**0.173 [ns]**

[ns] insignificant in 95 percent confidence level

Table 7: Results of significant analysis of percent canopy cover of Parrotia- carpinus forest type in managed and unmanged stands

Attribute	T statistics	Significant level
Percent canopy cover	-1.176	0.244 [ns]

[ns] insignificant in 95 percent confidence level

Table 8: Results of significant analysis of percent canopy cover of Carpinus-parrotia forest type in managed and unmanged stands

Attribute	T statistics	Significant level
Percent canopy cover	-1.403	0.164 [ns]

[ns] insignificant in 95 percent confidence level

Results of significant different of canopy structure metrics extracted by Lidar data between managed and unmanaged stands using independent T-test (table 9) and forest type of both stands (tables 10 and 11) showed that in some canopy hight attributes there is significant different, however in some cases there is not insignificant different.

Table 9: Results of significant analysis of canopy height in managed and unmanged stands

Attribute	Significant level	T statistics
Minimum canopy height	0.121 [ns]	-1.555
Maximum canopy height	0·00000**	-6.53
Range canopy height	0·00000**	-3.576
Mean canopy height	0·00000**	-5.652
Standard deviation canopy height	0.078 [ns]	-1.769
Variance canopy height	0.063 [ns]	-1.866

** significant in 99 percent confidence level., [ns] insignificant in 95 percent confidence level

Table 10: Results of significant analysis of canopy height of Carpinus-parrotia forest type in managed and unmanaged stands

Attribute	Significant level	T statistics
Minimum canopy height	0.549 [ns]	-0.601
Maximum canopy height	0·000**	-4.459
Range canopy height	0·021**	-2.347
Mean canopy height	0·000**	-4.329
Standard deviation canopy height	0.886 [ns]	-0.143
Variance canopy height	0.726 [ns]	-0.351

** significant in 99 percent confidence level., [ns] insignificant in 95 percent confidence level

Table 11: Results of significant analysis of canopy height of Parrotia-carpinus forest type in managed and unmanged stands

Attribute	Significant level	T statistics
Minimum canopy height	0.254^{ns}	-1.15
Maximum canopy height	0.157^{ns}	-1.432
Range canopy height	0.956^{ns}	-0.056
Mean canopy height	0.068^{ns}	-1.858
Standard deviation canopy height	0.368^{ns}	-0.907
Variance canopy height	0.244^{ns}	-1.176

**significant in 99 percent confidence level., ^{ns} insignificant in 95 percent confidence level

Conclusion:

. In this study, different of canopy cover density and crown height variances between managed and unmanaged forest stands were investigated using Lidar data and statistical significant analysis. The significant analyses were done by independent T-test between two stands in same dominant forest types. Results showed that there are no significant differences between canopy cover mean in two stands as well as forest types. Result of statistically analysis on height characteristics showed that there are a decreasing the forest height and its variance in the managed forest compared to unmanaged stands. In addition, there is a significant difference between maximum, range, and mean heights of two stands in 99 percent confidence level. However, there is no significant difference between standard deviation and canopy height variance of managed and unmanged stands. These results showd that accomplished management treatments and cuttings could lead to reducing of height variances and converting multi-layers stands to two or single layers. Results are also showed that the canopy cover densities in the managed forest stands are changing from high dense cover to dense cover.

References:

Anderson, E. S., Thompson, J.A., Austin, R.E. 2005. Lidar density and linear interpolator effects on elevation estimates. International Journal of Remote Sensing, 26 (18): 3889–3900.

-Axelsson, P. 1999. Processing of laser scanner data—algorithms and applications. ISPRS Journal of Photogrammetry and Remote Sensing. 54 (2–3): 138–147.

- Ackermann, F. 1999. Airborne laser scanning—present status and future expectations. ISPRS Journal of Photogrammetry and Remote Sensing, 54: 64–67.

- Baltsavias, E.P. 1999. A comparison between photogrammetry and laser scanning. ISPRS Journal of Photogrammetry and Remote Sensing, 54: 83–94

- Cartus, O., Kellndorfer, J., Rombach, M., and Walker, W. 2012. Mapping Canopy Height and Growing Stock Volume Using Airborne Lidar, ALOS PALSAR and Landsat ETM+. Remote Sensing, 4: 3320-3345

-Cuesta, J., Chazette, P., Allouis, T., Flamant, P. H., Durrieu, S., Sanak, J., Genau, P., Guyon, D., Loustau, D. and Flamant, C. 2010. Observing the Forest Canopy with a New Ultra-Violet Compact Airborne Lidar. Sensors, 2010, 10: 7386-7403

-Erdody, T. L., and Moskal, L. M. 2009. Fusion of lidar and Imagery for Estimating Canopy Fuel Metrics in Eastern Washington Forests, American Society for Photogrammetry & Remote Sensing 2009 Annual Conference, Baltimore, MD, March 2009.

- Gaveau, D. L. A., and Hill, R. A. 2003. Quantifying canopy height under estimation by laser pulse penetration in small-footprint airborne laser scanning data. Canadian Jornal of Remote Sensing, 29: 650-657.

- Haala, N., and Brenner, C. 1999. Extraction of buildings and trees in urban environments. ISPRS Journal of Photogrammetry and Remote Sensing, 54:130–137.

- Hawbaker, T.J., Gobakken, T., Lesak, A., Trømborg, E., Contrucci, K., and Radeloff, V. 2010. Light detection and ranging-based measures of mixed hardwood forest structure. Forest Science, 56(3): 313-326, 14.

- Hoffman-Wellenhof, B., Lichtenegger, H., Collins, J. 2001. Springer–Wien, New York. 371p.

- Heritage, G.L., and Large, A.R.G. 2009. Laser scanning for the environmental sciences. Wiley-Blackwell press. 278p.

- Heurich, M., and Thoma, F. 2008. Estimation of forestry stand parameters using laser scanning data in temperate, structurally rich natural European beech (*Fagus sylvatica*) and Norway spruce (*Picea abies*) forests. Forestry, 81(5): 645-661.

- Holmgren, J. 2004. Prediction of tree height, basal area and stem volume in forest stands using airborne laser scanning. Scandinavian Journal of Forest Research, 19(6): 543 – 553.

- Hyyppä, J., Hyyppä, H., Litkey, P., Yu, X., Haggrén, H., Rönnholm, P., Pyysalo, U.,. Pitkänen, J., and Maltamo, M. 2004. Algorithms and methods of airborne laser scanning for forest measurements. International Archives of Photogrammetry, Remote Sensing and Spatial Information Sciences XXXVI (8/W2), 82–89.

- IGI mbh. 2013. AERO office, GPS/IMU processing, Direct Georefrecing and intergrated sensor orientation, www.igi-systems.com, Germany, 1p.

- Kane, V.R., McGaughey, R. J., Bakker, J.D., Gersonde, R.F., Lutz, J.A., and Franklin, J.F. 2010. Comparisons between field- and LiDAR-based measures of stand structural complexity. Can. J. For. Res. 40(4): 761-773. doi:l0.1139/X10-024.

- Kraus, K., and Pfeifer, N. 1998. Determination of terrain models in wooded areas with airborne laser scanner data, ISPRS J. Photogrammetry remote sensing. 53, 193-203.

- Lemmens, M. 2011. Airborne lidar. In: Gatrell, J.D., Jensen, R.R. (Eds.), Geoinformation, Geotechnologies and the Environment, vol. 5. Springer, Netherlands, pp. 153–170.

- Lloyd, C.D., Atkinson, P.M. 2002. Deriving DSMs from LiDAR data with kriging. International Journal of Remote Sensing. 23, 2519–2524.

- Liu, X. 2008. Airborne LiDAR for DEM generation: some critical issues. Progress in Physical Geography. 32, 1, 31–49.

- Magnusson, M. 2006. Evelution of remote sensing techniques for estimation of forest variables at stand level. PhD, Thesis, Swedish University of agricultural sciences, Umea, 38p.

- Maguya, A.S., Junttila, V., and Kauranne, T. 2013. Adaptive algorithm for large scale dtm interpolation from lidar data for forestry applications in steep forested terrain. ISPRS Journal of Photogrammetry and Remote Sensing. 85, 74–83.

- Marks, K., Bates, P. 2000. Integration of high resolution topographic data with floodplain flow models. Hydrological processes. 14, 2109-2122.

- Meng, X., Currit, N., Zhao, K. 2010. Ground filtering algorithms for airborne LiDAR data: a review of critical issues. Remote Sensing. 2, 833–860.

- Mongus, D., Žalik, B. 2012. Parameter-free ground filtering of LiDAR data for automatic DTM generation. ISPRS Journal of Photogrammetry and Remote Sensing. 67, 1–12.

- Moskal, L. M., Erdody, T., Kato, A., Richardson, J., Zheng, G. and Briggs, D. 2009. Lidar applications in precision forestry. Silvilaser2009, Washington, USA, http://depts.washington.edu/rsgal/pubs/Moskaletal_Silvilaser2009.pdf.

- Næsset, E., and Bjerknes, K.O. 2001. Estimating tree heights and number of stems in young forest stands using airborne laser scanner data. Remote Sensing of Environment, 78: 328-340.

- Næsset, E. 2002. Predicting forest stand characteristics with airborne scanning laser using a practical two-stage procedure and field data. Remote Sensing of Environment. 80, 88-99.
- Næsset, E. 2004. Accuracy of forest inventory using airborne laser scaning: evaluating the first Nordic full-scale operational project. Scandinavian Journal of Forest Research. 19, 6, 554-557.
- Naesset, E. 2007. Airborne laser scanning as a method in operational forest inventory: Status of accuracy assessments accomplished in Scandinavia. Scandinavian Journal of Forest Research, 22 (5): 433 – 442.
-Næsset, E. 2011. Estimating above-ground biomass in young forests with airborne laser scanning. International Journal of Remote Sensing, 32, 473–501.
- Nelson, R., Krabill, W., and Tonelli, J. 1988. Estimating forest biomass and volume using airborne laser data. Remote Sensing of Environment, 24: 247–267.
- Price, W.F., and Uren, J. 1989. Laser surveying, London Van Nostrand reinhold international.
- Rutledge, A. M. and Popescu, S. C. 2006. Using Lidar in determining forest canopy parameters. ASPRS Annual Conference, Reno, Nevada, May 1-5.
- Shataee, Sh. 2013. Forest attributes estimation using aerial laser scanner and TM data, Forest systems journal, 22(3): 484-496.
- Smart, L.S., Swenson, J.J., Christensen, N.L., and Sexton, J. O. 2012. Three-dimensional characterization of pine forest type and red-cockaded woodpecker habitat by small-footprint, discrete-return lidar. Forest Ecology and Management, 281: 100-110.
- Smith, S. L., Holland, D. A., and Longley, P. A. 2005. Quantifying interpolation errors in urban airborne laser scanning models. Geographical Analysis, 37: 200–224.
- Smith-voysey, S. 2006. Laser scaning (Lidar): A tool for future data collection, Ordance survey research labs annual review 2005-06, Southampton: Ordnance survey.
- Tonolli, S., Dalponte, M., Neteler, M., Rodeghiero, M., Vescovo, L., and Gianelle, D., 2011b. Fusion of airborne LiDAR and satellite multispectral data for the estimation of timber volume in the Southern Alps. Remote Sensing of Environment, 115: 2486–2498.
- Wehr, A., Lohr, U. 1999. Airborne laser scanning an introduction an overview. Journal of Photogrammetry and Remote Sensing, 54: 68-82.
- Vosselman, G. 2000. Slope based filtering of laser altimetry data. International Archives of Photogrammetry, Remote Sensing and Spatial Information Sciences XXXIII (B3/2: Part 3), 935–942.
- Zhao, K., Popescu, S., Meng, X., Pang, Y., and Agca, M. 2011. Characterizing forest canopy structure with lidar composite metrics and machine learning. Remote Sensing of Environment, 115:1978–1996

26

QUASI-EPIPOLAR RESAMPLING OF HIGH RESOLUTION SATELLITE STEREO IMAGERY FOR SEMI GLOBAL MATCHING

Nurollah Tatar [a]*, Mohammad Saadatseresht[a], Hossein Arefi[a], Ahmad Hadavand[a]

[a] School of Surveying and Geospatial Information Engineering, College of Engineering, University of Tehran

Commission III, WG III/1

KEY WORDS: Epipolar resampling, SGM, High resolution satellite stereo images, Small image tiles.

ABSTRACT

Semi-global matching is a well-known stereo matching algorithm in photogrammetric and computer vision society. Epipolar images are supposed as input of this algorithm. Epipolar geometry of linear array scanners is not a straight line as in case of frame camera. Traditional epipolar resampling algorithms demands for rational polynomial coefficients (RPCs), physical sensor model or ground control points. In this paper we propose a new solution for epipolar resampling method which works without the need for these information. In proposed method, automatic feature extraction algorithms are employed to generate corresponding features for registering stereo pairs. Also original images are divided into small tiles. In this way by omitting the need for extra information, the speed of matching algorithm increased and the need for high temporal memory decreased. Our experiments on GeoEye-1 stereo pair captured over Qom city in Iran demonstrates that the epipolar images are generated with sub-pixel accuracy.

1. INTRODUCTION

Rectified images, resampled along the epipolar line to omit y-parallax, which are called epipolar images are considered as input to SGM algorithm (Hirschmüller, 2008). Using epipolar images increase the speed of stereo matching by reducing the search space from 2 dimensional to 1 dimensional. Knowing the properties of epipolar line is critical to build epipolar image. Epipolar line in images taken by traditional frame cameras appear as lines. In linear array scanners, image is formed during the sensor movement along its orbit. So as presented in Figure 1 there is multiple projection centres for one image scene. As we consider on image point in left scene, its conjugate location in right scene is determined as c, b, d and e, which are obtained by changing the correspondence object point on the ray of "a". These points creates the epipolar curve of point " a' " in right scene.

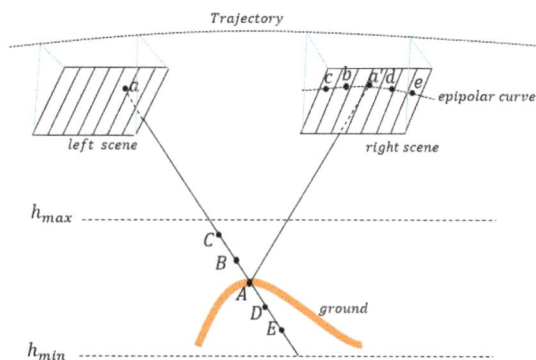

Figure 1. Epipolar geometry in linear array scanners

Epipolar curve as showed in Figure 1 is depends to motion equation of sensor. Studies on orbital parameters of sensors demonstrates that the epipolar line appears as quasi-hyperbola (Kim, 2000). To determine this curve one needs to access accurate orbital parameters of sensor. Image providers usually doesn't provide these information as public. Using ground control points (GCPs) is an alternative to determine epipolar curve (Lee and Park, 2002). Also terrain-independent rational polynomial coefficients (RPCs) are proposed to overcome the

problem of field work in measuring GCPs (Grodecki and Dial, 2003).

In high resolution satellite imagery, an image scene are taken just a few seconds. In this way we can assume constant altitude and constant velocity condition for each image scene. It is proved that under this condition and after parallel projection a simple 2D-affine transformation model suffices to determine epipolar images (Habib et al., 2005; Morgan et al., 2006; Morgan et al., 2004a, b; Ono, 1999). Determining the parallel projection transformation is based on RPCs and digital terrain model. RPCs also helps to create virtual control points and use 2D affine transformation to epipolar resampling (Oh et al., 2006).

Epipolar images could be generated in object space on a reference plan with constant height value equals to mean height of area. In this method the relation between the original image and generated epipolar image is established by RPCs (Wang et al., 2011). RPCs also could be used to find conjugate points between images and build epipolar image in image space (Oh, 2011; Wang et al., 2010; Zhao et al., 2008). The weak point of using RPCs is their need to correction for bias and drift (Fraser and Hanley, 2003).

In this paper a new solution is proposed to handle the epipolar resampling as the input of SGM algorithm. Our proposed algorithm omits the need for RPCs and improve the SGM performance by reducing the memory required by this algorithm.

2. SEMI-GLOBAL MATCHING ALGORITHM

SGM is known as a powerful pixel-wise stereo matching algorithm with reliable performance in photogrammetric applications. SGM comprised of four main steps including: matching cost computation, cost aggregation, disparity map generation and disparity refinement. Here we briefly discuss matching cost computation and cost aggregation.

Respect to disparity range, every pixel in left epipolar image could be matched by a pixel in right epipolar image. Correspondence degree of points is measured by a cost function. Cross correlation, sum of absolute differences, census and mutual information are some examples. The success of SGM depends on the cost function. After selection of cost function and minimum and maximum of disparity range, for each pixel in left epipolar image, a vector of matching costs are calculated. This process as shown in Figure 2 for whole image, results a matching costs cube which its dimension equals to left epipolar image size and its height equals to disparity range.

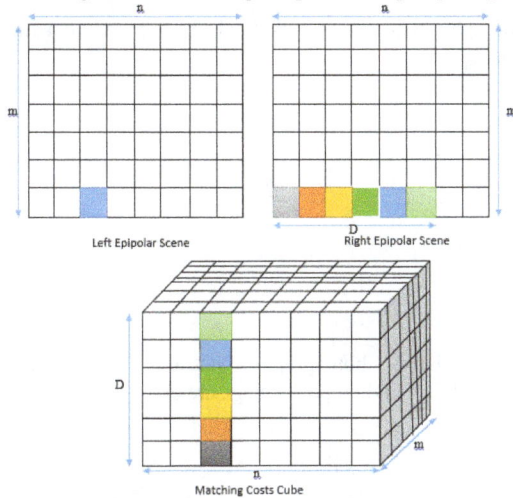

Figure 2. Matching cost computation and costs cube formation in SGM algorithm

Capability of matching based on cost computation, is reduced due to equal cost value for neighbour pixels, lack of texture in objects and deficiency of cost function. So the aggregation of matching cost from several paths have been proposed. In this step, based on matching cost and disparity in 8 or 16 paths matching costs aggregated. Figure 3 shows the aggregating matching costs in 8 paths.

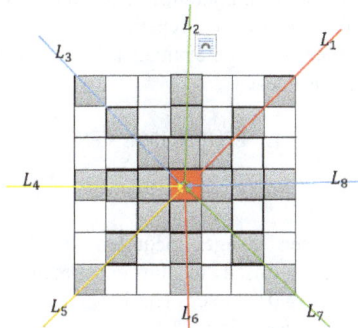

Figure 3. Matching cost aggregation in 8 paths

Equation 1 is used to aggregate the matching costs in each path. Then using equation 2 the costs in all paths are aggregated.

$$L_r(p,d) = C(p,d) + \min \begin{cases} L_r(p-r,d) \\ L_r(p-r,d\pm1) + p_1 \\ \min_k L_r(p-r,k) \\ \min_i L_r(p-r,i) + p_2 \end{cases} \quad (1)$$

Where;

p: image location of interest pixel
d: disparity value
$Lr(p,d)$: cost path toward the actual pixel of path
$C(p,d)$: pixel-wise matching cost
$P1$: a small value penalizing disparity changes between neighbouring pixels of one pixel

$P2$: a large value penalizing disparity changes between neighbouring pixels of one pixel
r: actual path
k: pixels in each path

$$S(p,d) = \sum_{r=1}^{8 \, or \, 16} L_r(p,d) \quad (2)$$

These equations provides aggregated cost values for each element in matching costs cube which is called aggregated cost cube. Now the disparity for each pixel could be calculated by minimizing aggregated cost values, using the following equation:

$$D = \min_d S(p,d) \quad (3)$$

To estimate the disparity in sub-pixel level, a quadratic curve is fitted to the neighbouring costs and the position of minimum is calculated.

SGM algorithm needs large amount of temporary memory for saving matching costs cube and aggregated costs cube. The size of temporary memory depends on the image size and the disparity range. The solution which has been proposed by SGM is to divide the epipolar images into small image tiles. This idea also followed in our proposed method.

3. PROPOSED METHOD

Based on flowchart of our proposed method have 4 main steps. In first step, due to the need of SGM algorithm to small tiles, a new method is proposed.

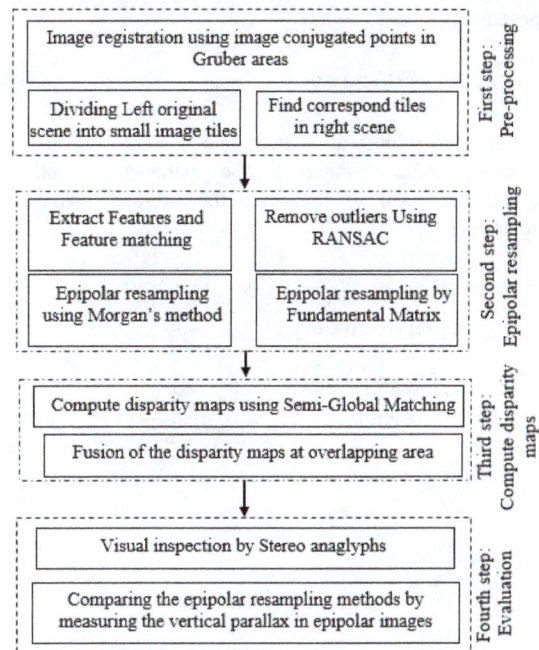

Figure 4. Flowchart of proposed method to create epipolar image as input of SGM algorithm

In the pre-processing step the left and right images are registered. Affine transformation is employed to model the geometrical relationship between images. An automated procedure is used to estimate affine transformation parameters. SURF operator (Bay et al., 2008) is used to extract reliable corresponding key features in both images. Then the RANSAC algorithm (Fischler and Bolles, 1981) is used to detect and remove outliers through the features. While the registration is solved, the left image is divided to small overlapping image tiles as presented in Figure 5.

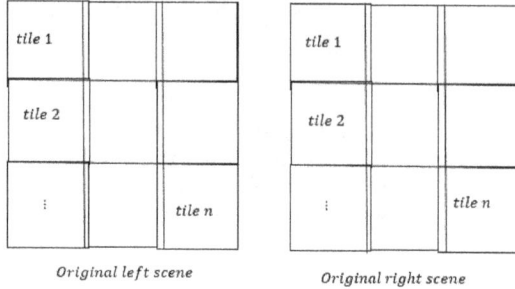

Figure 5. Dividing original images into small image tiles

For the image tiles in left image, correspond tile should be found in right image, using affine transformation.

In the second step the epipolar image is generated for each corresponding image tile. Epipolar geometry in small image tiles could be assumed as straight line. Fundamental matrix (Loop and Zhang, 1999) and epipolar line equation (Morgan et al., 2004b) is could be used to create epipolar image. To estimate the fundamental matrix or epipolar line parameters we need correspond points in each pair of image tiles. SURF feature extractor followed by RANSAC algorithm such as previous step is employed here.

Epipolar resampling method, used here is originally proposed by Morgan (Morgan et al., 2004b). He proved that for images with parallel projection, epipolar lines could be defined using an affine transformation. 2D affine equation as below is proposed to this end:

$$G_1 x + G_2 y + G_3 x' + G_4 y' + 1 = 0 \qquad (4)$$

Four conjugate points are required to estimate the unknown parameters. After computing the parameters, following equations are used to transform the original images into epipolar ones.

$$\begin{bmatrix} x \\ y \end{bmatrix}_{Epipolar} = \frac{1}{S} \begin{bmatrix} \cos(\theta) & \sin(\theta) \\ -\sin(\theta) & \cos(\theta) \end{bmatrix} \begin{bmatrix} x \\ y \end{bmatrix}_{original} + \begin{bmatrix} 0 \\ -\frac{\Delta y}{2} \end{bmatrix} \qquad (5)$$

$$\begin{bmatrix} x' \\ y' \end{bmatrix}_{Epipolar} = S \begin{bmatrix} \cos(\theta') & \sin(\theta') \\ -\sin(\theta') & \cos(\theta') \end{bmatrix} \begin{bmatrix} x' \\ y' \end{bmatrix}_{original} + \begin{bmatrix} 0 \\ \frac{\Delta y}{2} \end{bmatrix} \qquad (6)$$

Where,

$$\theta = \arctan(-\frac{G_1}{G_2}) \qquad (7)$$

$$\theta' = \arctan(-\frac{G_3}{G_4}) \qquad (8)$$

$$S = \sqrt{-\frac{G_4 \cos(\theta)}{G_2 \cos(\theta')}} \qquad (9)$$

$$\Delta y = -\frac{S \times \cos(\theta')}{G_4} \qquad (10)$$

Fundamental matrix is also used to produce epipolar images (Loop and Zhang, 1999). Fundamental matrix relates the corresponding points in stereo images. Fundamental matrix

solves this relationship without the need for interior and relative orientation parameters. Using this matrix, the relationship between image coordinates in stereo images is expressed by:

$$\begin{bmatrix} x & y & 1 \end{bmatrix} \begin{bmatrix} F_1 & F_2 & F_3 \\ F_4 & F_5 & F_6 \\ F_7 & F_8 & 1 \end{bmatrix} \begin{bmatrix} x' \\ y' \\ 1 \end{bmatrix} = 0 \qquad (11)$$

Homography matrices are derived from above equation as follows:

$$X^T H^T i_x H' X' = 0 \Rightarrow F = H^T [i]_x H' \qquad (12)$$

In which;

$$[i]_x = \begin{bmatrix} 0 & 0 & 0 \\ 0 & 0 & -1 \\ 0 & 1 & 0 \end{bmatrix} \qquad (13)$$

After calculating homography matrices, the relationship between original and epipolar image are expressed by:

$$\begin{bmatrix} x' \\ y' \\ 1 \end{bmatrix}_{Epip} = \begin{bmatrix} h_1 & h_2 & h_3 \\ h_4 & h_5 & h_6 \\ h_7 & h_8 & h_9 \end{bmatrix} \begin{bmatrix} x \\ y \\ 1 \end{bmatrix}_{Orig} \Rightarrow X_{epip} = H X_{orig} \qquad (14)$$

$$X_{orig} = H^{-1} X_{epip} \qquad (15)$$

Next step is to calculation of disparity map from epipolar images. SGM algorithm is employed in this step. Tuning of P1, P2 and disparity range is critical in SGM algorithm. P1 and P2 are necessary in calculating aggregated costs and disparity range is essential in matching cost cube generation. Higher disparity range results in high computation cost and also increase the chance of outliers caused by repetitive textures in the image scene. Finally the results are evaluated by comparing y-parallax values for epipolar images.

4. EXPERIMENTS

4.1 Dataset

Panchromatic stereo images acquired by GeoEye-1 high resolution sensor is used in the experiments. The spatial resolution of data is 0.5 meter. The image is captured over an urban area in Qom city in Iran.

4.2 Pre-processing

In pre-processing step, to avoid the high computational load of SURF algorithm, the image pyramid is built for original images. In lower levels of pyramid SURF algorithms is employed to detect correspond key features which are used to estimate an approximate affine transformation between images. Then on original left image, 9 Gruber areas are considered and their correspond areas on right image is delineated by affine transformation. SURF algorithm is used on correspond Gruber areas as showed in Figure 6 and the accurate affine transformation between images is determined.

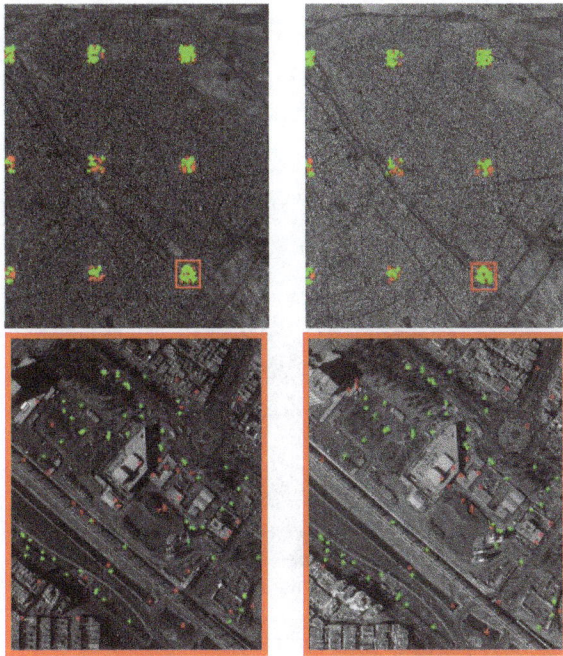

Figure 6. Key features extracted in 9 Gruber areas for corresponding original images

After co-registration of images, left image is divided into 3000*3000 tiles and their corresponding areas are found using the 2D affine transformation obtained in previous step. There would be 30 image tiles in each image used in the experiments.

4.3 Epipolar resampling and computing disparity map

Now in each correspond image tile key features are detected such as previous step and using epipolar line equation obtained by Morgan method and fundamental matrix the epipolarly resampled images are generated. Figure 7 shows the epipolar images obtained by both methods.

After creation of epipolar images for each corresponding tile, SGM algorithm is used to compute disparity map. As mentioned earlier, disparity range is a critical input for SGM algorithm. Its value depends on the height variation in the image scene. By tiling the original image, height variation in each image tile is decreased compared with the original one. So using optimum values for disparity range will produce good results, with lower computation cost and also prevents mismatches due to repetitive texture. Best disparity range value for each image tile obtained manually by visual inspection are listed in Table 1.

Table 1. Selected disparity range for each image tile

Tile number*	Disparity range
19, 25, 29, 30	[0, 64]
5, 12, 18, 24, 28	[0, 198]
Other tiles	[0, 128]

*The numbering strategy is presented in Figure 5

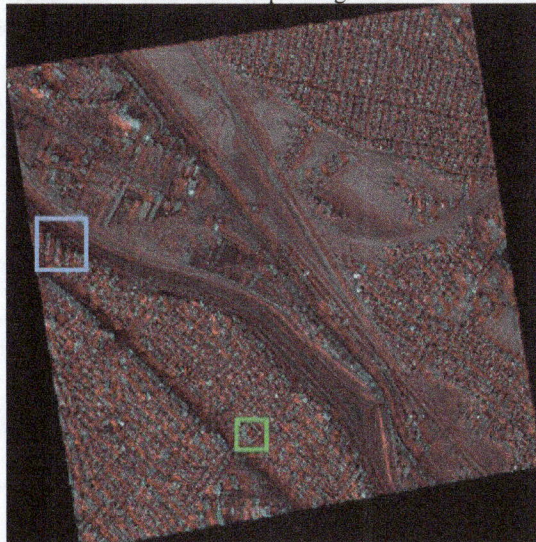

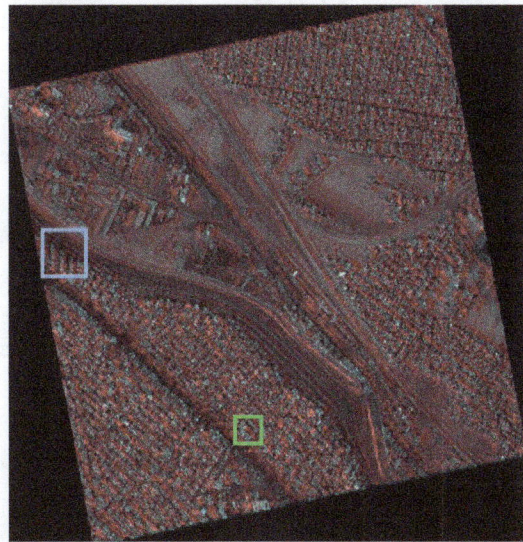

Figure 7. Stereo anaglyph of the generated epipolar images by fundamental matrix (left) and Morgan's method (right)

Figure 8 contains the computed disparity map for both epipolar resampling methods.

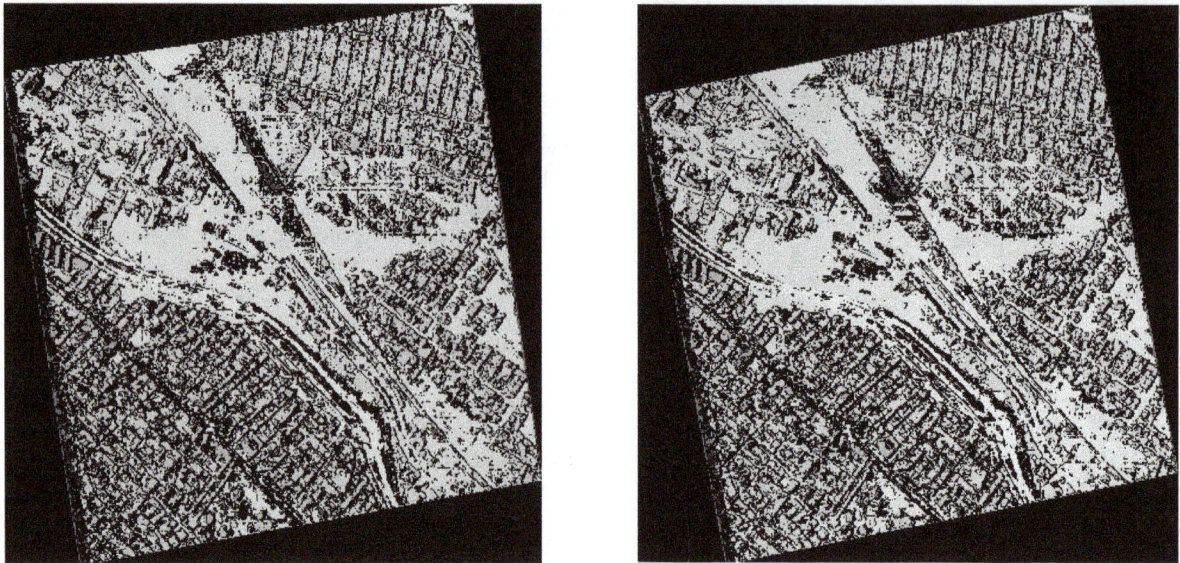

Figure 8. Disparity map computed by SGM algorithm for epipolar images produced by fundamental matrix (left) and Morgan's method (right)

4.4 Evaluation of results

Y-parallax values for correspond points in epipolar images are used to evaluate the results. Correspond points are generated through the SURF and RANSAC algorithm as proposed in previous steps. Generated correspond points and their y-parallax value are presented in Figure 9. Statistical analysis on y-parallax values are shown in Table 2.

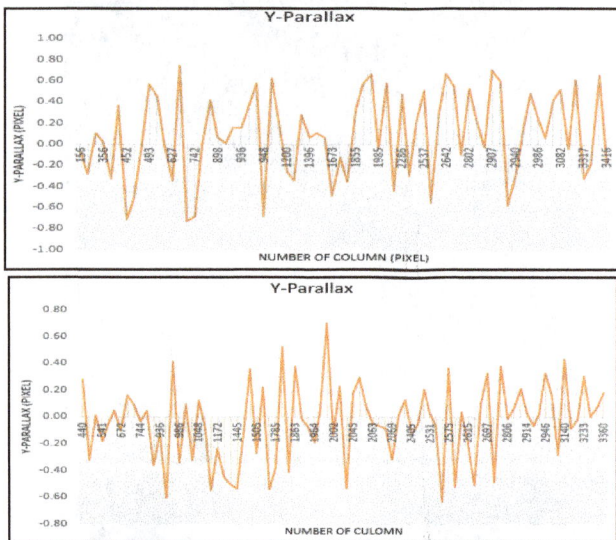

Figure 9. Y-parallax of conjugate image points on the generated epipolar images by fundamental matrix (top) and Morgan's method (bottom)

Table 2. Statistical analysis of y-parallax values

Epipolar resampling method	Maximum y-parallax (pix)	Mean y-parallax (pix)	Standard deviation of y-parallaxes (pix)
Morgan's method	0.78	0.25	0.34
Fundamental matrix	0.80	-0.33	0.33

To evaluate the performance of SGM algorithm in disparity computation, the disparity value for 80 correspond points are measured manually. Then these values are compared to the values computed by SGM algorithm. Evaluation results for both epipolar resampling methods are presented in Table 3.

Table 3. Statistical analysis of SGM algorithm

Epipolar resampling method	*Mean (Px-d)	Standard deviation (Px-d)
Morgan's method	0.02	0.44
Fundamental matrix	0.09	0.54

*Px: x-parallax; d: disparity value

5. CONCLUDING REMARKS

In this paper we propose and analyse new epipolar resampling methods for high resolution stereo images. By dividing original images into small image tiles, the epipolar geometry in pushbroom images is simplified to straight line. Also using small tiles in SGM algorithm omits the need for high disparity change values. Consequently it increases the speed of algorithm and decrease wrongly matched points due to repetitive texture. Result of experiments in Table 2 demonstrates that epipolar images are produced with sub-pixel accuracy.

SURF and RANSAC algorithms with computer vision basis are used in the proposed method to automate the process of finding corresponding points. These algorithms omit the need for RPCs in this process.

The effect of epipolar resampling method on the result of image matching is presented in Table 3. It also demonstrates that the proposed methods have acceptable performance in SGM image matching.

REFERENCES

Bay, H., Ess, A., Tuytelaars, T., Van Gool, L., 2008. Speeded-up robust features (SURF). Computer vision and image understanding 110, 346-359.

Fischler, M.A., Bolles, R.C., 1981. Random sample consensus: a paradigm for model fitting with applications to image analysis and automated cartography. Communications of the ACM 24, 381-395.

Fraser, C.S., Hanley, H.B., 2003. Bias compensation in rational functions for IKONOS satellite imagery. Photogrammetric Engineering & Remote Sensing 69, 53-57.

Grodecki, J., Dial, G., 2003. Block adjustment of high-resolution satellite images described by rational polynomials. Photogrammetric Engineering & Remote Sensing 69, 59-68.

Habib, A.F., Morgan, M.F., Jeong, S., Kim, K.-O., 2005. Epipolar geometry of line cameras moving with constant velocity and attitude. Electronics and Telecommunications Research Institute Journal 27, 172-180.

Hirschmüller, H., 2008. Stereo processing by semiglobal matching and mutual information. Pattern Analysis and Machine Intelligence, IEEE Transactions on 30, 328-341.

Kim, T., 2000. A study on the epipolarity of linear pushbroom images. Photogrammetric engineering and remote sensing 66, 961-966.

Lee, H.-Y., Park, W., 2002. A New Epipolarity Model Based on the Simplified Push-Broom Sensor Model, International Archives of the Photogrammetry, Remote Sensing and Spatial Information Sciences, Ottawa. Canada, pp. 631-636.

Loop, C., Zhang, Z., 1999. Computing rectifying homographies for stereo vision, Computer Vision and Pattern Recognition, IEEE Computer Society Conference on. IEEE, Fort Collins, CO, USA.

Morgan, M., Kim, K.-O., Jeong, S., Habib, A., 2006. Epipolar resampling of space-borne linear array scanner scenes using parallel projection. Photogrammetric Engineering & Remote Sensing 72, 1255-1263.

Morgan, M., Kim, K., Jeong, S., Habib, A., 2004a. Epipolar geometry of linear array scanners moving with constant velocity and constant attitude, International Archives of the Photogrammetry, Remote Sensing and Spatial Information Sciences, Istanbul, Turkey, pp. 52-57.

Morgan, M., Kim, K., Jeong, S., Habib, A., 2004b. Indirect epipolar resampling of scenes using parallel projection modeling of linear array scanners, International Archives of the Photogrammetry, Remote Sensing and Spatial Information Sciences, pp. 52-57.

Oh, J.-H., Shin, S.-W., Kim, K.-O., 2006. Direct epipolar image generation from IKONOS stereo imagery based on RPC and parallel projection model. Korean Journal of Remote Sensing 22, 451-456.

Oh, J., 2011. Novel Approach to Epipolar Resampling of HRSI and Satellite Stereo Imagery-based Georeferencing of Aerial Images. The Ohio State University, The Ohio State, USA.

Ono, T., 1999. Epipolar resampling of high resolution satellite imagery, Sensors and Mapping from Space, Hannover, Germany.

Wang, M., Hu, F., Li, J., 2010. Epipolar arrangement of satellite imagery by projection trajectory simplification. The Photogrammetric Record 25, 422-436.

Wang, M., Hu, F., Li, J., 2011. Epipolar resampling of linear pushbroom satellite imagery by a new epipolarity model. ISPRS Journal of Photogrammetry and Remote Sensing 66, 347-355.

Zhao, D., Yuan, X., Liu, X., 2008. Epipolar line generation from IKONOS imagery based on rational function model, International Archives of the Photogrammetry, Remote Sensing and Spatial Information Sciences, pp. 1293-1297.

27

UAV PHOTOGRAMMETRY:
A PRACTICAL SOLUTION FOR CHALLENGING MAPPING PROJECTS

M. Saadatseresht [a] *, A.H. Hashempour[b], M. Hasanlou[a]

[a] School of Surveying and Geospatial Engineering, University of Tehran, Tehran, Iran – (msaadat, hasanlou)@ut.ac.ir
[b] Robotic Photogrammetry Research Group, Close Range Photogrammetry Lab., University of Tehran, Tehran, Iran – hashempour_amir@yahoo.com

ICWG I/V-B

KEY WORDS: UAV Photogrammetry, Large Scale Mapping, Unmanned Aerial Vehicle, Close Range Photogrammetry.

ABSTRACT

We have observed huge attentions to application of unmanned aerial vehicle (UAV) in aerial mapping since a decade ago. Though, it has several advantages for handling time/cost/quality issues, there are a dozen of challenges in working with UAVs. In this paper, we; as the Robotic Photogrammetry Research Group (RPRG), will firstly review these challenges then show its advantages in three special practical projects. For each project, we will share our experiences through description of the UAV specifications, flight settings and processing steps. At the end, we will illustrate final result of each project and show how this technology could make unbelievable benefits to clients including 3D city realistic model in decimetre level, ultra high quality map production in several centimetre level, and accessing to a high risk and rough relief area for mapping aims.

1. INTRODUCTION

Since last decade ago, UAV Photogrammetry has been under attention as a new technology for topographic mapping (Blom, 2006, Oliver and Money, 2001). The main reasons are less cost, more safety, higher quality, more popularity and more adoptablity for mapping of relatively small distributed areas. To achieve these advantages in UAV photogrammetry, some challenges should be resolved masterly such as correct designing due to limited space, weight and component placement in UAV body, operational aspects in take-off, flight and landing, periodical pre/post flight tests, experienced and skilful pilot, flight and photography licences, huge and complex data processing issue, and flight limitations due to aerial and terrestrial direct insights, topography, wind, light, site conditions, telecommunication limits, flight height and duration.

The Robotic Photogrammetry Research Group (RPRG) has been established by authors in University of Tehran, School of Surveying and Geospatial Engineering since 2014 to design and develop the unmanned aircraft and terrestrial mobile mapping systems. Before that, since six years ago, we have developed several fixed and rotatory wing UAVs for mapping aims. Now, dealing with design and development of different fixed wing UAV photogrammetric systems through practical experiments, progressively enhance them. This paper focuses on the result of different mapping projects done by our developed UAV photogrammetric systems. Fortunately, all these mapping projects have faced to some challengeable aspects so that conventional surveying methods might be impractical such as time/cost limits, quality demands and rough working conditions and requested outputs.

In this paper, details of map production line of three different projects are described including project definition, special challenges of the project, UAV system specifications, ground station establishment, ground control point (GCP) and flight design, flight operations, raw data pre-processing, image feature extraction and matching, network formation, aerial triangulation,

dense point cloud generation, orthoimage-mosaic and 3D textured model generation, and linear map generation.

In continue, this paper will firstly review the advantages of using UAV photogrammetry relative to field surveying for mapping aims. Then it will explain various challenges in working with UAV. In the following, the details of three practical projects will be described including 3D modelling of Khorramabad's MASKAN-MEHR construction site, large scale mapping of the Siman Fars manufacture site, and mapping of very hard monotonous geological Koomeh area. At the end, a discussion and conclusion on how could UAV photogrammetry technology provide unbelievable result and unique solution for mapping projects.

2. UAV PHOTOGRAMMETRY

2.1 UAV Photogrammetry Advantages

There are two types of UAV for mapping applications including multi-rotor and fixed-wing drones (Fahlstrom and Gleason, 2012). The multi-rotor drones capable to have higher payload that make it possible to carry heavy precise camera and even LiDAR sensor but less endurance. This causes the multi-rotor drones are generally utilized for very high accuracy mapping from small areas less than 100 hectares such as architectural 3D fine reconstruction applications. In contrast, fixed-wing drones are lighter and faster with more endurance which make them more proper for aerial imaging from larger areas usually from several hundreds to thousands hectares. Therefore, here we list some advantages of fixed-wing UAV photogrammetry in comparison to field surveying:

Higher quality and reliability of spatial products: high overlapped vertical aerial images of UAV photogrammetry are capable to make higher dense and more reliable spatial products. For example, in field surveying to generate 1:2000 map scale, 3D

---

* Corresponding author

topographic points are collected each 20 meters but multi image matching gives us 3D points each 0.5 meters which means there are about 1500 times more data. In addition, UAV photogrammetry not only uses several redundant image coordinates to compute a 3D point but also, it is possible to evaluate the accuracy of reconstructed point in the office via stereoscopic vision that means the result is more reliable. In opposite, the field surveyor usually has no choice but to confide totally to collected data in office.

More diversity of spatial products: by using field surveying, only linear maps and DTM can be produced. However, UAV photogrammetry is capable to produce more additional outputs including ortho-image-mosaic, image-map, 3D textured realistic model, high density coloured point cloud, and 3D flight simulation video.

More user friendly: ortho-image-maps and 3D realistic models are more user friendly for public users because all objects and features are displayed by their realistic apparent which causes simpler map interpretation and user interaction. In contrast, conventional linear maps have symbolic cartography that makes them more usable only to professional users.

Speed up the mapping process: in mapping projects with very high map scales e.g. 1:200 or with very large mapping areas e.g. several thousands hectares, the time period needed for UAV photogrammetry is several times less than field surveying. The reasons are (1) positioning of camera stations could be done by PPK (post processing kinematic) method without any field activities and (2) collection of object details rapidly is done by vertical aerial imaging via an drone, instead of manual sighting and observing of each object one by one.

Legal validity and consistency checks: imaging is a direct automatic data collection method without human interaction which makes it a legal credit and reference. If an image is manipulated and defaced, then non-consistency to other overlapped images will reveal it. Additionally, stereoscopic vision on raw images sharply illustrates the ground truth that is more reliable than linear maps produced by field surveyor.

More reasonable: although, UAV photogrammetry is a faster method for making higher quality, more reliable, and more diversity of products with legal validity, it costs several times less than field surveying due to its process automation without any huge field workings.

Less interruptions in operations: field surveying generally suffers from field interruptions in population areas due to continuously owner allowances to access the private and governmental regions. Since drones fly over such areas, only a general license from administrative security office is enough, therefore any interruption would not be happened.

More accessibility to rough areas: mapping from very high mountainous or swampy areas is difficult and even dangerous for field surveyors. In contrast, drones can rapidly access to these areas without any danger for operators.

These advantages of UAV photogrammetry cause the fixed-wing drones becomes a serious alternative for conventional field surveying in topographic map production for areas more than 100 hectares.

2.2 Challenges in UAV Photogrammetry

Although UAV photogrammetry has many advantages relative to field surveying, it faces to several challenges that limits its usage popularity. We classify these challenges in six categories:

Making drone as aerial robot:
- Space limitations: installation of several electronic boards, wirings, batteries, sensors, antennas and so on in a limited space in the drone.
- Weight limitations: limited payload for better performance and endurance.
- Cooling limitations: batteries, electronic boards and motor engine need a fluent air circulation.
- Design limitations: keeping gravity centre of UAV in a correct position, setting equipment parts in correct locations, manual accessing to switches and connectors, correct placement of blades, communication antennas, GPS insight, OSD video camera and so on.

Making an operational system:
- UAV taking-off by hands, a launcher or on an airstrip.
- UAV landing by skate, parashoot, into a net or by wheels.
- Being resistant to dust, stroke, water, and high/low temperatures.
- Safety aspects including return-to-home (RTH) strategy in communication loss states, parashooting in crash, crashed drone finding by GSM messages.
- System reparations and services during field operation.

Pilot and Autopilot:
- UAV flight needs experienced and skilful pilot in theory and practice. Pilot should be capable to manually control UAV especially during drone landing and take-off.
- Concentration and awareness of every moment flight conditions and navigation data and solving the software and hardware problems and interruptions during flight.
- Periodical pre/post flight tests for checking correct functionality of drone including flight mechanical parts, navigation and communication parts, and data collection sensors.
- Stopping drone flight while observing any evidence of part failures.
- Data collection, grouping, and backup of raw data
- Writing the operation process and problems in a diary booklet.

Flight limitations:
- Study area reconnaissance for ground station placement. It needs aerial and terrestrial direct insights to UAV without any obstacle such as electric wires and towers for communication needs and landing/take-off.
- Considering flight limitations in flight design such as topography, wind, light, site conditions, telecommunication limits, flight height and endurance.

Flight security licence and allowance:
- No clear rules and known instructions for UAV flight in comparison to manned planes.
- Attention to ICAO maps for knowing safe areas, heights, and period of flight.
- High security issues in population and military areas.
- Aerial imaging license is different from flight license and both should be taken.

Data processing considerations:
- Huge number of aerial images due to small sensor format, low flying height, and high overlap and sidelap.

- Low image quality in comparison of professional photogrammetric systems due to low image resolution, image motion (no FMC), noise in shadows (low radiometric resolution) and high image tilt and distortions.

- Need to powerful computers due to complex data processing issue in feature extraction and matching, bundle adjustment, DEM generation, and so on.

2.3 UAV Photogrammetry Process

Before describing the mentioned three practical projects, the nine steps production line of UAV photogrammetry are outlined.

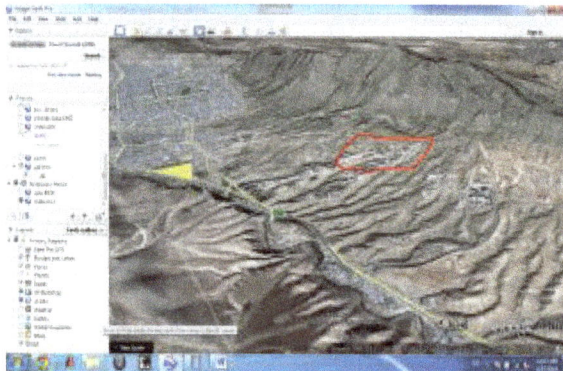

Step1: To define the study area by client in KMZ format in Google Earth

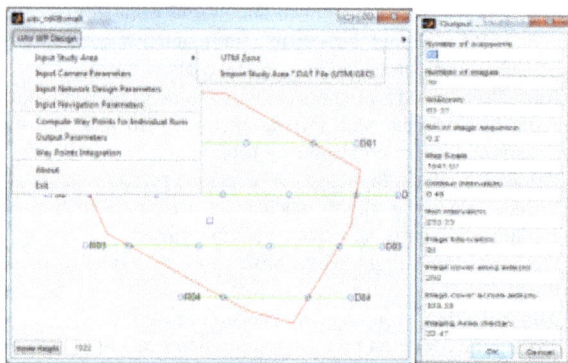

Step2: Initial reconnaissance, design GCP stations, flight schedule and design relative to circumstance conditions and map quality request.

Step3: Operation of making BMs and signalized GCPs based on network design and their positioning by GPS.

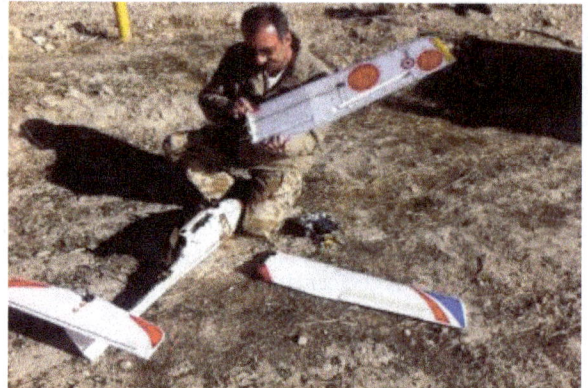

Step4: Flight preparation, setting ground station and UAV equipment installation.

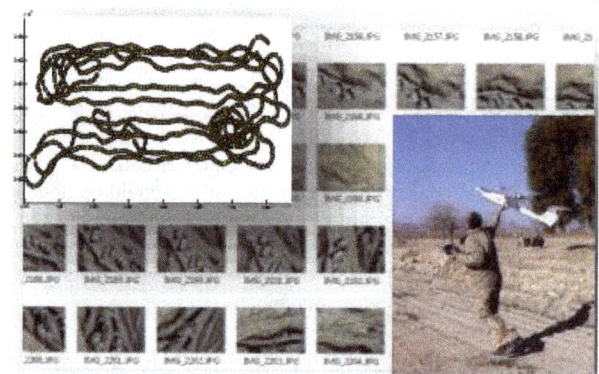

Step5: UAV Flight in the defined heights and regions and data recording including aerial images and navigation data.

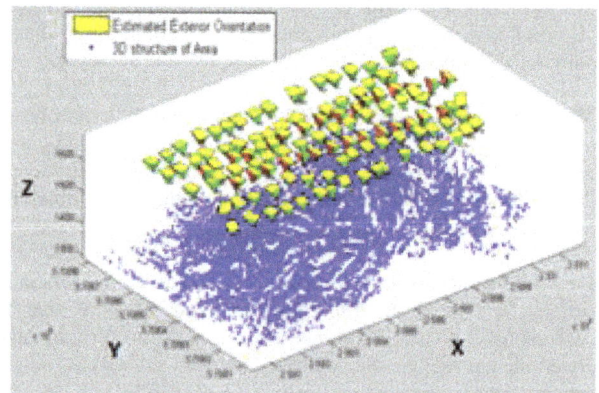

Step6: Data preparation, image network formation, automatic feature extraction and matching, entering GCPs coordinates, aerial triangulation by bundle adjustment and self calibration, distortion-free image generation, and making photogrammetric block project file for stereo-plotting operations.

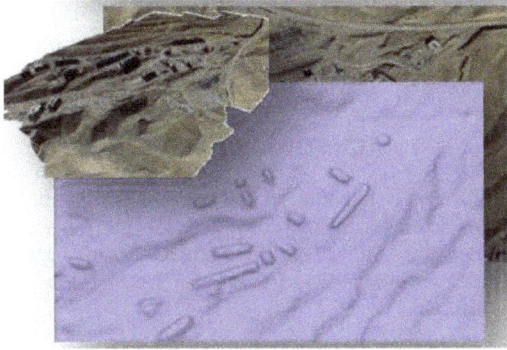

Step7: Initial DSM generation for ortho-image-mosaic and 3D textured realistic model generation.

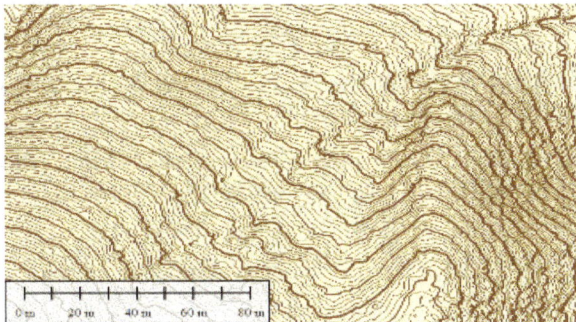

Step8: Dense coloured point cloud generation by multi-image-matching process (comparable to aerial LiDAR), DSM generation, filtering and editing, DTM extraction from DSM, Contour line generation from DTM and cartographic editing.

Step9: Stereo plotting of 3D features, map cartography, ortho-image-map generation, raster and vector data generalization, and spatial database generation.

3. PRACTICAL PROJECTS

In this section, specifications of three practical mapping projects that are accomplished by our fix-wing UAV photogrammetry are explained.

3.1 UAV Photogrammetry for Making 3D Realistic City Model

The aim of first project was 3D modelling of Khorramabad's MASKAN-MEHR construction site to show project development to Iran president at year of 2010. The topography of site was mountainous and full of tall buildings under construction in 40 hectares area. No other methods else UAV photogrammetry could produce a high quality 3D realistic model of site (Figure 1)

in a force time for a vital presentation. The client required an effective 3D presentation in order to get more budget for MASKAN-MEHR constructions.

Figure 1: 3D realistic modelling of MASKAN MEHR construction site by UAV Photogrammetry

For this project, we utilized a fix-wing drone without any autopilot and manually control it by RC (radio control) in designed flight lines. About 1000 images was taken under flight height of 200 meters with GSD of 5-10 cm. Then, 150 images was selected for rapid processing and after bundle adjustment and DEM and ortho-image generation, a 3D realistic model was generated in a simulation computer graphic software for presentation.

3.2 UAV Photogrammetry for Making Very Large Scale Map

The second project was 1:200 large scale mapping of a manufacture site named Siman Fars. The site was 100 hectares and had somehow flat bare topography including a large niches and small fenced equipment area. As the client doubted on the quality of result, a ground surveying also was done in a part of area to compare the results. The 10 cm dense coloured 3D point cloud making by UAV photogrammetry showed that quality of 20cm contour line output is clearly better than direct ground mapping due to density of generated 3D points that was impossible for ground surveying (Figure 2).

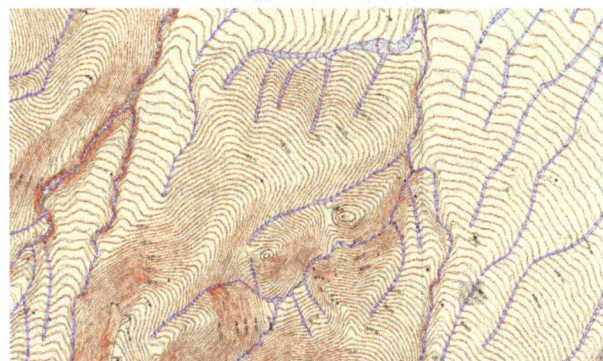

Figure 2: consistency of one meter contour lines computed by field surveying (blue and black lines) and UAV photogrammetry (red lines).

The GSD of images in this projects was about 2 cm which could possible by using SLR camera in 80 meters flying height (Figure 3). The drone was automatically navigate by autopilot and all images were taken from predesigned positions without any human interaction. To have better result, images with 90% overlap and 60% sidelap were acquired that caused the occluded areas decrease dramatically.

Figure 3: 2.5cm GSD of images taken from 80m flying height makes possible to produce high quality maps.

3.3 UAV Photogrammetry for Mapping from an Inaccessible High Risk Rough Area

The last project was mapping from very hard monotonous geological area that was dangerous for surveyors to climbing and impossible for conventional surface feature collection. It is 400 hectares area with 400 meters height difference having 25-75 degrees local slopes covered by sharp cutting salt stones surrounded with 100m tranches. Its topography was unique in difficulty of working conditions and full of occluded areas so that utilizing total station, RTK GPS, or terrestrial laser scanner for surface reconstruction in 1:1000 map scale are impractical (Figure 4).

Figure 4: UAV Photogrammetry is only solution for surface mapping of hazardous area. The corner figure shows the 3D (not 2D) flight lines of aerial imaging.

UAV photogrammetry is only solution that is accomplished by our fixed wing ultralight system. We did some enhancement in our UAV and designing 3D flight lines 250m height up over area. As local high slops make hard shadows, we decided to schedule duplicate flights in different hours due to proper sun directions. The final result was 25cm dense point cloud and 1m contour lines on 10cm orthoimage mosaic.

4. CONCLUSION

We observe huge attention to application of unmanned aerial vehicle (UAV) in aerial mapping since a decade ago. Though UAV photogrammetry has several advantages for handling time/cost/quality issues, there are a dozen of challenges in working with UAVs. In this paper, we reviewed these advantages as well as challenges in details. Then, the process of map production line for UAV photogrammetry was outlined. In continue, we described three special practical projects accomplished by our fix-wing UAV photogrammetry systems in order to show above challenges and advantages in practice.
The mentioned operational examples show that UAV photogrammetry is a new solution for surveyors with its special advantages according to other methods. It means they should be familiar with this technology in practice and benefits from its lower cost, higher quality, less time, and more safety in their

practical mapping projects. The end of this paper finished with the topic: UAV photogrammetry is a practical solution for challenging mapping projects in terms of cost, time, accuracy and safety.

ACKNOWLEDGEMENTS

Authors of paper thank to private companies supported RPRG team to do research and development on fix-wing UAV photogrammetry utilized for three projects in this paper. These companies are Siman-Fars Co., Madan-Koomeh-Pars Co., and Bonyan-Sazan Co.

REFERENCES

Blom, J.D., 2006. Unmanned Aerial Systems: A Historical Perspective, Combat Studies Institute Press, USA.

Fahlstrom, P.G., Gleason, T.J., 2012. Introduction to UAV Systems, John Wiley & Sons Ltd. Publications. UK.

Oliver, D.R., Money, A.L., 2001. Unmanned Aerial Vehicle Roadmap 2000-2015, Office of Secretary of Defence, USA.

28

INTRA-URBAN MOVEMENT FLOW ESTIMATION USING LOCATION BASED SOCIAL NETWORKING DATA

A. Kheiri [a], F. Karimipour [b,*], M. Forghani [b]

[a] Faculty Technical and Engineering, Eslamic Azad University of Larestan, Iran - asma.kheiri@ymail.com
[b] Faculty of Surveying and Geospatial Engineering, College of Engineering, University of Tehran, Tehran, Iran - (fkarimipr, mo.forghani)@ut.ac.ir

KEY WORDS: Location Based Social Network, Check-in Data, O/D Matrix, Mobility

ABSTRACT

In recent years, there has been a rapid growth of location-based social networking services, such as Foursquare and Facebook, which have attracted an increasing number of users and greatly enriched their urban experience. Location-based social network data, as a new travel demand data source, seems to be an alternative or complement to survey data in the study of mobility behavior and activity analysis because of its relatively high access and low cost. In this paper, three OD estimation models have been utilized in order to investigate their relative performance when using Location-Based Social Networking (LBSN) data. For this, the Foursquare LBSN data was used to analyze the intra-urban movement behavioral patterns for the study area, Manhattan, the most densely populated of the five boroughs of New York city. The outputs of models are evaluated using real observations based on different criterions including distance distribution, destination travel constraints. The results demonstrate the promising potential of using LBSN data for urban travel demand analysis and monitoring.

1. INTRODUCTION

An important part of human activities in an urban environment is their mobility behavior. Nowadays, measuring the movement of people is a fundamental activity in modern societies. First insight regarding the mobility within a region can be captured by extracting the origin–destination (O/D) matrix, which specifies the travel demands between the origin and destination nodes on a network. This matrix could be on different scales including macroscopic scales, e.g., at the inter-urban level, or at microscopic scales, e.g., at the intra-urban level. In the intra-urban level, OD matrix is indicative of the movement of people between different areas of the city. Many methods have been suggested for OD-matrix estimation, which can be classified into three main categories (Jin et al., 2013): survey-based methods, traffic counts, and methods based on the positioning technology. Survey-based methods as traditional OD estimation methods such as telephone, in-person interview, mail or email survey are always time-consuming and costly tasks and they have limited sample sizes and lower frequencies. Moreover, the survey data cannot provide up-to-date information to reflect the rapid changes in travel demand pattern. Traffic count based methods calibrate an OD matrix based on traffic detector data (Jin et al., 2014). These methods are based on an existing metering infrastructure, which may be expensive to install or maintain. Moreover, estimation of OD matrices from this data is extremely challenging because the data is very often limited in extent, which can lead to multiple plausible non-unique OD matrices (Jin et al., 2013; Igbal et al., 2014). Using LBSN data for OD matrix estimation is a method based on the positioning technology which is raised in recent literatures as a new travel demand data source and has attracted an increasing number of users and significantly raised their urban experience. Location based social networking sites such as Foursquare allow a user to "check in" at a real-world POI (point of interest, e.g., a hotel, coffee shop, art gallery, etc.), leave tips about the POI, and share the check-in with their online friends.

Compared to traditional GPS data, location based social networks data have unique features with many information to reveal human mobility, i.e., "when and where a user (who) has been to for what," and the temporal check-in sequence of a specific person can be considered as his/her trajectory. Therefor, these information provides opportunities to better understand human mobility from spatial, temporal, social, and content aspects. Additionally, with the rapid growth of smartphones, the LBSN application can be easily built in personal mobiles and tablets without concerning the maintenance and update issues in the traditional traffic monitor infrastructure. The sample size can be much larger than other methods because the penetration rate of social networking service is growing at a rapid pace. Moreover this data has the potential to provide origin-destination movement estimation with significantly higher spatial and temporal resolution at a much lower cost in comparison with traditional methods.

On the other hand, despite the long history of modeling human mobility, predicting mobility patterns in cities has been a challenging task until now, and the lack of an accurate approach with low data requirements for predicting mobility patterns in cities can still be felt. So far, many of researchers have tried to predict intra-urban movement using mobile positioning technologies but the results show that there is not significant achievements (Calabrese et al., 2011).

This study examines the efficiency of LBSN check-in data provided by Foursquare in the estimation of the intra-urban OD matrix. The remainder of this paper is organized as follows: Section 2 introduces the used dataset and an initial analysis conducted for the characteristics of check-ins collected. The data-filtering steps are presented in Section 3. Section 4 presents OD estimation models and our proposed approach, which are evaluated using a MTA OD matrix in Section 5. Finally, Section 6 presents some discussions and conclusions.

2. DATASET AND PRELIMINARY ANALYSIS

2.1 Review of the Dataset

In this paper, Manhattan, the most densely populated of the five boroughs of New York city, is selected as the study area.

Manhattan had a population of 1.6363m people in July 2014 according to the U.S. Census Bureau estimate and encompasses an area of 87.46 km². The data used in this study for analysis on the intra-urban movement can be categorized into three parts: the 2010 census tracts data of New York city from U.S. Census Bureau, New York Customer Travel Survey data from the Metropolitan Transportation Authority (MTA)[1] and the check-ins data from Foursquare.

The census tract data is used as spatial resolution to estimate movement distribution. There are 288 identified tracts within the borough of Manhattan, which will serve as the study area for this paper. MTA origin-destination survey data serves as the ground truth data used for comparison. The Survey provided a rich source of information about the tract to tract travel behavior of New York city residents from May through November 2008. This survey captured detailed information on the travel of 4,014 residents in 3,433 households in the borough of Manhattan. In this paper, Foursquare data set was chosen to study the human movement behavioral patterns based on geo-social networks because Foursquare is the most popular LBSN, with 31% of mobile users active on social networks using it. The check-ins data related to study area is extracted from source data. The resulted dataset contains 100,879 check-ins of 1083 users for about 10 month (from 12 April 2012 to 16 February 2013).

2.2 Preliminary Analysis of the Check-ins Data

In this sub-section, a preliminary analysis is conducted on the characteristics of the check-ins occurrence by investigating of the spatial pattern of the check-ins data. The spatial distribution of the 100,879 check-ins are represented using scatter dots in Figure 1(a). As shown in Figure 1(b), a heat map also represents the geographic density of check-ins features on study area by using graduated color areas to represent the quantities of those points. Also, according to the primary analysis, we found that the average number of check-ins per user is around 99 and the median is approximately 83.

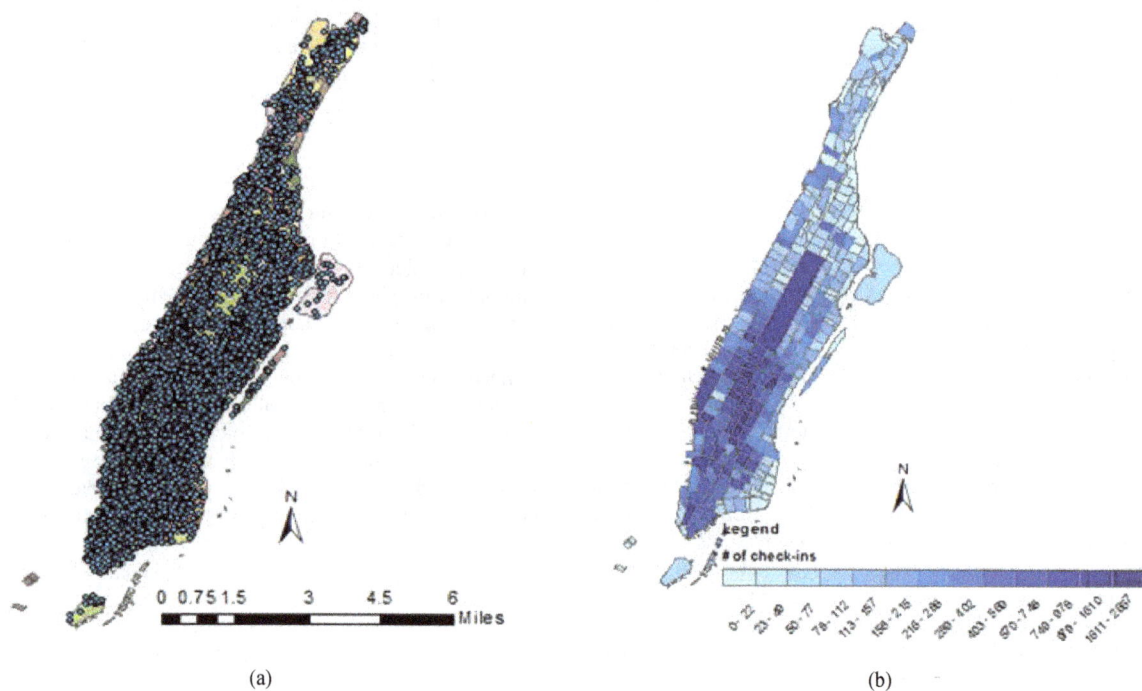

(a)   (b)

Figure 1. Foursquare check-ins locations and their spatial distribution among tracts

3. FILTERING CHECK-IN RECORDS

Although most social media services provide some mechanism to prevent the emergence of fake check-ins, invalid check-ins and trips still exist. Invalid check-ins prevent the efficiency of data for exploring intra-urban human mobility patterns and must be eliminated. This section presents how the check-in records and trips were filtered out and extracts daily trajectories from check-in data.

1) Filtering the duplicate check-ins: Some users check-in several times at the same location upon or after arrival, which must be filtered.

2) Filtering the user with only one check-in: The user with only one check-in were filtered out as they do not allow any movement analysis.

Moreover, having individual spatio-temporal trajectories extracted by connecting the consecutive check-ins, we filtered out trips according to the following rules:

3) Removing the trips with time intervals less than 1 minute and more than 12 hours: Very short trips (<1 minute) are not so sueful ans so filtered out. In addition, if time intervals between previous check-in is more than 12h, this means that some trips are very likely to be missing, and may deviate the results, so they also be removed.

---

[1] http://web.mta.info/mta/planning/data-nyc-travel.html

4) Removing trips with speed greater than 200 km/h. If the speed of the trip is more than 200 km/h, it means the user is traveling at a speed extremely higher than any urban transportation modes, including bus, subway, car, etc. Such trips were considered fake trips and were removed.

After applying the above filtering process, 69675 check-ins and 31978 trips were finally obtained.

4. METHODOLOGY

Since the 1940s, various models such as gravity model (Zipf, 1946), intervening opportunity model (Stouffer, 1940), radiation model (Simini et al., 2012), rank-based model (Noulas et al., 2012) and population-weighted opportunities (PWO) model (Yan et al., 2014) have been proposed for estimation of OD matrix. In this study, the three following models have been utilized to compare their performance for LBSN data analysis:

4.1 Radiation Model

The "radiation model" (Simini et al., 2012) defines a commuting flux from location i to j as:

$$T_{ij} = T_i \frac{m_i n_j}{(m_i+s_{ij})(m_i+n_j+s_{ij})} \tag{1}$$

where m_i and n_j are the population of locations i and j, T_i is the number of trips starting from i, and s_{ij} is the total population in the circle of radius r_{ij} (the distance between the origin i and destination j) centered at location i (excluding the source and destination population). This model needs only the spatial distribution of population as an input, without any adjustable parameters. Therefore, as an advantage, this model can be applied in areas where the previous mobility measurements are not available.

4.2 Rank-based Model

The "rank-based model" is inspired by the theory of intervening opportunities (Stouffer, 1940), which presents the rank distance as a key component, accounting in the number of places between origin and destination, rather than the pure physical distance. This model assumes that the probability of an individual travelling from an origin to a destination depends (inversely) only upon the rank-distance between the destination and the origin. The model is described as:

$$T_{ij} = T_i \frac{R_i(j)^{-\gamma}}{\sum_{k \neq i}^{N} R_i(j)^{-\gamma}} \tag{2}$$

where $R_i(j)$ is the rank-distance from location j to i (e.g., if j is the closest location to i, $R_i(j) = 1$; if j is the second closest location to i, $Ri(j) = 2$, and so on) and is an adjustable parameter. This model belongs to the category of parameterized models. However, γ presents minor variations from city to city. This model needs very low input information to reproduce some key characteristics of human mobility patterns.

4.3 Population-weighted Opportunities Model

Finally, the "population-weighted opportunities model" (Yan et al., 2014) assumes that the attraction of a destination is inversely proportional to the population S_{ji} in the circle centered at the destination with radius r_{ij} (the distance between the origin i and destination j) minus a finite-size correction $1/M$:

$$A_j = o_j \left(\frac{1}{S_{ji}} - \frac{1}{M} \right) \tag{3}$$

where A_j is the relative attraction of destination j to travelers at origin i, o_j is the total opportunities of destination j and M is the total population in the city. Hence, probability of travel from i to j is proportional to the attraction of j. Moreover, with the assumption that the number of opportunities o_j is proportional to the population m_j, the travel from i to j is defined as:

$$T_{ij} = T_i \frac{m_j(\frac{1}{S_{ji}} - \frac{1}{M})}{\sum_{k \neq i}^{N} m_k(\frac{1}{S_{ki}} - \frac{1}{M})} \tag{4}$$

where T_i is the trips starting from i and N is the number of locations in the city.

4.4 Proposed Approach

In order to evaluate the efficiency of LBSN data in the estimation of the intra-urban OD matrix using three mentioned models, firstly, Foursquare check-ins data for Manhattan, was collected and the collected raw data was filtered according to the approach described in section 3. Secondly, as each LBSN data record has a time attribute, check-ins were sorted based on time and individuals' trajectories were extracted using consecutive check-ins. The study area was then partitioned based on census tracts and using aggregated trajectories between these areas, the movement flow intensity between each pair of tracts was estimated by each of the mentioned models. Although the mentioned models have no adjustable parameters, they require information on variables such as population distribution and location attraction as inputs. In order to extract these inputs from location-based social networking data and make the proposed models compatible with this kind of data, it is essential to renew and develop the models, considering specific characteristics and limitations of LBSN data. Therefore, in radiation and population-weighted opportunities, the number of check-ins were used instead of the population in the models. And in rank-based model, the number of venues that are closer in terms of distance to origin than destination were used instead of the rank-distance from location j to i. In this paper, γ was assigned the value 0.84 as proposed by (Noulas et al., 2012).

5. EVALUATION OF MODELS

This section compares the estimated Foursquare OD matrix using mentioned models with the MTA ground truth matrix. To evaluate the performance of the models we investigated the travel distance distribution by three models based on ground truth data. Travel distance distribution is an important statistical property to capture human mobility behaviours (Yan et al., 2014).

As shown in Figure (2) it can be found that the distribution of travel distance estimated by the rank-based model has a good accordance with the ground truth data compared to the other models. Moreover, we computed the probability of travel towards a location with population m, $P_{dest}(m)$, for both ground truth data and the models. As shown in Figure (3), the results of rank-based model have a better agreement with the ground truth data than those of the other models. Furthermore, in order to evaluate the similarity between the data obtained from the mentioned models and the ground truth data, we used Cosine and Sørensen similarity indices. Therefore, the O/D matrices were converted to vector form and cosine similarity method was used as follow:

$$Similarity(i,j) = cos(\vec{i},\vec{j}) = \frac{\vec{i}.\vec{j}}{\|\vec{i}\| \times \|\vec{j}\|} \qquad (5)$$

The Sørensen similarity index is a statistic tool to identify the similarity between two samples (Yan et al., 2014). In this paper, we were used the index to measure the degree of agreement between reproduced travel matrices and empirical observations. The index is defined as:

$$SSI \equiv \frac{1}{N^2}\sum_i^N \sum_j^N \frac{2\min(T'_{ij}, T_{ij})}{T'_{ij}+T_{ij}} \qquad (6)$$

where T'_{ij} is the travels from location i to j predicted by the model and T_{ij} is the observed number of trips. The results of evaluation are presented in Table 1.

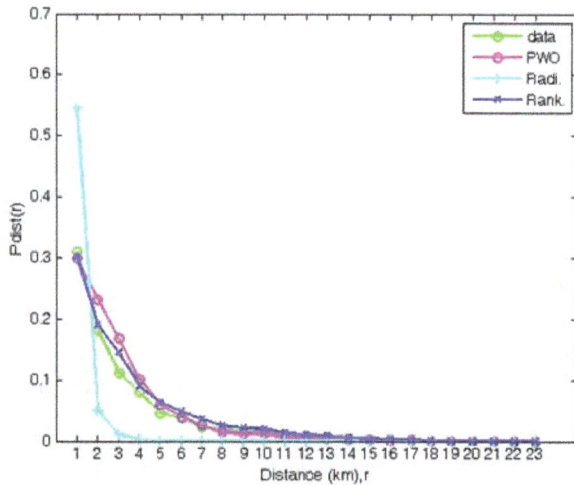

Figure 2. Travel distance distribution $P_{dist}(r)$ produced by the mentioned models in comparison with ground truth data. Here, $P_{dist}(r)$ is defined as the probability of travel between locations at distance r.

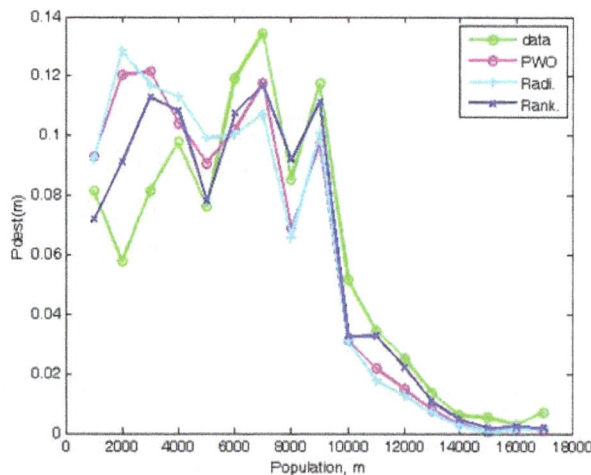

Figure 3. Comparing the destination travel constraints mentioned models with ground truth data. $P_{dest}(m)$ is the probability of travel to a location with population m.

Table 1: Cosine and Sørensen similarities between estimated OD matrices and ground truth data

Model	Cosine Similarity	Sørensen Similarity
PWO	0.688	0.449
RANK-BASED	0.674	0.49
RADIATION	0.707	0.417

6. CONCLUSION

This study examines the efficiency of LBSN data in the estimation of the intra-urban OD matrix in Manhattan, the one of the five boroughs of New York city. In this paper, the check-ins data from the leading LBSN provider, Foursquare, were used to analyze the intra-urban movement, and MTA origin-destination survey data served as the ground truth data to evaluate the performance of the proposed methodology. With respect to traditional and emerging travel demand data collection technologies, LBSN data has potential to investigate better spatial and temporal coverage, have real-time updating capability and much lower data collection cost. In this paper, three OD estimation models have been utilized to compare their relative performance for LBSN data. To evaluate the outputs of models, the travel distance distribution and destination travel constraints were investigated for three models based on ground truth data. The results show that the rank-based model has a better agreement with the ground truth data than those of the other models. Moreover, two indices, Cosine and Sørensen similarities, were used to measure the degree of similarity between reproduced travel matrices and empirical observations. According to the results, we found that the rank-based model has a better performance when using LBSN data than the other models. These results prove the assumption that intervening opportunity is more promising than geographical distance. In other words, in the development of future models for predicting intra-urban movement, it is essential to pay more attention to effects of the number of the closer places.

REFERENCES

Jin, P. J., Yang, F., Cebelak, M., Ran, B., & Walton, C. M., 2013. Urban travel demand analysis for Austin TX USA using location-based social networking data. In *TRB 92nd Annual Meeting Compendium of Papers*.

Jin, P., Cebelak, M., Yang, F., Zhang, J., Walton, C., & Ran, B., 2014. Location-Based Social Networking Data: Exploration into Use of Doubly Constrained Gravity Model for Origin-Destination Estimation. *Transportation Research Record: Journal of the Transportation Research Board*, (2430), pp. 72-82.

Iqbal, M. S., Choudhury, C. F., Wang, P., & González, M. C., 2014. Development of origin–destination matrices using mobile phone call data. *Transportation Research Part C: Emerging Technologies*, 40, pp. 63-74.

Calabrese, F., Di Lorenzo, G., Liu, L., & Ratti, C., 2011. Estimating origin-destination flows using mobile phone location data. *IEEE Pervasive Computing*, 4(10), pp. 36-44.

Zipf, G. K., 1946. The P1 P2/D hypothesis: On the intercity movement of persons. *American sociological review*, pp. 677-686.

Stouffer, S. A., 1940. Intervening opportunities: a theory relating mobility and distance. *American sociological review*, 5(6), pp. 845-867.

Simini, F., González, M. C., Maritan, A., & Barabási, A. L., 2012. A universal model for mobility and migration patterns. *Nature*, 484(7392), pp. 96-100.

Noulas, A., Scellato, S., Lambiotte, R., Pontil, M., & Mascolo, C., 2012. A tale of many cities: universal patterns in human urban mobility. *PloS one*, 7(5), e37027.

Yan, X. Y., Zhao, C., Fan, Y., Di, Z., & Wang, W. X., 2014. Universal predictability of mobility patterns in cities. *Journal of The Royal Society Interface*, 11(100), 20140834.

29

A NEW OBJECT-BASED FRAMEWORK TO DETECT SHADOWS IN HIGH-RESOLUTION SATELLITE IMAGERY OVER URBAN AREAS

Nurollah Tatar [a]*, Mohammad Saadatseresht[a], Hossein Arefi[a], Ahmad Hadavand[a]

[a] School of Surveying and Geospatial Information Engineering, College of Engineering, University of Tehran

Commission VII, WG VII/4

KEY WORDS: Shadow Detection, Spectral Index, High resolution satellite imagery, Segmentation, Object-based, Majority Voting.

ABSTRACT

In this paper a new object-based framework to detect shadow areas in high resolution satellite images is proposed. To produce shadow map in pixel level state of the art supervised machine learning algorithms are employed. Automatic ground truth generation based on Otsu thresholding on shadow and non-shadow indices is used to train the classifiers. It is followed by segmenting the image scene and create image objects. To detect shadow objects, a majority voting on pixel-based shadow detection result is designed. GeoEye-1 multi-spectral image over an urban area in Qom city of Iran is used in the experiments. Results shows the superiority of our proposed method over traditional pixel-based, visually and quantitatively.

1. INTRODUCTION

From 1999 by the launch of IKONOS, known as the first high resolution satellite imaging system, new applications in photogrammetry and remote sensing are emerged such as: producing high resolution digital surface model, high precision land cover mapping, change detection and hazard management. Appearance of small urban objects e.g. buildings, single trees and cars build the possibility to detect and analyse these objects. Diversity of features in high resolution images from the perspective of spectral and geometrical properties make some difficulties in analysing these images. Variation of height in urban areas, coincide with the sun elevation angle makes shadows in the image scene. Shadow has both constructive and destructive role in the processing of the images. It helps in well detection of different objects visually and also automated detection of collapsed buildings after natural disasters e.g. earthquakes and 3D reconstruction of buildings (Huang and Kwoh, 2007; Tong et al., 2013). While it ruins the contrast of objects casted by shadow and consequently classification, objects detection and automated stereo image matching (Shahtahmassebi et al., 2013; Tsai, 2006).

Shadows are produced when an opaque objects prevent the light rays to shine on a surface. It makes overacted areas appear darker than their surroundings. As presented in Figure 1 shadow consists of two parts: cast shadow and self-shadow. In aerial and satellite imagery the cast shadow affects more and in the following in this paper, everywhere we mention shadow it means cast shadow.

Knowing the accurate position of the sun and the sensor's platform in the imaging time and accurate 3D model of the imaged scene, location of shadows in the image could be simulated geometrically (Nakajima et al., 2002; Zhan et al., 2005). This method demands deep and expensive information.

There is a plenty of methods which proposed to detect shadows in the remotely sensed imagery. Spectral indices obtained by simple computations over spectral bands is of the simplest methods (Song and Civco, 2002; Tsai, 2006). Although these indices unable to discern water bodies, asphalt roads and clouds. More advanced indices are proposed to solve this issue, for example thresholding on near infrared band to discriminate clouds or using spectral bands in visible parts of the spectrum to discriminate water bodies (Shahtahmassebi et al., 2013). Modelling blackbody radiator model is another approach which integrates physical properties of shadows to design adaptive index. This method needs ground truth and metadata over the sensor and the imaged scene (Makarau et al., 2011).

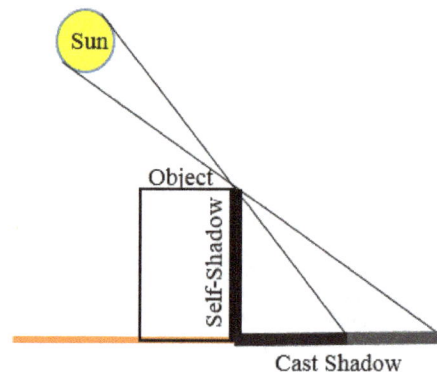

Figure 1. Shadow formation and its components

Using spectral indices in shadow detection, usually is integrated to thresholding in pixel level. In this way, the correlation and contextual information of neighbouring pixels is neglected. In high resolution images, shadow regions also could be detected, analysing edge information and image segmentation (Arévalo et al., 2008; Dare, 2005; Elbakary and Iftekharuddin, 2014; Sarabandi et al., 2004). Some researchers also employed region growing concept to grow shadow seed points. This solution also has some deficiencies. Morphological filtering, gap filling methods and edge information are used to fulfil deficiencies (Arévalo et al., 2008; Song et al., 2014).

Supervised machine learning algorithms are widely used in shadow detection problems. Support vector machine and artificial neural networks are employed to detect shadows in pixel level (Liu et al., 2011; Lorenzi et al., 2012). In high resolution images single pixels are not meaningful independently. Object-based methods, integrating similar neighbouring pixels provides powerful tool to analyse high resolution image data. Object based image classification paradigm is also used to detect shadows (Liu and Yamazaki, 2012; Zhang et al., 2014).

In this paper we propose a new shadow detection framework which integrates new spectral indices, machine learning algorithm and object-based image analysis principals. In the following the basic shadow indices are described and then the detail of proposed method is discussed. Then the result of experiments on high resolution satellite images is provided and the paper will ends with discussion and conclusion.

2. SHADOW DETECTION INDICES

Among different indices which are proposed to detect shadows, C_3 component and blue/near infrared ratio are used in our proposed method. The computational aspects of these indices will be described in the following.

2.1 Invariant colour Model

Visible spectral bands includes blue, green and red are employed to calculate these components using the following equations (Gevers and Smeulders, 1999):

$$C_1 = tan^{-1}(\frac{R}{max(G, B)}) \tag{1}$$

$$C_2 = tan^{-1}(\frac{G}{max(R, B)}) \tag{2}$$

$$C_3 = tan^{-1}(\frac{B}{max(R, G)}) \tag{3}$$

Among these components, C_3 is useful in shadow detection (Tsai, 2006).

2.2 Blue/near infrared ratio

Absorption of electromagnetic waves is a function of their wavelength and expressed by Rayleigh equation as below:

$$I(\lambda)_{scattering} \propto \frac{I(\lambda)_{incident}}{\lambda^4} \tag{4}$$

In equation (4), $I(\lambda)_{scattering}$ is the grey value after absorption, $I(\lambda)_{incident}$ is the grey value before absorption and λ is the wavelength. Based on this equation, for lower wavelengths the absorption is more. In addition, reflectivity in shadowed area have low value. So the variation of grey values in shadow and non-shadow areas is different and could be used as a measure in shadow detection. Following normalized index uses this fact:

$$Ratio_{B\_NIR} = \frac{B - NIR}{B + NIR} \tag{5}$$

3. DATASET

A panchromatic and 4 band multi-spectral image acquired by GeoEye-1 high resolution sensor is used in the experiments. The spatial resolution of data is 0.5 and 2 meter for panchromatic and spectral bands respectively. The image is captured over an urban area in Qom city in Iran.

4. METHOD

As presented in Figure 2, our proposed method consists of 4 main steps. These steps are introduced in detail in the following.

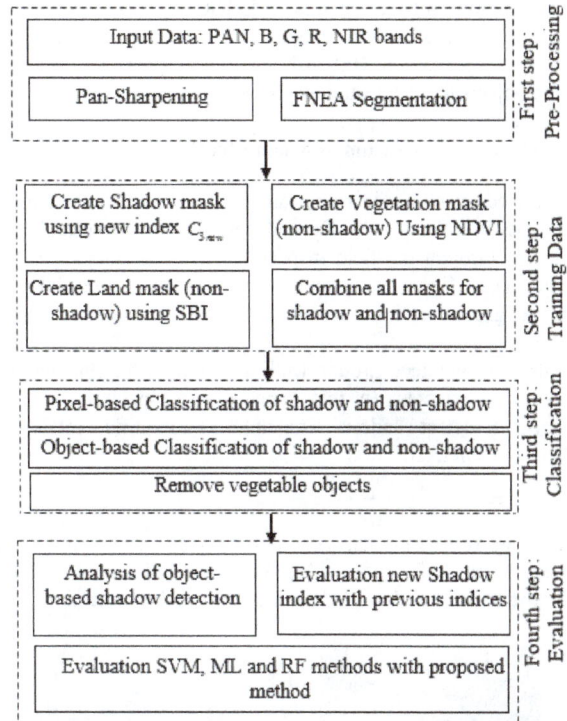

Figure 2. flowchart of proposed shadow detection method

4.1 1st step: Pre-processing

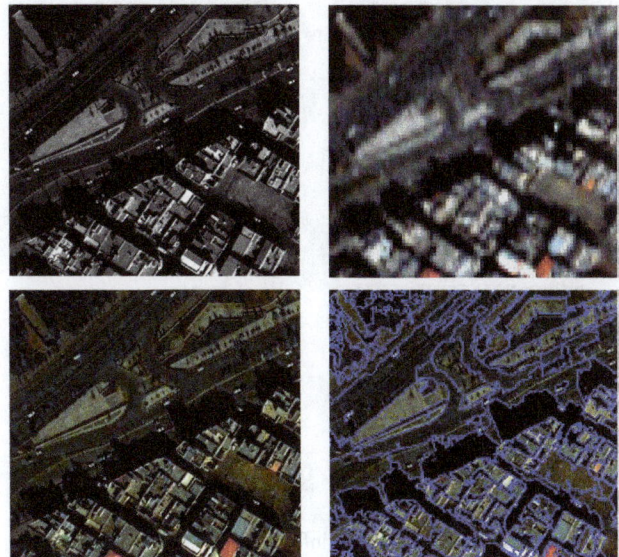

Figure 3. Original panchromatic and spectral bands (top), result of image fusion and image segmentation (bottom)

To prepare the data for further analysis, in the beginning fusion of panchromatic and spectral bands is used to enhance the spatial resolution of spectral bands. Due to its superiority in maintaining spatial accuracy, IHS algorithm (Strait et al., 2008) is employed here. Also in this step image objects are produced to use in further object-based process. Fractal Net Evolution Approach (FNEA) segmentation algorithm (Benz et al., 2004), implemented in eCognition software is used in this step. Figure

3 contains an overview of results of image fusion and segmentation.

We used FNEA segmentation algorithm which is implemented in eCognition software to this end. This algorithm gets scale parameter, shape and compactness weights as inputs. This parameters are set to 100, 0.1 and 0.9 respectively to build image objects to detect shadow areas.

4.2 2nd step: Generation of ground truth data

Supervised machine learning algorithms need some ground truth data in training step. Ground truth information usually are collected through the field inspection or visual analysis of the image data. Here we propose an automatic procedure to generate ground truth information.

In our work we need ground truth information for shadow and non-shadow classes. In this paper a new spectral index is designed to detect shadows. This index is a modified version of C3 component as follows:

$$C_{3new} = tan^{-1}(\frac{B}{PAN})$$ (6)

Following histogram analysis and comparison for proposed index versus C_3 and blue/near infrared simple ratio demonstrates the ability of C_{3new} index in shadow detection.

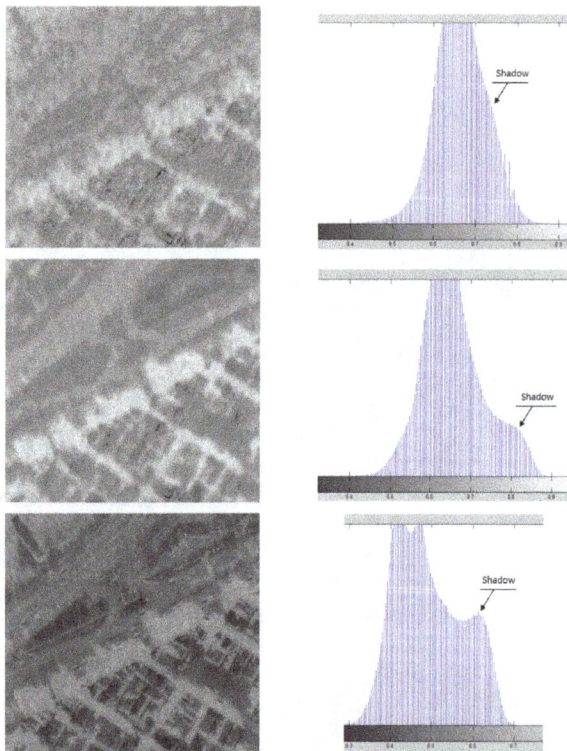

Figure 4. C3 (top), blue/near infrared ratio (middle) and C3new (bottom) indices and following histograms

After calculating index values for the image pixels, Otsu thresholding algorithm is employed to automatically find the best threshold to detect shadow pixels. Morphological erosion filter the result to increase reliability of detected shadow pixels. Ground truth information for non-shadow class prepared by analysing normalized difference vegetation index (NDVI) and soil brightness (SBI) indices. Otsu thresholding and morphological erosion is also employed in a similar process to get the final ground truth in non-shadow class. In Figure 5 the result of computing indices and selected pixels after Otsu thresholding could be seen.

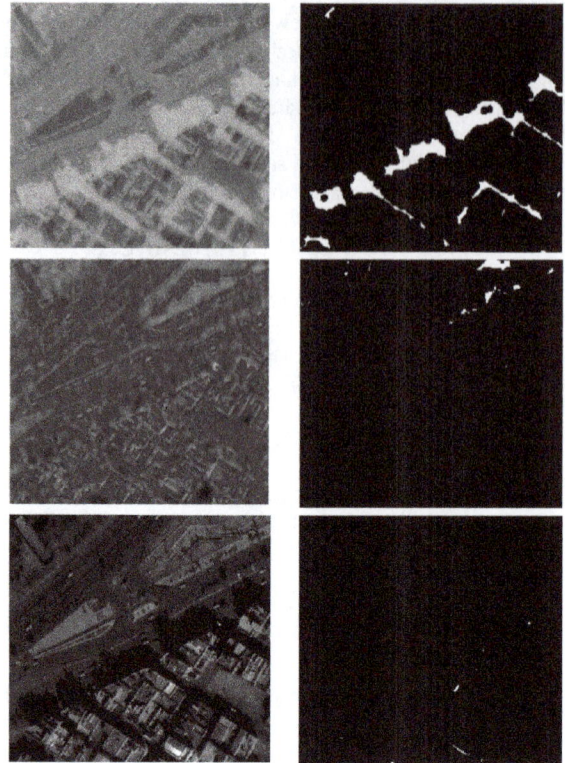

Figure 5. Result of computing indices (left) and Otsu thresholding (right) for C3New (top), NDVI (middle) and SBI (bottom)

4.3 3rd step: Object-based and pixel-based shadow detection

Ground truth information are used by supervised machine learning algorithms to separate shadow and non-shadow pixels. To detect shadow areas in object level, a majority voting analysis is used on number of shadow pixels in each image objects. To solve the ambiguity between vegetated and shadow objects an extra condition is checked to confirm that an object belongs to shadow class. This condition uses the mean NDVI value of pixels in each image object. For shadow objects the NDVI should have low values. The result of detecting shadows on pixel level using SVM algorithm, shadow detection on object level and overlay of shadow areas on original image is presented in Figure 6.visually the superiority of the object-based shadow detection results is clear and in evaluation step it will be approved.

Figure 6. Shadow detection result in pixel and object level

4.4 4th step: Evaluation

Results of shadow map, evaluates using a two class confusion matrix. As presented in Figure 7 there are 4 measures in this matrix includes: true positive (TP) number correctly classified shadow pixels, false positive (FP) number of wrongly classified non-shadow pixels as shadow, false negative (FN) number of shadow pixels which detected as non-shadow, and true negative (TN) number of non-shadow pixels classified correctly.

		Ground truth information	
		Shadow	Non-shadow
Detection result	Shadow	TP	FP
	Non-shadow	FN	TN

Figure 7. Confusion matrix for shadow detection

Compactness, correctness and F-measure are calculated based on confusion matrix and are used to evaluate and compare the results. These measures are computed as below:

$$Compactness = 100 \times \frac{TP}{TP + FN} \qquad (7)$$

$$Correctness = 100 \times \frac{TP}{TP + FP} \qquad (8)$$

$$F - measure = 100 \times \frac{2 \times TP}{2 \times TP + FN + FP} \qquad (9)$$

Details on evaluation of results will be expressed in the next section.

5. EVULATION OF RESULTS

To evaluate the results, 408 image objects in shadow class and 487 image objects in non-shadow class are selected manually. The confusion matrix is calculated for these objects and compactness, correctness and F-measure are calculated for the result. Table 1 contains the result of shadow detection respect to the result of SVM, random forest (RF) and maximum likelihood pixel-based classifiers. To assess the sensitivity of threshold in majority voting process, different thresholds are also selected and the results are compared.

Table 1. The result of object-based shadow detection using majority voting on different classifiers (Best result appears in bold face)

Classifier	Majority voting threshold	Completeness	Correctness	F-measure
SVM	30	85	96	90
	40	**90**	**93**	**91**
	50	93	86	89
	60	95	74	83
	70	98	58	72
RF	**30**	**93**	**91**	**92**
	40	95	83	88
	50	97	72	82
	60	98	58	73
	70	98	40	57
ML	**30**	**95**	**86**	**90**
	40	97	76	85
	50	98	62	76
	60	98	49	65
	70	98	30	95

Table 2 contains the best object-based result versus pixel-based result for each classifier. This enables us to compare the result of pixel-based and object-based shadow detection.

Table 2. Comparison on object-based and pixel-based result of shadow detection

Method	Completeness	Correctness	F-measure
SVM Pixel-based	91	89	90
SVM Object-based	90	93	91
RF Pixel-based	98	85	90
RF Object-based	93	91	92
ML Pixel-based	100	72	84
ML Object-based	95	86	90

6. CONCLUSION

Accurately detection of shadows is a critical pre-processing step in many remote sensing image processing applications. Here we proposed a new object-based shadow detection method and take several experiments to compare our method with the traditional pixel-based one. Also the sensitivity of our algorithm against the selection of classifier and majority voting threshold is examined.

Result of our experiments in Table 2 demonstrates the superiority of proposed object-based over the pixel-based method respect to correctness and F-measure for different classifiers. This superiority could be seen in Figure 6. It is also evident that object-based method have well behaviour on the edge of shadow areas and perfectly detect shadows.

Figure 4 shows the ability of our proposed index to detect shadows. Bisection shape of C_{3New} index makes it possible to detect shadow by thresholding. So the Otsu algorithm expected to work well with this index.

The sensitivity of object-based method to the threshold of majority voting is examined and the results are presented in Table 1. It seems that choosing higher thresholds increase the misclassified shadow areas and worsen the result. Comparison on different classifiers show that their performance doesn't have meaningful difference.

Main deficiency of our proposed method is its failure in detecting the shadow of small objects. The fusion of pixel-based and object-based result is proposed to solve this issue and will follow by the authors.

REFERENCES

Arévalo, V., González, J., Ambrosio, G., 2008. Shadow detection in colour high-resolution satellite images. International Journal of Remote Sensing 29, 1945-1963.

Benz, U.C., Hofmann, P., Willhauck, G., Lingenfelder, I., Heynen, M., 2004. Multi-resolution, object-oriented fuzzy analysis of remote sensing data for GIS-ready information. ISPRS Journal of photogrammetry and remote sensing 58, 239-258.

Dare, P.M., 2005. Shadow Analysis in High-Resolution Satellite Imagery of Urban Areas. Photogrammetric Engineering & Remote Sensing 71, 9.

Elbakary, M.I., Iftekharuddin, K.M., 2014. Shadow detection of man-made buildings in high-resolution panchromatic satellite images. IEEE Transactions on Geoscience and Remote Sensing 52, 5374-5386.

Gevers, T., Smeulders, A.W.M., 1999. Color-based object recognition. Pattern Recognition 32, 453-464.

Huang, X., Kwoh, L.K., 2007. 3D building reconstruction and visualization for single high resolution satellite image, IEEE International Geoscience and Remote Sensing Symposium (IGARSS), p. 4.

Liu, J., Fang, T., Li, D., 2011. Shadow detection in remotely sensed images based on self-adaptive feature selection. IEEE Transactions on Geoscience and Remote Sensing 49, 5092-5103.

Liu, W., Yamazaki, F., 2012. Object-based shadow extraction and correction of high-resolution optical satellite images. IEEE Journal of Applied Earth Observations and Remote Sensing 5, 1296-1302.

Lorenzi, L., Melgani, F., Mercier, G., 2012. A complete processing chain for shadow detection and reconstruction in VHR images. IEEE Transactions on Geoscience and Remote Sensing 50, 3440-3452.

Makarau, A., Richter, R., Muller, R., Reinartz, P., 2011. Adaptive shadow detection using a blackbody radiator model. IEEE Transactions on Geoscience and Remote Sensing 49, 2049-2059.

Nakajima, T., Tao, G., Yasuoka, Y., 2002. Simulated recovery of information in shadow areas on IKONOS image by combing ALS data, Proc. Asian Conference on Remote Sensing.

Sarabandi, P., Yamazaki, F., Matsuoka, M., Kiremidjian, A., 2004. Shadow detection and radiometric restoration in satellite high resolution images, IEEE International Geoscience and Remote Sensing Symposium (IGARSS), pp. 3744-3747.

Shahtahmassebi, A., Yang, N., Wang, K., Moore, N., Shen, Z., 2013. Review of shadow detection and de-shadowing methods in remote sensing. Chinese Geographical Science 23, 403-420.

Song, H., Huang, B., Zhang, K., 2014. Shadow Detection and Reconstruction in High-Resolution Satellite Images via Morphological Filtering and Example-Based Learning. IEEE Transactions on Geoscience and Remote Sensing 52, 2545-2554.

Song, M., Civco, D.L., 2002. A knowledge-based approach for reducing cloud and shadow, Proceedings of the American Society of Photogrammetry and Remote Sensing (ASPRS-ACSM), Washington.

Strait, M., Rahmani, S., Merkurev, D., 2008. Evaluation of pan-sharpening methods, UCLA Department of Mathematics.

Tong, X., Lin, X., Feng, T., Xie, H., Liu, S., Hong, Z., Chen, P., 2013. Use of shadows for detection of earthquake-induced collapsed buildings in high-resolution satellite imagery. ISPRS Journal of Photogrammetry and Remote Sensing 79, 15.

Tsai, V.J., 2006. A comparative study on shadow compensation of color aerial images in invariant color models. IEEE Transactions on Geoscience and Remote Sensing 44, 10.

Zhan, Q., Shi, W., Xiao, Y., 2005. Quantitative analysis of shadow effects in high-resolution images of urban areas, International Archives of Photogrammetry and Remote Sensing.

Zhang, H., Sun, K., Li, W., 2014. Object-Oriented Shadow Detection and Removal From Urban High-Resolution Remote Sensing Images. IEEE Transactions on Geoscience and Remote Sensing 52, 11.

30

DETERMINATION OF OPTIMUM CLASSIFICATION SYSTEM FOR HYPERSPECTRAL IMAGERY AND LIDAR DATA BASED ON BEES ALGORITHM

F. Samadzadegan [a], H. Hasani [a], *

[a] School of Surveying and Geospatial Engineering, College of Engineering, University of Tehran, Tehran, Iran –
(samadz, hasani)@ut.ac.ir

Commission I, ICWG III/VII

KEY WORDS: Hyperspectral, LiDAR, Fusion, SVM Classifier, Optimization, Parameter Determination, Feature Selection, Bees Algorithm

ABSTRACT

Hyperspectral imagery is a rich source of spectral information and plays very important role in discrimination of similar land-cover classes. In the past, several efforts have been investigated for improvement of hyperspectral imagery classification. Recently the interest in the joint use of LiDAR data and hyperspectral imagery has been remarkably increased. Because LiDAR can provide structural information of scene while hyperspectral imagery provide spectral and spatial information. The complementary information of LiDAR and hyperspectral data may greatly improve the classification performance especially in the complex urban area. In this paper feature level fusion of hyperspectral and LiDAR data is proposed where spectral and structural features are extract from both dataset, then hybrid feature space is generated by feature stacking. Support Vector Machine (SVM) classifier is applied on hybrid feature space to classify the urban area. In order to optimize the classification performance, two issues should be considered: SVM parameters values determination and feature subset selection. Bees Algorithm (BA) is powerful meta-heuristic optimization algorithm which is applied to determine the optimum SVM parameters and select the optimum feature subset simultaneously. The obtained results show the proposed method can improve the classification accuracy in addition to reducing significantly the dimension of feature space.

1. INTRODUCTION

Recently, with the progress in remote sensing technologies, it is possible to measure different characteristics of objects on the earth such as spectral, height, amplitude and phase information by multispectral/hyperspectral, LiDAR and SAR respectively (Debes et al. 2014). Availability of different types of data, provides means of detecting and discriminating of land use land cover in complex urban area (Ramdani, 2013). Classification of urban area has been used in wide range of application, such as mapping and tracking, risk management, social and ecological problems (Fauvel, 2007).

Hyperspectral remote sensing data is characterized by a very high spectral resolution that usually results in hundreds of observation bands. According to spectral richness, it plays very important role in discrimination of land-cover with similar spectral reflectance (Chang, 2013). Although hyperspectral imagery provides comprehensive spectral information but classification of complex urban area based on just spectral information has some limitations: same objects with different spectral characteristic don't classify in a class (e.g. buildings with different roof material/color don't classify in one class) and different objects with same spectral appearance may classify in same class (e.g. tree and grass/ roof and road). On the other hand, LiDAR sensor provides 3D information from surfaces and mapping with LiDAR data depend on the ability to detect objects with different height. There is a complementary relationship between passive hyperspectral images and active LiDAR data, as they contain very different information (Khodadazadeh et al. 2015).

According to availability, robustness and accuracy of spectral and structural information of hyperspectral images and LiDAR data, fusion of hyperspectral images and LiDAR data in a joint classification system, may yield more reliable and accurate classification results. While there have been numerous investigations that have reported on the use of other multisensory data (e.g. LiDAR and Multispectral), very few results are available about simultaneously integration of these two data sources in classification tasks (Debes et al. 2014, Latifi et al. 2012).

This paper presents an optimum hybrid classification system by simultaneous determination of the SVM parameters and the selection of features through swarm optimization process in order to fuse hyperspectral imagery and LiDAR data.

2. RELATED WORK

During last years, some investigations were carried out on fusion/integration of hyperspectral images and LiDAR data in different application, such as forest structure analysis, urban area mapping, identification of tree species, forest fire management, etc. (Alonzo et al. 2014, Brook et al., 2010, Dalponte et al. 2008, Koetz et al. 2008, Latifi et al. 2012). In some research works, LiDAR data is used for separation of 2D and 3D objects and then hyperspectral images are applied to discriminate among different species of an object, such as roofing material (Niemann, et al., 2009; Zhang and Qiu, 2012). Sugumaran and Voss (2007), apply the object based classification where LiDAR data is used for segmentation and hyperspectral image to classify the segments.

Dalponte et al. (2008) merge a subset of hyperspectral bands with two LiDAR imaging data (intensity and nDSM), then fuse it with

---

* Corresponding author

results of the image classified by SVM and Gaussian Mixture Model. Liu et al. (2011) compute Canopy Height Model (CHM) from first LiDAR return and Minimum Noise Fraction (MNF) transformation is executed based on the pixel-level fusion of hyperspectral imagery and CHM channels. Then the first 26 eigenvalue bands are kept as input data for SVM classifier. Latifi et al. (2012) fuse hyperspectral bands and LiDAR features using Genetic Algorithm (GA) and apply this to select the feature subset in order to model forest structure.

In order to optimize the classification performance of high dimensional data, several methods are proposed in literatures which can be categorized into three groups: parameter determination of classifier (Liu et al. 2014), feature selection (Rashedi and Nezamabadi-pour, 2014) and simultaneously consider both of them (Samadzadegan, et al. 2012). The parameters of classifier has significant effect on its performance where grid search is common way to determine them (Hsu et al. 2003). Moreover, the selection of the feature subset may affect several classification aspects, including classification accuracy, computation time, training sample size, and the cost associated with the features (Lin et al., 2008). Several studies are focused on optimization these two issues which show that according to the dependency of parameters and features, simultaneous parameter determination and feature selection yield the most accurate results (O'Boyle et al., 2008). Recently Liu et al. applied PSO to determine the SVM kernel and margin parameters in classification of hyperspectral imagery and the results compared with grid search method which show the superiority of the proposed method (Liu et al. 2014). Feature selection is another essential step in classification of high dimension data. Rashedi and Nezamabadi-pour (2014) proposed an improved version of the binary gravitational search algorithm as a tool to select the best subset of features with the goal of improving classification accuracy. As parameter values effect on feature subset selection and vice versa, Samadzadegan et al. (2012) show that the best performance of classification is obtained by simultaneously classifier determination and feature selection by Ant Colony Optimization.

In this paper an optimum hybrid classification system is presented that simultaneously determines SVM classifier parameters and selects the feature subset to optimize classification performance for combined hyperspectral imagery and LiDAR data.

3. PROPOSED METHOD

In order to fuse hyperspectral imagery and LiDAR data, a hybrid feature space consisting of spectral and structural features is generated. Spectral feature space composed of original hyperspectral bands, vegetation indices and principle components. On the other hand, textural analysis on normalized DSM (nDSM), roughness and its textures, slope descriptors are extracted from LiDAR data which make the structural feature space. By combining spectral and structural feature space, the hybrid feature space is defined. Then normalization is used to transform data into the range [0, 1], in order to reduce numerical complexity.

According to the stability of SVM in high dimensional space [5], SVM is selected as classifier. There are two important challenges in classification of high dimensional data by a SVM classifier: SVM parameter determination (kernel and regularization parameters) and feature subset selection. In order to optimize classification of this hybrid feature space based on SVM, optimized SVM parameters values and appropriate feature

subsets should be selected. For this purpose the Binary Bees Algorithm, as a powerful population based optimization algorithm, is applied to determine SVM parameters and selection of features subset simultaneously. Figure 1 presents the flowchart of the proposed.

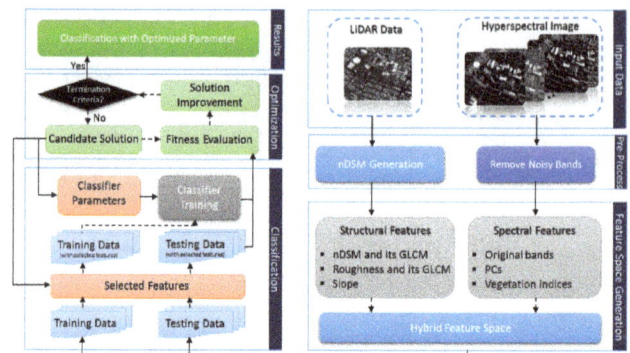

Figure 1. Flowchart of the proposed method

3.1 Feature Space Generation

Hyperspectral original bands include rich sources of spectral information but some indicators such as PCA components and vegetation indices may give additional information. Therefore PCA transformation is applied to the hyperspectral images and first three PCs are extracted additionally for use in the feature space. Then 30 vegetation indices are computed to discriminate vegetation classes from other classes (Table 1).

LiDAR-derived DSM provides height information, however more structural features should be generated to improve its ability in discrimination between classes. The nDSM is generated from DSM by geodesic morphological operation. In order to analyse the nDSM accurately, several types of features such as texture analysis, roughness and slope descriptors are extracted.

Grey Level Co-occurrence Matrices (GLCM) approach is used in this paper to extract second order statistical textural features from nDSM. In this paper, 16 features (Variance, Homogeneity, Contrast, Entropy, Dissimilarity, Sum Average, Angular Second Moment, Maximum Probability, Inverse Difference Moment, Sum Entropy, Sum Variance, Difference Variance, Correlation, Difference Entropy and two Information Measure of Correlation) are extracted from the GLCM matrix (Haralick et al. 1973).

Roughness is another structural feature which is extracted from nDSM. For this purpose, the terrain roughness is parameterized by the standard deviation of the detrended z-coordinates of the neighborhood. The plane is fitted to each neighborhood by the least square method and then the standard deviation of detrended height is determined. Texture analysis on the roughness map is also performed to better analysis of roughness. Moreover the slope of each neighbourhood in the nDSM is computed by applying the normal vector for the obtained plane which leads to a contribution of the slope feature to the structural feature space.

Finally, by stacking the spectral features from hyperspectral imagery and structural features from LiDAR data, the hybrid feature space is generated.

Table 1. Spectral indices from hyperspectral image, R_x is the reflectance at x nm

Name	Equation
Normalized Difference Vegetation Index (NDVI)	$(R_{800} - R_{670})/(R_{800} + R_{670})$
Simple Ratio (SR)	R_{800}/R_{670}
Enhanced Vegetation Index (EVI)	$2.5((R_{800} - R_{670})/(R_{800} + 6R_{670} - 7.5R_{475} + 1))$
Atmospherically Resistant Vegetation Index (ARVI)	$(R_{800} - 2R_{670} + R_{475})/(R_{800} + 2R_{670} - R_{475})$
Sum Green Index (SGI)	$mean(R_i), i = 500, \dots, 600$
Red Edge Normalized Difference Vegetation Index (RENDVI)	$(R_{750} - R_{705})/(R_{750} + R_{705})$
Modified Red Edge Simple Ratio Index (MRESRI)	$(R_{750} - R_{445})/(R_{750} + R_{445})$
Modified Red Edge Normalized Difference Vegetation Index (MRENDVI)	$(R_{750} - R_{705})/(R_{750} + R_{705} - 2R_{445})$
Vogelmann Red Edge Index 1, 2 (VREI 1)	$(R_{734} - R_{747})/(R_{715} + R_{726}), (R_{734} - R_{747})/(R_{715} + R_{720})$
Red Edge Position Index (REPI)	$wavelength\ of\ steepest\ slope\ within\ the\ range\ 690\ to\ 740\ nm$
Photochemical Reflectance Index (PRI)	$(R_{531} - R_{570})/(R_{531} + R_{5700})$
Structure Insensitive Pigment Index (SIPI)	$(R_{800} - R_{445})/(R_{800} + R_{680})$
Red Green Ratio Index (RGRI)	$mean(red\ bands)/mean(green\ bands)$
Plant Senescence Reflectance Index (PSRI)	$(R_{680} - R_{500})/R_{750}$
Carotenoid Reflectance Index 1, 2 (CRI 1,2)	$(1/R_{510}) - (1/R_{550}), (1/R_{510}) - (1/R_{700})$
Anthocyanin Reflectance Index 1, 2 (ARI 1,2)	$(1/R_{550}) - (1/R_{700}), R_{800}[(1/R_{550}) - (1/R_{700})]$
Modified Simple Ratio (MSR)	$(R_{800}/R_{670} - 1)/\sqrt{R_{800}/R_{670} + 1}$
Renormalized Difference Vegetation Index (RDVI)	$(R_{800} - R_{670})/\sqrt{R_{800} + R_{670}}$
Soil Adjusted Vegetation Index (SAVI)	$(1.5)(R_{800} - R_{670})/(R_{800} + R_{670} + 0.5)$
Improved SAVI (MSAVI)	$1/2[2R_{800} + 1 - \sqrt{(2R_{800} + 1)^2 - 8(R_{800} - R_{670})}]$
Modified Chrophyll Absorption Ration Index (MCARI)	$[(R_{700} - R_{670}) - 0.2(R_{700} - R_{550})](R_{700}/R_{670})$
MCARI1	$1.2[2.5(R_{800} - R_{670}) - 1.3(R_{800} - R_{550})]$
MCARI2	$\dfrac{1.5[2.5(R_{800} - R_{670}) - 1.3(R_{800} - R_{550})]}{\sqrt{(2R_{800} + 1)^2 - (6R_{800} - 5\sqrt{R_{670}}) - 0.5}}$
Triangular Vegetation Index (TVI)	$0.5[120(R_{750} - R_{550}) - 200(R_{670} - R_{550})]$
Modified TVI (MTVI)	$1.2[1.2(R_{800} - R_{550}) - 2.5(R_{670} - R_{550})]$
MTVI2	$\dfrac{1.5[1.2(R_{800} - R_{550}) - 2.5(R_{670} - R_{550})]}{\sqrt{(2R_{800} + 1)^2 - (6R_{800} - 5\sqrt{R_{670}}) - 0.5}}$
Water Band Index (WBI)	R_{900}/R_{970}

3.2 SVM Classifier

SVM is a learning technique derived from statistical learning theory. It is calculating an optimally separating hyperplane that maximizes the margin between two classes. If samples are not separable in the original space, kernel functions are used to map data into a higher dimensional space with a linear decision function (Abe et al. 2010).

Given a dataset with n samples $\{(x_i, y_i) \mid i = 1, \dots, n\}$ where $x_i \in \Re^k$ is a feature vector with k components and $y_i \in \{-1, 1\}$ denotes the label of x_i. The SVM looks for a hyperplane $w.\phi(x) + b = 0$ in a high dimensional space, able to separate the data from classes 1 and -1 with a maximum margin. w is a weight vector, orthogonal to the hyperplane, b is an offset term and ϕ is a mapping function which maps data into a high dimensional space to separate data linearity with a low training error. Maximizing the margin is equivalent to minimizing the norm of w. thus by solving the following minimization problem, SVM will be trained:

$$\text{Minimize: } \frac{1}{2}\|w\|^2 + C\sum_{i=1}^{n} \xi_i$$
$$\text{Subject to: } y_i(w.\phi(x) + b) \geq 1 - \xi_i \text{ and } \xi_i \geq 0, for\ i = 1, \dots, n \tag{1}$$

where C is a regularization parameter that imposes a trade-off between the number of misclassification in the training data and the maximization of the margin and ξ_i are slack variables. The decision function obtained through the solution of the minimization problem in Equation (1) is given by:

$$f(x) = \sum_{x_i \in SV} y_i \alpha_i \phi(x_i).\phi(x) + b \tag{2}$$

where the constants α_i are called Lagrange multipliers determined in the minimization process. SV corresponds to the set of support vectors, training samples for which the associated Lagrange multipliers are larger than zero. The kernel functions compute dot products between any pair of samples in the feature space. Gaussian Radial Basic Function (RBF) is a common kernel which is used in this paper and it is defined by (3).

$$K_{Gaussian}(x_i, x_j) = e^{\frac{-|x_i - x_j|}{2\sigma^2}} \tag{3}$$

In the proposed method, the classification module plays an important role in evaluation of the fitness function where SVM is trained by training data and trained SVM is evaluated by testing (unseen) data.

3.3 Bees Algorithm Optimization

Bees Algorithm is a meta-heuristic optimization algorithm that model the foraging behaviour of honey bee colony. The foraging process of honey bee colony in the nature begins in nature by scout bees which move randomly to search for promising flower patches. Flower patches with large amounts of nectar visited by more bees in neighbourhood of that site, whereas patches with less nectar receive fewer bees and other bee fly randomly for discovering new food source (Pham et al. 2006).

Bees Algorithm starts with the n scout bee move randomly in the search space. The quality of the sites visited by the scout bees (each bee represents a candidate solution) are evaluated by fitness function. Then bees that have the highest fitnesses are selected and sites visited by them are chosen for neighborhood search (m). In the next step, algorithm conducts searches in the neighborhood of the selected sites, assigning more bees to search near to the best e sites. The bees are chosen directly according to the fitnesses associated with the sites they are visiting. Searches in the neighborhood of the best e sites which represent more promising solutions are made more detailed by recruiting more bees to follow them than the other selected bees (m-e). Together with scouting, this differential recruitment is a key operation of the Bees Algorithm.

However for each patch only the bee with the highest fitness will be selected to for the next bee population. In nature, there is no such a restriction. This restriction is introduced here to reduce the number of points to be explored. Then, the remaining bees in the population are assigned randomly around the search space scouting for new potential solutions. These steps are repeated until a stopping criterion is met (Pham et al. 2006).

3.4 Determination of Optimum Classification System for Classification of Hyperspectral Imagery and LiDAR Based on Bees Algorithm

In order to determine the SVM parameters values and feature subset simultaneously based on Bees Algorithm, binary coding is applied. In the proposed method, binary string composed of three main parts is considered: features, regularization parameter and kernel parameter. The first part of binary string consist of n_f bits equal to dimension of feature space. Where '0' and '1' in the ith bit means that i^{th} feature should be discard and considered, respectively. Regularization and kernel parameters are real-valued and transform to binary coding for consistency with the binary nature of the feature selection process. The length of regularization (n_c) and kernel parameters (n_k) depends on the range of the parameters and the required precision.

Evaluation of the candidate solution is done by using a fitness function. The first part of the binary of the solution define which feature should be selected. For the determination of the SVM parameters, the binary format of the second and third parts of the solution converts to a real-value, expressed by Equation (4).

$$p = min_p + \frac{max_p - min_p}{2^l - 1} \times d \qquad (4)$$

where p is the real value of the bit string, min_p and max_p are minimum and maximum values of the parameter p, determined by the user. l is the length of the bit string (for each parameter) and d is a decimal value of the bit string.

Results may have fewer selected features and a higher classification accuracy. The combination of classification accuracy and the number of selected features constitutes the evaluation function. Multiple criteria problems can be solved by creating a single objective fitness function that combines the two goals into one. The objective function is defined by Equation (5).

$$f = \rho \times (1 - accuracy) + (1 - \rho) \times \frac{1}{N_f} \qquad (5)$$

where f is the fitness value, ρ is a constant parameter in [0,1], accuracy obtained by Kappa coefficient and N_f is the number of selected features.

The proposed method starts with generation of the candidate solutions which are formed randomly at the first iteration. Then each bee (represent by candidate solution) is evaluated by Equation (5) and the bee with higher classification accuracy and the lowest selected feature subset is selected as promising solution for the population (with maximum fitness value). Neighbourhood search around the best solutions is performed by changing the value of a random bit of that solutions. Other bees are search randomly by generating the random binary string which show the candidate solutions. This process is iterated till the termination criterion (maximum iteration) is satisfied.

4. EXPERIMENTAL RESULTS

To evaluate the performance of the proposed method, experiments are performed on Compact Airborne Spectrographic Imager (CASI) hyperspectral imagery and LiDAR derived DSM acquired by the NSF-funded Center for Airborne Laser Mapping (NCALM), both at the same spatial resolution (2.5 m). The hyperspectral imagery consists of 144 spectral bands in the spectral range between 380 nm to 1050 nm and the corresponding co-registered DSM consists of elevation in meters above sea level (Geoid 2012A model).

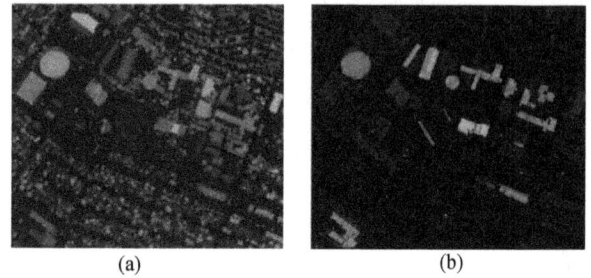

| (a) | (b) |

Figure 2. (a) LiDAR derived DSM (b) Hyperspectral imagery

Land cover classes consist of three types of grass (healthy, stressed and synthetic), road, soil, residential and commercial buildings. Spectral features (144 spectral bands, 30 vegetation indices and 3 PCs) have ability to discriminate different grass types and 2D objects; however referring to similar geometrical structure and height, LiDAR data cannot provide more information. On the other hand, spectral similarity of tree and grass/ roof and road may cause hyperspectral encounter some challenges but according to the height difference, fusion of hyperspectral imagery and LiDAR data may improve discrimination of complex urban objects.

4.1 Feature Space Generation

Generation of feature space is performed by processing both hyperspectral imagery and LiDAR data. Hyperspectral image was acquired by the CASI sensor and it has 144 bands. Moreover 30 vegetation indices are computed (Table 1). PCA transformation is applied on hyperspectral imagery and 3 first PCs with more than 99% eigenvalues are selected to complete spectral feature space. Consequently, the spectral feature space compose of 177 descriptors.

DSM derived from LiDAR data is a source of structural information. Geodesic morphological operation with circular structural element is applied on the DSM to create nDSM. Texture analysis of nDSM is performed based on GLCM features that 16 descriptors are extracted. Then roughness map and its 16 textural descriptors are also computed. Slope is further descriptor which is useful in classification, extracted from nDSM. Therefore the structural feature space is generated by merging all these 35 features.

By merging, spectral and structural feature space, a hybrid image is generated that contains rich information content for each pixel and forms our feature space with 212 features for pixel-based classification.

4.2 Classification Based on SVM

SVM classifier is applied to evaluate the quality of hybrid feature space. The SVM classification was done by using the LIBSVM through its Matlab interface (Chang and Lin, 2001). The Kappa coefficient and the overall accuracy are commonly used to determine the classification accuracy. These criteria were used to compare classification results and were computed by using the confusion matrix.

Ground truth samples are divided into training and testing data sets. The SVM classifier is trained based on training data and the best parameters are tuned and the classification performance is evaluated by unseen data (testing data). Among 7 classes, the classes tree, residential and commercial are placed in the "3D objects group", where fusion of LiDAR and hyperspectral data may improve classification results. For 2D objects, hyperspectral data are an efficient tool for discrimination among them. However 2D objects are commonly grouped as ground level in LiDAR data but the data are also useful in separating 2D and 3D objects.

Table 2 present the results of SVM classification along with determined parameters (based on grid search) for hyperspectral, LiDAR, spectral, spatial and hybrid feature space.

Table 2. Classification accuracy and SVM parameters

Dataset	C	Gamma	Kappa	OA
Hyperspectral	128	1	0.82	84.78%
LiDAR	1028	8	0.29	33.65%
Spectral	64	0.25	0.84	86.74%
Structural	64	0.25	0.47	52.35%
Hybrid Feature	4	0.25	0.87	89.13%

Obtained results show that LiDAR data are not accurate enough to classify the dataset, however by extracting the structural features, the classification accuracy improves significantly. On the other side hyperspectral data show comparable results with respect to the hybrid feature space. However the hybrid image still exhibits a superior performance through the fusion of two datasets with different information content.

4.3 Results of the Proposed Method

Although the hybrid image improve the classification accuracy, but there are several correlated and redundant features which degrade classification performance. On the other side the SVM parameters is another important elements in classification. SVM parameters influence on feature subset selection and vice versa, therefore in this section simultaneous SVM parameters tuning and feature subset selection based on Bees Algorithm is performed. Table 3 contains important values for the Bees Algorithm.

Table 3. Parameters values of Bees Algorithm

Parameters	Values
Number of bee (n)	30
Number of best bees (m)	15
Number of elite (e)	5
Neighbourhood (N_e)	4
Neighbourhood best (N_{m-e})	2
Iteration (t)	100

Figure 3 shows the convergence plots for the Bees Algorithm procedures in spectral and structural features and hybrid feature space. The fitness value for the best individual in each generation is shown. The weight parameter in objective function (Equation 5) is set to $\rho = 0.8$ which considers 80% of fitness to accuracy and 20% to dimensionality of feature space.

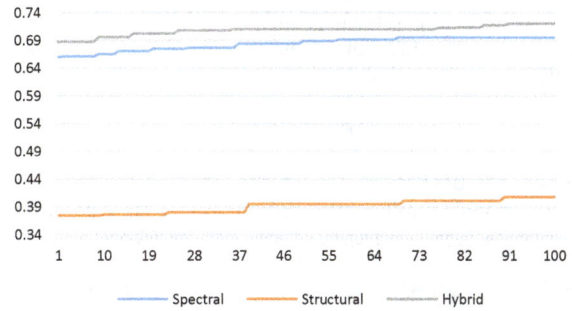

Figure 3. Convergence plot of the fitness value

Table 4 contains the number of selected features, as well as the values of regularization and kernel parameters and the classification accuracy for testing dataset, determined with the proposed method for spectral and structural and hybrid feature space.

Table 4. Results of the proposed method

Dataset	# Feature	C	Gamma	Kappa	OA
Spectral	81	131.5	0.175	0.87	89.65%
Structural	20	57.3	0.265	0.51	53.89%
Hybrid	101	76.34	0.274	0.901	92.53%

Analysing Table 4 reveals that applying the proposed method on hybrid image yields the best performance with respect to each dataset separately.

Comparing the results of hyperspectral imagery classification with the optimized classification system of hybrid images show that using DSM beside hyperspectral imagery and optimization of the SVM parameter and selection of feature subset, improve classification system approximately 8%, moreover it eliminate 111 redundant features.

5. CONCLUSION

This study investigates the framework for optimization of a hybrid classification system to fuse hyperspectral and LiDAR data based on Bees Algorithm. Experiments were carried out using CASI hyperspectral image data and a DSM derived from LiDAR data. Several spectral and structural features were extracted from hyperspectral and LiDAR data respectively. Although SVM is an appropriate classifier for this high dimensional space, its performance is optimized by simultaneously determination of parameters and selection of feature subsets.

The obtained results show that utilizing 3D information from LiDAR data in addition to high spectral information of hyperspectral data, improves the classification performance. Optimization of the hybrid classification system based on Bees Algorithm improves classification accuracy about 3.5% along with the elimination of 111 redundant features. Therefore the optimum hybrid classification system reaches more accurate results in a less complex space.

ACKNOWLEDGEMENTS

The authors would like to thank the Hyperspectral Image Analysis group and the NSF Funded Center for Airborne Laser

Mapping (NCALM) at the University of Houston for providing the data sets used in this study, and the IEEE GRSS Data Fusion Technical Committee for organizing the 2013 Data Fusion Contest.

REFERENCES

Alonzo, M., Bookhagen, B., Roberts, D.A., 2014. Urban tree species mapping using hyperspectral and lidar data fusion. *Remote Sensing of Environment*, Vol. 148, pp. 70-83.

Brook, A., Ben-Dor, E., Richter, R., 2010. Fusion of Hyperspectra Images and LiDAR Data for Civil Engineering Structure Monitoring. *2nd workshop on hyperspectral image and signal processing: Evolution in Remote Sensing (WHISPERS)*, pp. 1-5.

Chang, C. I. *Hyperspectral Data Processing: Algorithm Design and Analysis*, JohnWiley & Sons, Inc., USA, 2013.

Dalponte, M., Bruzzone, L., and Gianelle, D., 2008. Fusion of Hyperspectral and LIDAR Remote Sensing Data for Classification of Complex Forest Area. *IEEE Transaction on Geoscience and Remote Sensing*, Vol. 46, No. 5, pp. 1416-1427.

Debes, C., Merentitis, A., Heremans, R., Hahn, J., Frangiadakis, N., van Kasteren, T., Liao, W., Bellens, R., Pizurica, A., Gautama, S. and Philips, W., Prasad, S., Du, Q., Pacifici, F., 2014. Hyperspectral and LiDAR data fusion: Outcome of the 2013 GRSS Data Fusion Contest. *IEEE Journal of Selected Topics in Applied Earth Observations and Remote Sensing*.

Fauvel, M., (2007). Spectral and Spatial Method for Classification of Urban Remote Sensing Data, Ph.D thesis.

Haralick, R.M., Shanmugam, K., Dinstein, I., 1973. Textural Features for Image Classification. *IEEE Transaction on Systems, Man and Cybernetics*, Vol. SMC-3, No. 6, pp. 610-621.

Hsu, C.W., Chang, C.C., and Lin, C.J., 2003. A Practical Guide to Support Vector Classification, *National Technical Report*, Taiwan University.

Khodadadzadeh, M., Li, J., Prasad, S., Plaza, A., (2015). Fusion of Hyperspectral and LiDAR Remote Sensing Data Using Multiple Feature Learning. *IEEE Journal of Selected Topics in Applied Earth Observations and Remote Sensing*, Vol. 8, No. 6.

Koetz, B., Morsdorf, F., van der Linden, S., Curt, T., Allgöwer, B., 2008. Multi-source Land Cover Classification for Forest Fire Management based on Imaging Spectrometry and LiDAR data. *Forest Ecology and Management*, Vol. 256, pp. 263-271.

Latifi, H., Fassnacht, F., Koch, B., 2012. Forest Structure Modeling with Combined Airborne Hyperspectral and LiDAR Data. *Remote sensing of Environment*, Vol. 121, pp.10-25.

Lin, S.W., Ying, K.C., Chen, S.C., and Lee, Z.J., 2008. Particle Swarm Optimization for Parameter Determination and Feature Selection of Support Vector Machines. *Expert Systems with Applications*, Vol. 35, No. 4, pp. 1817-1824.

Liu, Q.J., Jing, L.H., Wang, L.M., Lin, Q.Z., 2014. A Method of Particle Swarm Optimized SVM Hype-spectral Remote Sensing Image Classification. *35th International Symposium on Remote Sensing of Environment*.

Niemann, O.O., Gordon, W.F., Rafael, L., Fabio, V., 2009. LiDAR-Guided Analysis of Airborne Hyperspectral Data. First Workshop on Hyperspectral Image and Signal Processing: Evolution in Remote Sensing, pp. 1-4.

O'Boyle, N.M., Palmer, D.S., Nigsch, F., and Mitchell, J.B.O., 2008. Simultaneous Feature Selection and Parameter Optimisation using an Artificial Ant Colony: Case Study of Melting Point Prediction. *Chemistry Central Journal*, Vol. 2, pp.21-25.

Ramdani, F., 2013. Urban Vegetation Mapping from Fused Hyperspectral Image and LiDAR Data with Application to Monitor Urban Tree Heights. *Journal of Geographic Information System*, 5(4).

Rashedi, E., Nezamabadi-pour, H., 2014. Feature Subset Selection Using Improved Binary Gravitational Search Algorithm. *Journal of Intelligent and Fuzzy Systems*, Vol. 26, No. 13, pp. 1211-1221.

Samadzadegan, F., Hasani, H., Schenk, T., 2012. Determination of optimum classifier and feature subset in hyperspectral images based on ant colony system. *Photogrammetric Engineering & Remote Sensing*, Vol. 78, No. 12, pp. 1261-1273.

Sugumaran, R., Voss, M., 2007. Object-Oriented Classification of LiDAR-Fused Hyperspectral Imagery for Tree Species Identification in an Urban Environment. *Urban Remote Sensing Joint Event*, pp. 1-6.

Zhang, C., Qiu, F., 2012. Mapping Individual Tree Species in an Urban Forest using Airborne LiDAR Data and Hyperspectral Imagery, *Photogrammetric Engineering and Remote Sensing*. Vol. 78, No. 10, pp. 1079-1087. (AAG Remote Sensing Specialty Group 2011 Award Winner).

2013 IEEE GRSS Data Fusion Contest, Online: http://www.grss-ieee.org/community/technical-committees/data-fusion/

31

THE EFFECT OF LAND USE CHANGE ON LAND SURFACE TEMPERATURE IN THE NETHERLANDS

S. Youneszadeh [a, *], N. Amiri [b,c], P. Pilesjo [a]

[a] Dept. of Physical Geography and Ecosystem Science, Lund University, Lund, Sweden - s.youneszadeh@gmail.com, Petter.Pilesjo@gis.lu.se
[b] Dept. of Geoinformatics, Munich University of Applied Sciences, Munich, Germany - n.amiri@hm.edu
[c] Faculty of Geoinformation science and Earth observation, ITC, University of Twente, Enschede, The Netherlands – n.amiri@utwente.nl

Commission VI, WG VI/4

KEY WORDS: GIS, remote sensing, MODIS, land surface temperature, LST

ABSTRACT

The Netherlands is a small country with a relatively large population which experienced a rapid rate of land use changes from 2000 to 2008 years due to the industrialization and population increase. Land use change is especially related to the urban expansion and open agriculture reduction due to the enhanced economic growth. This research reports an investigation into the application of remote sensing and geographical information system (GIS) in combination with statistical methods to provide a quantitative information on the effect of land use change on the land surface temperature. In this study, remote sensing techniques were used to retrieve the land surface temperature (LST) by using the MODIS Terra (MOD11A2) Satellite imagery product. As land use change alters the thermal environment, the land surface temperature (LST) could be a proper change indicator to show the thermal changes in relation with land use changes. The Geographical information system was further applied to extract the mean yearly land surface temperature (LST) for each land use type and each province in the 2003, 2006 and 2008 years, by using the zonal statistic techniques. The results show that, the inland water and offshore area has the highest night land surface temperature (LST). Furthermore, the Zued (South)-Holland province has the highest night LST value in the 2003, 2006 and 2008 years. The result of this research will be helpful tool for urban planners and environmental scientists by providing the critical information about the land surface temperature.

1. INTRODUCTION

This Earth system is a complicated cycle with many interconnected components like the Earth's surface and it's interior. Naturally the Earth surface is covered by different land cover types which are mainly distributed based on the environmental and climatically patterns. By adding the rapidly increasing human population and his needs to this balanced system, we will face with many disturbances from the concept of how we change the use of land due to our needs based on its capacity or environmental impact.

Land use is defined as "the arrangements, activities and inputs people undertake in a certain land cover type to produce a change or maintain it" (FAO/UNEP, 1999). Land use is a change over the time and the most important and primary factor in land use changes is the human need. Human population as settlements and especially large urban and industrial areas could significantly modify their sounding environment. Therefore, it is critical to have a detailed information of temporal and spatial land use changes and their rate. Land use should be matched with land capability and at the same time it should respect to the environment, and global climate systems (UNEP, 1996).

The land surface temperature (LST) is the temperature of the skin surface of a land which can be derived from the satellite information or direct measurements in the remote-sensing terminology. LST is the surface radiometric temperature emitted by the land surfaces and observed by a sensor at instant viewing angles (Prata et al., 1995). This is an accurate measurement tool for indicating the energy exchange balance between the atmosphere and the Earth. The degree of land surface temperature (LST) is affected by the surface attributes, which are significantly influenced by the elevation, slope and aspect. However, topography is one of the factors that controls the soil moisture distribution and exerting an additional influence on the land surface temperature.

The Netherlands has an almost flat topography, so it can be a proper case to separate the effect of topographic factors from the land use properties on land surface temperature (LST) behaviour analysis. The combination of the land use analysis result with the mean land surface temperature (LST) can offer useful information to study the urban land use change effects in the Dutch cities.

In this study, we investigate the effects of land use change on the land surface temperature in the Netherlands provinces based on the remote sensing analysis and zonal statistic methods.

2. DATA COLLECTION

In this research, two different data sources were used for our analysis, which are presented in the Figure 1. The land surface temperature were available from the NASA website by MODIS Terra (temporal interval of 8 days) and the land use data

---

* Corresponding author

provided from the land use base of the statistics Netherlands (temporal interval of 2 or 3 years).

Data type	Data source	Product name	Platform	Projection system	Type	Res(m)/Scale	Temporal interval	Temporal coverage
Land surfaces temperature	NASA website	MODIS LST/MOD11A2	Terra	sinusoidal projection	Raster/HDF	1000m	8 Day	2000 onward
Land use	Land Use Base of Statistics Netherlands	BBG land use map	-	RD-New	Vector	1:10,000	2 or 3 years	1996,2000,2003,2006,2008

Figure 1. Overview of the datasets

3. METHODOLOGY

In the global MODIS LST tiling system, the Netherlands is located in tile h18v03. This tile is gridded in a network of approximately 1 km by 1 km. The BBG land use maps (updated detailed national land use map through visual analysis of aerial photography, called "Bestand Bodemgebruik") of the 2003, 2006 and 2008 years were used to assess the land use types. Multiple land uses can be present in 1 km^2 LST pixel size and each of the land use classes can affect the LST mixed value. So it was needed to have the land use change maps with the same resolution as the LST images (1 km^2) are available, including the areal proportions of each land use within each LST pixel.

Dutch BBG land use map contains 8 main classes: Traffic, Build-up areas, Semi Build-up areas, Recreational, Agricultural, Forest and natural open land, Inland waterway and offshore areas. These classes were reclassified to new categories in which different classes had different heat capacity, emissivity and reflection characteristics. The BBG land use maps were reclassed into 6 classes, using the dissolve function of Arc Map. New classes are open agriculture, build-up areas, recreational areas, greenhouse farming, inland waterway, offshore areas and forest. To summarize the values of the LST within each province and land use type and to provide tangible and practical information for decision making, zonal statistic methods were used. The spatial mean LST was then computed for all Dutch provinces and land use types. To achieve this, yearly mean LST raster files of the 2003, 2006 and 2008 years were used. Nearly 45 images for each year. To calculate the average LST of each province, the LST image values were converted to integer values, using the Int. function of the spatial analyst tool in the ArcGIS 10 software. Afterwards, the zonal statistic function was applied to calculate the mean LST.

4. RESULTS AND DISCUSSION

The Thermal characteristics, conductivity, albedo, surface roughness and heat capacity are among the fundamental factors which affect the amount of the LST of different land uses (Brovkin et al., 2006; Davin & de Noblet-Ducoudré, 2010). The spatial arrangement, area, adjacent land uses and connectivity of different land uses also have impact on the mixed value of LST for each pixel. Reference (van Leeuwen et al., 2011) argued that the LST is regulated by the several parameters e.g. surface conductance, amount of water available for evaporative cooling, wind speed, and surface roughness which regulates the power of sensible and latent heat fluxes.

Year	2003	2006	2008
Mean LST for All the Pixels with Open agriculture/ Coverage percentage of the land use in the pixels>95%	4.45	5.57	5.49
Mean LST for All the Pixels with Inland waterway and offshore area/ Coverage percentage of the land use in the pixels >95%	7.90	8.30	8.21
Mean LST for All the Pixels with Forest/ Coverage percentage of the land use in the pixels >95%	4.59	5.02	4.85
Mean LST for All the Pixels with Greenhouse farming/ Coverage percentage of the land use in the pixels >80%	4.93	4.31	5.27
Mean LST for All the Pixels with Build-up area/ Coverage percentage of the land use in the pixels >95%	6.48	6.98	6.93

Table 1. The average of mean yearly night LST for different land use types in the 2003, 2006 and 2008 years

Table 1 shows that the lowest LST in 2003 year is observed in open agriculture, followed by forest, greenhouse farming, build-up areas, inland waterway and offshore areas. The 2006 year pattern is slightly different, where the lowest LST is found in greenhouse farming, followed by forest, open agriculture, build-up areas, inland waterway and offshore areas. In 2008 year, the lowest LST is for forest followed by greenhouse farming, open agriculture, build-up areas, inland waterway and offshore area.

These results are consistent with the study of Weng et al., in 2004 (Weng, Lu, & Schubring, 2004). They suggest that the higher biomass/vegetation abundance a land cover has, the lower the land surface temperature is. Furthermore, vegetative land uses possessed a smaller mean value than inland water and build-up area. Forest has the least LST. This result may be explained by the fact that forests strong vegetation can decrease the amount of heat stored in the soil and surface structures through the transpiration (Weng & Lu, 2008). A possible explanation for this is argued by Qian et al., in 2006 (Qian, Cui, & Chang, 2006). They discussed that these changes can stem from the discrepancy in solar illumination, atmospheric influences, and soil moisture content in the different study years.

Table 2 indicates the annual mean value of the LST for different provinces for the 2003, 2006 and 2008 years. Zued-Holland has the highest LST value in the 2003, 2006 and 2008 years. The range of LST for the year 2003 is from 3.39 to 6.39 °C. In the 2006 year, it is ranging from 4.02 to 7.21 °C. The range for 2008 year is from 4.15 to 7.05 °C. The mean LST value for each province can be an appropriate decision-making factor and an environmental warning tool for urban and environmental planners.

Province	Mean LST (°C), 2003	Mean LST (°C), 2006	Mean LST (°C), 2008
Groningen	4.38	5.99	5.93
Drenthe	4.37	5.81	5.48
Overijssel	4.42	5.39	5.46
Gelderland	4.78	5.39	5.40
Nord-Brabant	5.83	6.64	6.62
Limburg	5.84	6.68	6.57
Friesland	4.21	5.49	5.39
Nord-Holland	5.83	6.78	6.68
Zued-Holland	6.30	7.21	7.05
Zeeland	6.24	6.97	6.79
Utrecht	5.00	5.74	5.76
Flevoland	3.39	4.02	4.15

Table 2. The mean night LST value for each province

The zonal map of the 2006 year is provided. See Figure 1 which shows the average of the mean yearly night LST for each province. The values are the LST original raw values. To convert to Celsius, each value should be multiplied by 0.02 and subtracted from the 273.15 (Zhengming, 2007). The Zued-

Holland for all the three years has the largest mean yearly LST. Nord-Holland and Nord-Brabant are among the high LST provinces. The most urbanized parts of the country is Randstad which comprises the major cities of Amsterdam, Rotterdam, Utrecht and The Hague has around 5 million inhabitants (de Nijs et al., 2004). It can be noted that in the hot urbanized spots, the amount of mean LST is higher. Reference Zhou & Wang in 2011, (Zhou & Wang, 2011) argued that "apart from land-use change, urbanization with increased human population also contributes to the urban thermal environment change with the rising anthropogenic heat discharge."

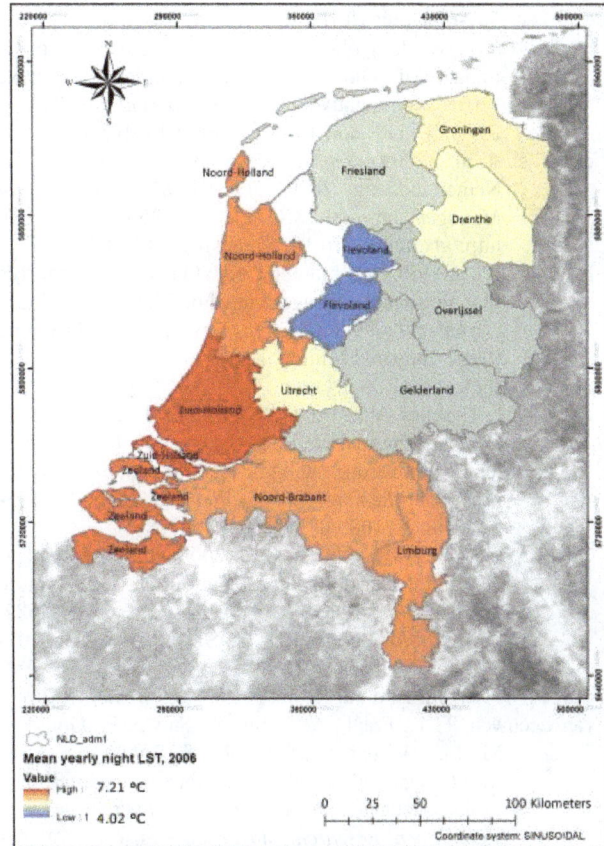

Figure 1. The up-scaled mean yearly night LST mean to the Dutch provinces scale in 2006 year

5. CONCLUSIONS

The above discussion has evaluated the effect of different land use types on the land surface temperature. The mean yearly value of LST is calculated. This value can serve as a suitable managerial tool for urban planner and environmental scientists. Combining the result of land use analysis and mean LST can offer useful information to study urban heat island in the Dutch cities. Zued-Holland has the highest amount of LST from 2003 to 2008 years. The amount of LST for this province has increased from 2003 to 2008 years by 0.75 °C which could be due to urbanization, population increase and also possible variations in weather temperature. In the same period, Flevoland which has the minimum amount of LST is also increased from the 3.39 °C in the 2003 year to the 4.15 °C in the 2008 year. The amount of LST for different land use types can be used for constructing the thermal environment of the Dutch provinces in relation to the land use types.

ACKNOWLEDGEMENTS

Our thanks to ACM SIGCHI for allowing us to modify templates they had developed.

REFERENCES

Brovkin, V., Claussen, M., Driesschaert, E., Fichefet, T., Kicklighter, D., Loutre, M. F., . . . Sokolov, A. (2006). Biogeophysical effects of historical land cover changes simulated by six Earth system models of intermediate complexity. *Climate Dynamics, 26*(6), 587-600. doi: 10.1007/s00382-005-0092-6

Davin, E. L., & de Noblet-Ducoudré, N. (2010). Climatic Impact of Global-Scale Deforestation: Radiative versus Nonradiative Processes. *Journal of Climate, 23*(1), 97-112. doi: 10.1175/2009JCLI3102.1

de Nijs et al. (2004). Constructing land-use maps of the Netherlands in 2030. *Journal of Environmental Management, 72*(1–2), 35-42. doi: http://dx.doi.org/10.1016/j.jenvman.2004.03.015

FAO/UNEP. (1999). Terminology for Integrated Resources Planning and Management. Food and Agriculture Organization/United Nations Environmental Program, Rome, Italy and Nairobi, Kenya

Table 3 -FAO.

Prata, A. J., Caselles, V., Coll, C., Sobrino, J. A., ., & Ottlé, C. (1995). Thermal remote sensing of land surface temperature from satellites: Current status and future prospects. *Remote Sensing Reviews, 12*(3-4), 175-224. doi: 10.1080/02757259509532285

Qian, L.-X., Cui, H.-S., & Chang, J. (2006). Impacts of Land Use and Cover Change on Land Surface Temperature in the Zhujiang Delta1. *Pedosphere, 16*(6), 681-689. doi: http://dx.doi.org/10.1016/S1002-0160(06)60103-3

UNEP. (1996). 0ur land our future. *FAO/AGLS.Rome. 48. P.*

van Leeuwen, T. T., Frank, A. J., Jin, Y., Smyth, P., Goulden, M. L., van der Werf, G. R., & Randerson, J. T. (2011). Optimal use of land surface temperature data to detect changes in tropical forest cover. *Journal of Geophysical Research: Biogeosciences, 116*(G2), n/a-n/a. doi: 10.1029/2010JG001488

Weng, Q., & Lu, D. (2008). A sub-pixel analysis of urbanization effect on land surface temperature and its interplay with impervious surface and vegetation coverage in Indianapolis, United States. *International Journal of Applied Earth Observation and Geoinformation, 10*(1), 68-83. doi: http://dx.doi.org/10.1016/j.jag.2007.05.002

Weng, Q., Lu, D., & Schubring, J. (2004). Estimation of land surface temperature–vegetation abundance relationship for urban heat island studies. *Remote Sensing of Environment, 89*(4), 467-483. doi: http://dx.doi.org/10.1016/j.rse.2003.11.005

Zhengming, W. (2007). Collection-5-MODIS Land Surface Temperature Products-Users' Guide. *ICESS, University of California, Santa Barbara.*

Zhou, X., & Wang, Y.-C. (2011). Dynamics of Land Surface Temperature in Response to Land-Use/Cover Change. *Geographical Research, 49*(1), 23-36. doi: 10.1111/j.1745-5871.2010.00686.x

32

ANALYSING THE EFFECTS OF DIFFERENT LAND COVER TYPES ON LAND SURFACE TEMPERATURE USING SATELLITE DATA

A. Şekertekin [a, *], Ş. H. Kutoglu [a], S. Kaya [b], A. M. Marangoz [a]

[a] BEU, Engineering Faculty, Geomatics Engineering Department 67100 Zonguldak, Turkey - (aliihsan_sekertekin, kutogluh, aycanmarangoz)@hotmail.com
[b] ITU, Civil Engineering Faculty, 80626 Maslak Istanbul, Turkey - (kayasina)@itu.edu.tr

SMPR 2015

KEY WORDS: Land Surface Temperature, Urbanization, Climate Change

ABSTRACT

Monitoring Land Surface Temperature (LST) via remote sensing images is one of the most important contributions to climatology. LST is an important parameter governing the energy balance on the Earth and it also helps us to understand the behavior of urban heat islands. There are lots of algorithms to obtain LST by remote sensing techniques. The most commonly used algorithms are split-window algorithm, temperature/emissivity separation method, mono-window algorithm and single channel method. In this research, mono window algorithm was implemented to Landsat 5 TM image acquired on 28.08.2011. Besides, meteorological data such as humidity and temperature are used in the algorithm. Moreover, high resolution Geoeye-1 and Worldview-2 images acquired on 29.08.2011 and 12.07.2013 respectively were used to investigate the relationships between LST and land cover type. As a result of the analyses, area with vegetation cover has approximately 5 °C lower temperatures than the city center and arid land., LST values change about 10 °C in the city center because of different surface properties such as reinforced concrete construction, green zones and sandbank. The temperature around some places in thermal power plant region (ÇATES and ZETES) Çatalağzı, is about 5 °C higher than city center. Sandbank and agricultural areas have highest temperature due to the land cover structure.

1. INTRODUCTION

In recent years, climate change has been one of the most important problems that the ecological system of the world has been encountering. Global warming and climate change have been studied frequently by all disciplines all over the world and Geomatics Engineering also contributes to such studies by means of remote sensing, Global Navigation Satellite System (GNSS) etc.

Retrieving LST is crucial for climate change, especially for understanding urban heat islands and local climate changes (Voogt & Oke, 2003; Kaya et al., 2012). The most commonly LST retrieval algorithms are split-window algorithm (Sobrino et al., 1996), temperature/emissivity separation method (Gillespie et al., 1998), mono-window algorithm (Qin et al., 2001) and single channel method (Jimenez-Munoz & Sobrino, 2003). Generally three algorithms are used to obtain LST by using Landsat 5 TM data. These algorithms are radiative transfer equation method, single channel method and mono-window algorithm. Radiative transfer equation method is not applicable because during the satellite pass, atmospheric parameters must be measured in-situ. Single channel method and mono-window algorithm both present satisfying results. However, mono-window algorithm can be implemented simply and practically because of using simulated linear transformation equations for some parameters.

2. STUDY AREA

The study area, Zonguldak is located on the coast of Western Black Sea region of Turkey (Figure 1). The city is also one of the most forested cities in Turkey. However, with the exploitation of the coal reserves, progressive deforestation and massive loss of wetlands followed on. As a study area, Zonguldak has some characteristic areas close to each other such as industrial areas, sandbank and forested land.

Figure 1. Study Area Zonguldak, Turkey

3. MATERIAL AND METHOD

In this research, Landsat 5 TM image acquired on 28.08.2011 was used to retrieve LST. High resolution satellite images, taken from Geoeye and Worlview-2, acquired on 29.08.2011 and 12.07.2013, respectively were used for regional analyses. Furthermore, meteorological data such as humidity and temperature were used in the algorithm.

---

* Corresponding author

LST retrieval algorithms using remote sensing data are dependent on data specifications such as spectral features and number of thermal bands. Considering LST retrieval from Landsat 5 TM sensor data, it is usually preferred one of the algorithms, namely radiative transfer equation method, mono-window algorithm and single-channel algorithm.

Radiative transfer equation method reveals the best results, but it requires in-situ radiosonde measurement of the atmospheric parameters during the satellite passes. Mono-window algorithm is also an effective method which provides satisfying values of root mean square deviation (Sobrino et al., 2004). Therefore, mono-window algorithm can be preferred in the absence of radiosonde data, and so mono-window algorithm has been chosen in this research.

Mono-window algorithm, developed by Qin et al. (2001) for Landsat 6 TM data, was used in this research. It is recommended for LST retrieval of the images with one thermal band. The method includes three main parameters, namely emissivity, atmospheric transmittance and effective mean atmospheric temperature (Qin et al., 2001). In order to apply mono-window algorithm, the required parameters are computed in a step-wise fashion which is elucidated below.

- Converting Digital Numbers (DNs) To Spectral Radiance Values
- Converting Spectral Radiance to Reflectance Values
- Converting Spectral Radiance to Brightness Temperature Values
- Estimation of Emissivity Values (Depends on Normalized Difference Vegetation Index (NDVI))
- Estimation of Atmospheric Transmittance (Depends on water vapour content)
- Calculation of Mean Atmospheric Temperature
- Calculation of LST

Firstly the images were prepared for processing. They were resampled and clipped as including the study area. Secondly radiometric corrections like converting digital numbers to spectral radiance values and converting spectral radiance to reflectance values were conducted. Then spectral radiance values of thermal bands were converted to brightness temperature by means of equation (1).

$$T = \frac{K_2}{\ln\left(\frac{K_1}{L_\lambda} + 1\right)} \qquad (1)$$

where T = effective at-satellite temperature in Kelvin
L_λ = spectral radiance at the sensor's aperture
K_1, K_2 = calibration constants

The other step is the estimation of Land Surface Emissivity (LSE) by using NDVI. A detailed estimation of LSE from NDVI was proposed by Zhang et al. (2006).

After that the estimation of mean atmospheric temperature (T_a) via atmospheric temperature at ground (T_o) was proposed by Qin et al. (2001). Estimation of the last parameter, atmospheric transmittance (τ_i), could be estimated from water vapor content (w_i) as demonstrated in Table 1 (Qin et al., 2001).

Profiles	Water Vapor (w_i)(g/cm^2)	Transmittance estimation equation (τ_i)	Squared correlation	Standard Error
High Air Temperature	0.4-1.6	0.974290-0.08007×w_i	0.99611	0.002368
	1.6-3.0	1.031412-0.11536×w_i	0.99827	0.002539
Low Air Temperature	0.4-1.6	0.982007-0.09611×w_i	0.99563	0.003340
	1.6-3.0	1.053710-0.14142×w_i	0.99899	0.002375

Table 1. Estimation of atmospheric transmittance from water vapour

Finally LST values could be obtained from equation (2), the equation of mono-window algorithm.

$$T_s = \{a \cdot (1-C-D) + [b \cdot (1-C-D) + C + D] \cdot T_i - D \cdot T_a\} \div C \qquad (2)$$

where $a = -67.355351$
$b = 0.458606$
$C = \varepsilon_i \times \tau_i$
$D = (1 - \tau_i)[1 + (1 - \varepsilon_i) \times \tau_i]$
T_s = LST in Kelvin
T_i = brightness temperature in Kelvin
T_a = effective mean atmospheric temperature
τ_i = atmospheric transmittance
ε_i = land surface emissivity
a, b = algorithm constants

4. RESULTS

A model of the mono-window algorithm is illustrated in Figure 2. The whole process in this figure is programmed in Erdas Imagine Spatial Modeler for processing the data automatically. After obtaining LST, the image was classified using threshold method (Figure 3). Regional LST changes were analyzed using high resolution Geoeye and Worldview-2 images (Figure 4). Sample points over different land cover types were chosen and the results were presented in the graph below (Figure 5).

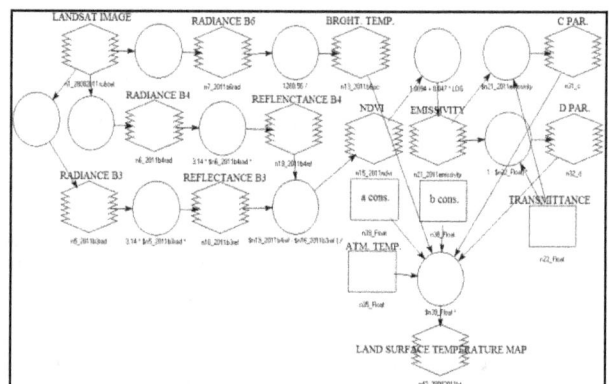

Figure 2. LST image model created in Erdas Imagine Spatial Modeler

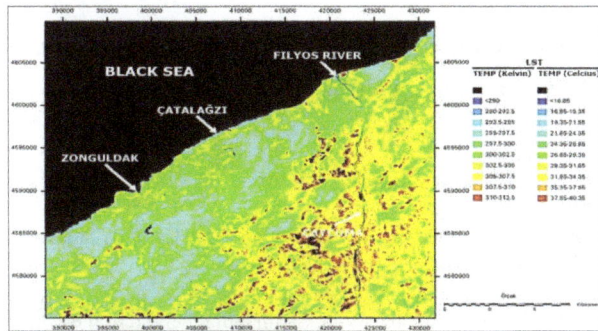

Figure 3. Classified LST image of study area

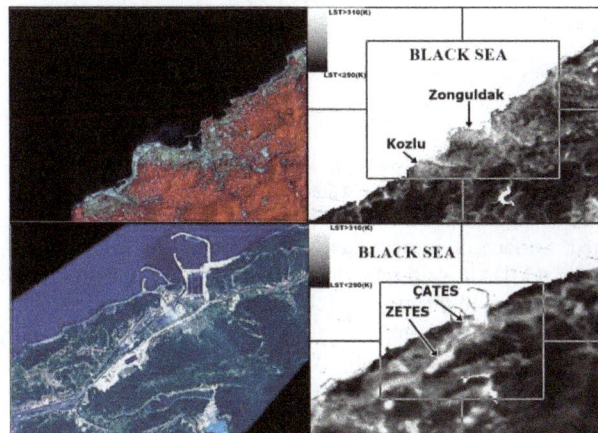

Figure 4. Association of high resolution images (left) and LST image (right)

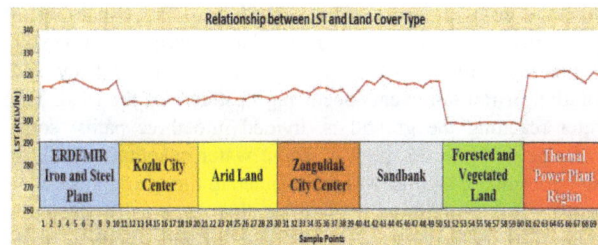

Figure 5. Relationship between LST and land cover types

Accuracy assessment of LST image was conducted using the linear correlation coefficient between the radiance and LST values. In general, the temperature values of LST image and local meteorological stations are compared for this evaluation. Because there is only one meteorological station in Zonguldak, the radiance and LST values were compared with each other for the accuracy assessment and the result was 86%.

In this study, some characteristic areas like industrial regions, city centers, sandbank, arid land, vegetated and forest land were evaluated as different land cover types with regard to LST. As a result of the analyses, area with vegetation cover has approximately 5 °C lower temperatures than the city center and arid land., LST values change about 10 °C in the city center because of different surface properties such as reinforced concrete construction, green zones and sandbank. The temperature around some places in thermal power plant region Çatalağzı, is about 5 °C higher than city center. Sandbank and agricultural areas have highest temperature due to the land cover structure. Furthermore, two plants ÇATES and ZETES also have high LST values compared the other land cover types.

Satellite imagery is an effective method to retrieve LST maps for large areas. Land cover types and the materials used as surface structure affect LST directly. Thus, it should be considered not to use materials in city centres that absorb the sun radiation too much for urban heat island effect. LST maps can be generated periodically by means of satellite images. Thus, it can not only be useful for agricultural activities but also for preparing a substructure especially for regional climate change researches.

ACKNOWLEDGEMENTS

Geoeye and Worldview-2 satellite images have been provided by BEUN Scientific Research Project: 2012-17-12-03.

REFERENCES

Gillespie, A. R., Rokugawa, S., Matsunaga, T., Cothern, J. S., Hook, S. J., & Kahle, A. B. 1998. A temperature and emissivity separation algorithm for advanced space borne thermal emission and reflection radiometer (ASTER) images. IEEE Transactions on Geoscience and Remote Sensing, 36, 1113-1126. doi:10.1109/36.700995

Jimenez-Munoz, J. C., & Sobrino, J. A. 2003. A generalized single-channel method for retrieving land surface temperature from remote sensing data. Journal of Geophysical Research, 108(D22), 4688. doi:10.1029/2003JD003480

Kaya, S., Basar, U. G., Karaca, M., & Seker, D. Z. 2012. Assessment of Urban Heat Islands Using Remotely Sensed Data. Ekoloji, 21(84), 107-113. doi: 10.5053/ekoloji.2012.8412

Qin, Z., Karnieli, A., & Berliner, P. 2001. A mono-window algorithm for retrieving land surface temperature from Landsat TM data and its application to the Israel-Egypt border region. International Journal of Remote Sensing, 22(18), 3719-3746. doi:10.1080/01431160010006971

Sobrino, J. A., Li, Z. L., Stoll, M. P., & Becker, F. 1996. Multi-channel and multi-angle algorithms for estimating sea and land surface temperature with ATSR data. International Journal of Remote Sensing, 17, 2089-2114. doi:10.1080/01431169608948760

Sobrinoa, J. A., Jimenez-Munoz, J. C., & Paolini, L. 2004. Land surface temperature retrieval from LANDSAT TM 5. Remote Sensing of Environment, 90, 434–440. doi:10.1016/j.rse.2004.02.003

Voogt, J. A., & Oke, T. R. 2003. Thermal remote sensing of urban climates. Remote Sensing of Environment, 86, 370–384. doi:10.1016/S0034-4257(03)00079-8

Zhang, J., Wang, Y., & Li, Y. 2006. A C++ program for retrieving land surface temperature from the data of Landsat TM/ETM+ band6. Computers & Geosciences, 32, 1796–1805. doi:10.1016/j.cageo.2006.05.001

33

INFLUENCE OF THE PRECISION OF LIDAR DATA IN SURFACE WATER RUNOFF ESTIMATION FOR ROAD MAINTENANCE

H. González-Jorge [a, *], L. Díaz-Vilariño [a], S. Lagüela [b], J. Martínez-Sánchez [a], P. Arias [a]

[a] Applied Geotechnologies Group, Dept. Natural Resources and Environmental Engineering, University of Vigo, Campus Lagoas-Marcosende, CP 36310 Vigo, Spain (higiniog, lucia, joaquin.martinez, parias)@uvigo.es

[b] Department of Cartographic and Terrain Engineering, University of Salamanca, 05003 Ávila, Spain (sulaguela)@usal.es

Commission III, WG III/2

KEY WORDS: Mobile LiDAR, Point Cloud, runoff, road maintenance.

ABSTRACT

Roads affect the natural surface and subsurface drainage pattern of a hill or a watershed. Road drainage systems are designed with the objective of reducing the energy generated by the flowing water and the presence of excess water or moisture within the road. A poorly designed drainage may affect to road maintenance causing cut or fill failures, road surface erosion and degrading the engineering properties of the materials with which it was constructed. Surface drainage pattern can be evaluated from Digital Elevation Models typically calculated from point clouds acquired with aerial LiDAR platforms. However, these systems provide low resolution point clouds especially in cases where slopes with steep grades exist. In this work, Mobile LiDAR systems (aerial and terrestrial) are combined for surveying roads and their surroundings in order to provide complete point cloud. As the precision of the point clouds obtained from these mobile systems is influenced by GNSS outages, Gaussian noise with different standard deviation values is introduced in the point cloud in order to determine its influence in the evaluation of water runoff direction. Results depict an increase in the differences of flow direction with the decrease of cell size of the raster dataset and with the increase of Gaussian noise. The last relation fits to a second-order polynomial Differences in flow direction up to 42° are achieved for a cell size of 0.5 m with a standard deviation of 0.15 m.

1. INTRODUCTION

The World Bank (2015) states that road construction includes design, contracting, implementation, supervision, and maintenance. Proper road maintenance contributes to reliable transport at reduced cost. An improperly maintained road can also represent an increased safety hazard to the user, producing more accidents.

The activities of road maintenance can be divided into three main categories: routine works, periodic works, and special works. Routine works are undertaken each year and funded by a recurrent budget. Examples are verge cutting, culvert cleaning, and patching, which is carried out in response to the appearance of cracks of pot-holes. Periodic works include activities undertaken at intervals of several years to preserve the structural integrity of the road such as resealing and overlay works. Special works are the activities for which demand cannot be estimated with in advance, and with reasonable certainty. These activities include emergency works to repair landslides and washouts removal or salting. Too much water flowing in too narrow channels over destabilized soil can produce washouts. Washouts that occur on road surfaces are generally a result of inadequate grading that allows water to channelize rather than staying spread over the whole surface. To avoid this, roads should be properly crowned, road shoulder false berms should be removed or never allowed to form, and cross drainage should be kept free and clear of debris or deposited soil. Roads need to be good quality stable gravel that resists the forces of water and traffic loads.

Drainage installations are sized according to the probability of occurrence of an expected peak discharge during the lifecycle of the installation. This fact is related to the intensity and duration of rainfall events occurring upstream of the road. The water reaching the ground is divided into three paths: some water percolates into the soil, some water evaporates back to the atmosphere, and the rest contributes to the overland flow or runoff. The proportion of rainfall that eventually becomes streamflow is dependent on the size of the drainage area, topography, and soil (Barreiro et al, 2014; Caine 1980).

A popular method to determine flow direction is the D8 algorithm that defines the flow direction in any raster cell through the evaluation of the cell along with its eight neighbouring cells. Digital Elevation Models (DEM) are the input data for the D8 calculation (Douglas 1986; Wang et al, 2014) and water is supposed to follow the steepest descent.

DEM are typically calculated from point clouds that contain geometric information from the environment. Point clouds for these applications are usually provided by means of mobile platforms, aerial and terrestrial. Aerial LiDAR platforms provide results with lower spatial resolution, while terrestrial platforms give higher spatial resolution (Puente et al, 2013; Meesuk et al, 2015). Mobile LiDAR systems combine global navigation satellite systems (GNSS) and inertial measurement units (IMU) for positioning and orientation, with LiDAR systems for range measurements (Petri 2010). All systems are time stamped and boresighted to provide a geo-referenced point cloud.

---

* Corresponding author.

The quality of the point cloud is affected by external factors that contribute to decrease the precision. One example is the dilution of precision in the GNSS that typically occurs in mountain roads with high slopes, urban areas, and forested roads. The surveying methodology tries to control these aspects, although sometimes it is difficult and the precision of measurements decreases affecting the quality of the point cloud and the derived DEM.

The present work focuses on two aspects. On one hand, it combines the use of airborne LiDAR and terrestrial mobile LiDAR for the evaluation of road runoff. In this way the aerial LiDAR provides information from the top of the mountains surrounding the road and the terrestrial mobile LiDAR provides information from the road slopes and pavement. On the other hand, the influence of LiDAR precision in the evaluation of runoff direction is calculated taking into account the cell size of the DEM and the noise of the LiDAR data. Noise of LiDAR data is artificially introduced according to a Gaussian distribution.

2. AREA OF STUDY

Figure 1 depicts the area of study (42°17'41''N and 7°35'19'' W) that corresponds to a mountain road (OU-536) connecting the city of Ourense and the village of A Rúa in Spain. The road presents several slopes, causing major runoff over the road during rainy weather, very common in this region.

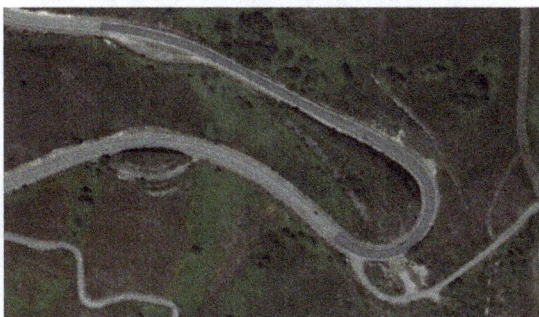

Figure 1. Area of study

3. MOBILE LIDAR SYSTEMS

3.1 Aerial LiDAR survey

The National Geographic Institute of Spain (IGN) started a campaign for the acquisition of aerial LiDAR data in 2009, finished in 2012. The sensor used in this area was the LMS-Q680 from Riegl. The scanned field of view is 50° with a scanning frequency of 45 kHz. Point density is 0.5 points per m^2, with position precision below 0.2 m. The coordinate system used is UTM – WGS84. Figure 2 shows an example of the aerial point cloud from the area of study.

Figure 2. Aerial point cloud from the area of study.

3.2 Terrestrial LiDAR survey

A mobile LiDAR surveying from the area of study was done using and Optech Lynx Mobile Mapper system during 2012. Figure 3 shows the survey van and Figure 4 the resulting point cloud. The Lynx contains an Applanix POS-LV 520 GNSS/IMU for navigation and two Optech LiDAR sensors for range measurement. The main technical specifications are shown in Table 1. It can be observed how GNSS outages decrease precision over 0.1 m.

Figure 3. Survey van with Optech Lynx Mobile Mapper.

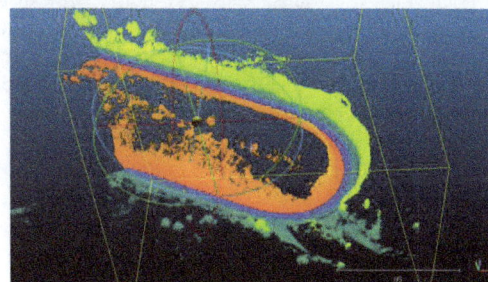

Figure 4. Point cloud obtained from the Mobile Terrestrial LiDAR System.

	With GNSS	GNSS Outage (1 minute)
GNSS X,Y precision (after post-processing)	0.020 m	0.100 m
GNSS Z precision (after post-processing)	0.050 m	0.070 m
IMU Roll and Pitch precision	0.005 °	0.005 °
IMU Heading precision	0.015 °	0.015 °
IMU measurement rate	200 Hz	200 Hz
LiDAR range	200 m	200 m
LiDAR precision	0.008 m	0.008 m
Absolute X, Y precision (GNSS + LiDAR)	0.022 m	0.100 m
Absolute Z precision (GNSS + LiDAR)	0.051 m	0.070 m
LiDAR pulse repetition rate	500 kHz	500 kHz
LiDAR scan frequency	200 Hz	200 Hz
LiDAR echoes	≤ 4	≤ 4
Operation temperature	-10 °C to + 40 °C	-10 °C to + 40 °C

Table 1. Technical specifications of Optech Lynx Mobile Mapper.

The survey was performed at 200 Hz with a pulse repetition rate of 500 kHz for each sensor. The resultant point cloud shows 26,591,026 points with an average point density of 2,084 points/m^2. The coordinate system used is UTM – WGS84.

4. DATA PROCESSING

This section deals with the proposed methodology (Figure 5), aiming to evaluate the influence of the precision of point clouds in the estimation of surface drainage pattern in roadsides. The method starts with the registration of aerial and terrestrial point clouds (step 1). Afterwards, Gaussian noise with different standard deviation values is introduced in the point cloud to simulate GNSS outages (step 2). Step 3 consists on generating DEM from point clouds with different cell resolution, which are used as the input to analyse flow directions (Step 4).

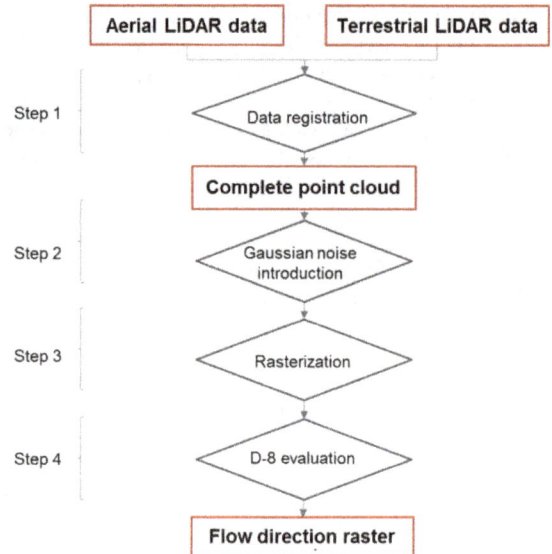

Figure 5. Schema of the proposed methodology.

4.1. Aerial and terrestrial data fusion

First step in data processing consists on fusing the terrestrial and aerial mobile LiDAR point clouds. Coarse registration was done by selecting three points from the pavement at the corners of horizontal traffic signs. Fine registration was done using Iterative Closest Point (ICP) algorithm. Cloud Compare software was used for this operation. The final point cloud shows 27,056,827 points (Figure 6).

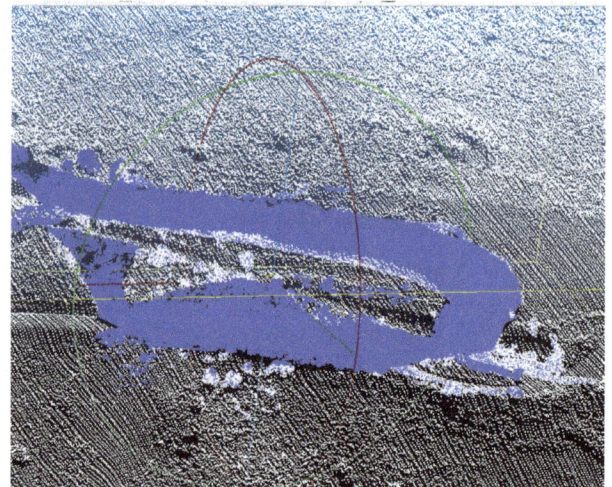

Figure 6. Terrestrial (blue) and aerial (white) mobile LiDAR point clouds registered in one dataset.

4.2. Generation of Gaussian noise

As the main aim of this work is to evaluate the influence of the precision of point clouds in the estimation of water runoff, point clouds are not submitted to pre-processing operations such as noise removal or filtering.

Gaussian noise is directly introduced in the point cloud in order to simulate the decrease in precision related with GNSS outages. Noise is equally introduced in X, Y and Z coordinates.

Random numbers are generated in MatLAB according different standard deviations and a Gaussian distribution to simulate the noise. The values of the standard deviations taken into account are 0.003 m, 0.005 m, 0.007 m, 0.010 m, 0.020 m, 0.030 m, 0.040 m, 0.050 m, 0.060 m, 0.080 m, 0.100, and 0.150 m. Figure 7 shows a section of the point cloud where points from the original point cloud are represented in white while points from a noised point cloud are showed in black.

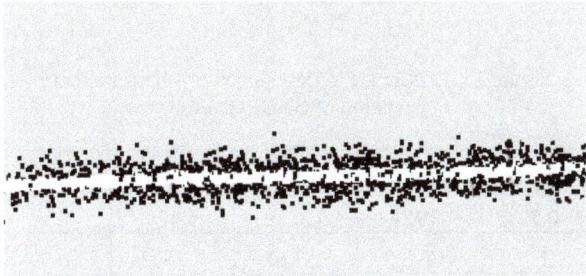

Figure 7. Original road – slope profile (white) and noised profile (black)

4.3. Generation of DEM from the point cloud

Third step consists on rasterizing the previous noised point clouds to obtain Digital Elevation Models, which are the input to the evaluation of the influence of LiDAR data precision in surface water runoff estimation.

The rasterization process starts by organizing the point clouds into a uniform XY grid. Next, a nearest neighbour algorithm is used to determine the points belonging to each cell. The algorithm calculates the Euclidean distance between each node of the matrix and the neighbourhood points. The height assigned to each grid cell corresponds to the mean height of the points belonging to each cell. A nearest neighbour Figure 8 shows an example of a raster layer generated from the case study.

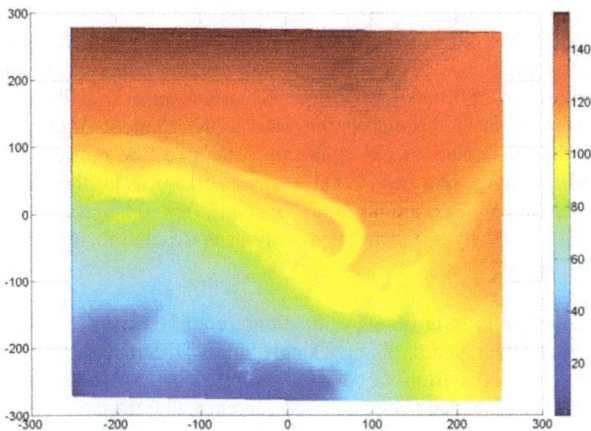

Figure 8. DEM with cell resolution of 2 m.

Cell sizes between 0.5 m and 5 m, with an interval of 0.5m, are generated to evaluate the influence of cell resolution. Figure 9 exhibits the relation between the number of cells and the resolution.

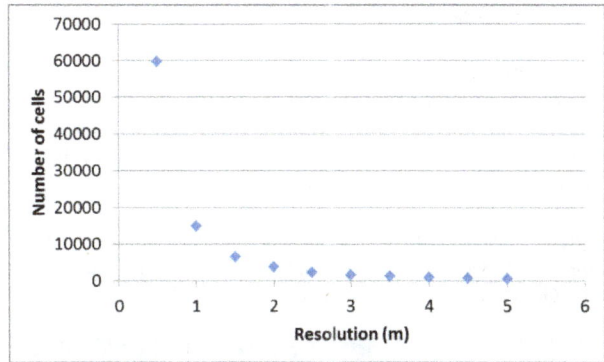

Figure 9. Relation between the number of cells of the DEM and cell resolution.

4.4. D8 evaluation

The D8 algorithm is based on the evaluation of the maximum terrain gradient for each cell of a DEM to approximate the primary flow direction (O`Callaghan 1984). It is a grid based algorithm extensively used in Geographic Information Systems (GIS) due to its simplicity and reliability.

For a given grid cell, the D8 algorithm approximates the primary flow direction by choosing the direction to the neighbor with maximal 2D gradient. Figure 10 schematizes the flow calculation with D8 algorithm. For example, the flow direction in the central pixel, with value 19, is down ward. The gradient is maximum towards the pixel directly below with value 17.

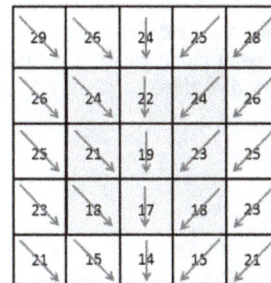

Figure 10. Flow directions evaluated with D8 algorithm.

The implementation of the algorithm is performed in MatLAB. Figure 11 shows the results of a runoff evaluation with a cell size of 3 m. Each direction is codified from 1 to 8 counter clockwise beginning at (0, 0).

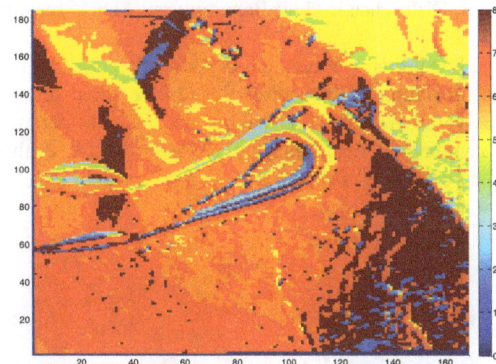

Figure 11. Runoff directions from a raster with a grid size of 3 m from an original point cloud.

4.5. Differences in runoff evaluation with D8

Final step of the methodology consists on evaluating the differences of flow directions between raster datasets from the original point cloud and those with Gaussian noise. The comparison is carried out for each cell size considered in this study and differences range between 0º to 180º.

Figure 12 shows de differences in the evaluation of runoff direction between the original data and a simulated point cloud with Gaussian noise with standard deviation of 0.1 m.

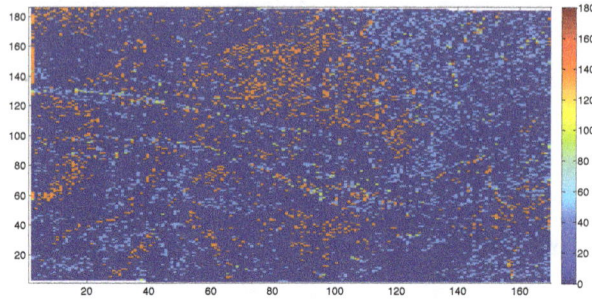

Figure 12. Differences in D8 evaluation.

5. RESULTS AND DISCUSSION

The difference in flow direction versus the size of the DEM cell for different standard deviations is shown in Figure 13. According to the results, differences in flow direction decrease with the increase of cell size. In addition, differences in flow direction increase with the increasing of standard deviation. For a resolution of 0.5 m, flow differences range between approximately 5º for standard deviations of 0.003 m to 42º for 0.150 m. A noisy point cloud clearly contributes negatively to the precision of flow direction.

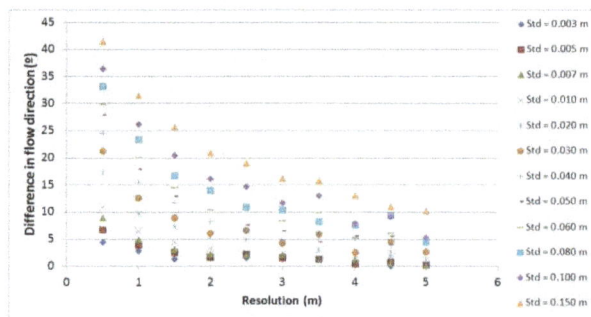

Figure 13. Difference in flow direction versus DEM resolution.

Figure 14 exhibits the relation in flow direction versus the standard deviation of the noise. Results are fitted to second-order polynomial (Table 2), making necessary to parametrize the behaviour and determine the precision of D8 evaluation depending on the cell size and precision of surveyed data.

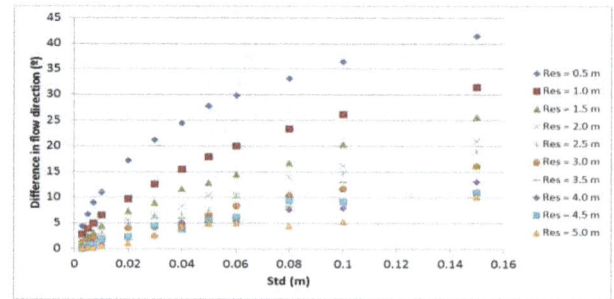

Figure 14. Difference in flow direction versus standard deviation of Gaussian noise.

Resol	a (m^{-1})	b	c (m)	R^2
0.5	-1957	526.39	5.428	0.98
1	-1093.2	353.61	2.6508	0.99
1.5	-641.47	254.87	1.615	0.99
2	-407.87	193.06	109.37	0.99
2.5	-221.12	150.92	1.3408	0.99
3	-141.72	122.92	1.0105	0.99
3.5	-232	137.12	0.5587	0.94
4	-104.69	99.58	0.1722	0.97
4.5	-433.16	139.46	-0.0902	0.97
5	-109.85	79.562	0.0454	0.92

Table 2. Fitting parameters to equation of the form:
$$difference = a{\cdot}std^2 + b{\cdot}std + c$$

CONCLUSIONS

In this work, the influence of the precision of LiDAR data is evaluated and parameterized for runoff estimation. The analysis is carried out for a real case study where terrestrial and aerial LiDAR data is combined in order to complete the available data.

The surface drainage pattern of the road and its surroundings is determined by using the D8 algorithm under different conditions of LiDAR precision. Gaussian noise is artificially generated in the original point cloud to determine its influence in the evaluation of D8 algorithm. Standard deviation of the noise ranges between 0.003 m and 0.15 m. Furthermore, different cell sizes are taken into account, from 0.5 m to 5 m, in order to evaluate the influence of raster resolution in runoff estimation.

The differences in runoff evaluation are analysed for each cell size and averaged to obtain a single value. Differences between the D8 results from the original point cloud and those point clouds with added Gaussian noise increase with the decrease of cell size and the increase of standard deviation. Differences range between 5º to 42º for standard deviations of 0.003 m and 0.150 m. Cell resolution in this case is 0.5 m. On the other hand, differences in flow direction change between 42º and 10º for 0.15 m standard deviation and cell size between 0.5 m and 5 m, respectively.

ACKNOWLEDGEMENTS

Authors want to give thanks to the Xunta de Galicia (CN2012/269; R2014/032) and Spanish Government (Grant No: TIN2013-46801-C4-4-R; ENE2013-48015-C3-1-R; FPU: AP2010-2969).

REFERENCES

Barreiro, A., Domínguez, J. M., Crespo, A. J. C., González-Jorge, H., Roca, D., Gómez-Gesteira, M. 2014. Integration of UAV photogrammetry and SPH modelling of fluids to study runoff on real terrains. *PlosONE*, 9(11), e 111031.

Caine, N. 1980. The rainfall intensity. *Geografiska Annaler. Series A, Physical*, 62(1/2), 23 – 24.

Douglas, D. H. 1986. Experiments to locate ridges, channels to create a new type of digital elevation model. *Cartographica*, 23(4), 29 – 61.

Meesuk, V., Vojinivic, Z., Mynett, A. E., Abdullah, A. F. 2015. Urban flood modeling combining top-view LiDAR data with ground-view SfM. *Advances in Water Resources*, 75, 105 – 117.

O'Callaghan, J. F. and Mark, D. M. 1984. The extraction of drainage networks from digital elevation data. *Computer Vision Graphics and Image Processing*, 28(3), 323 – 344.

Petri, G. 2010. Mobile mapping systems: An introduction to the technology. *Geoinformatics Magazine*, 13(1), 32 – 43.

Puente, I., González-Jorge, H., Martínez-Sánchez, J., Arias, P. 2013. Review of mobile mapping and surveying technologies. *Measurement*, 46(7), 2127 – 2145.

Wang, J., González-Jorge, H., Lindenbergh, R., Arias-Sánchez, P., Menenti, M. 2014. Geometric road runoff estimation from laser mobile mapping data. *ISPRS Annals of the Photogrammetry, Remote Sensing, and Spatial Information Sciences*, II-5, 385 – 391.

World Bank (2015); http://www.worldbank.org/transport/roads/con&main.htm#road maint

34

YOU DESCRIBE IT, I WILL NAME IT: AN APPROACH TO ALLEVIATE THE EFFECT OF USERS' SEMANTICS IN ASSIGNING TAGS TO FEATURES IN VGI

S. Hassany Pazoky, F. Karimipour, F. Hakimpour

Faculty of Surveying and Geospatial Engineering, College of Engineering, University of Tehran, North Kargar St., Tehran, Iran
(shpazooky, fkarimipr, fhakimpour)@ut.ac.ir

Commission II, WG II/4

KEY WORDS: Uncertainty, VGI, Users' Semantics, Tagging Features, OpenStreetMap, Perception

ABSTRACT

As an important factor of VGI quality, this paper focuses on uncertainty arisen in assigning tags to features by VGI users. The VGI portals ask their users to assign (or tag) one or more data types to features, from a set of pre-defined types, whose meanings may be vague for the user, or distinctions between some of them are not clear, i.e. depend on the users' semantics. This research believes such uncertainties are the results of perceptual issues arising in serial communication between the system and the user. We categorize the problem, and then utilize semantic modelling to reduce such uncertainties. A hierarchy of feature types is produced. At each step, users are asked a simple question with clear distinct answers, which gradually directs the user to the right type. We will describe the approach and present the initial results for the hierarchy produced for major linear features of OpenStreetMap.

1. INTRODUCTION

Among the technological advances of the third millennium, Web 2.0 in line with themes such as crowdsourcing, collaboration, wikis, and the GeoWeb have thoroughly changed the World Wide Web. Neogeography was a pioneer term introduced by Turner (2006) to convey the idea of participation of untrained users in map production process. In other words, the distinction between map producer, communicator and user loses clarity (Goodchild, 2009b). Goodchild (2007a) took advantage of this research and proposed a new term namely Volunteered Geographic Information (VGI), which is widely used in the literature and various research are being conducted on different aspects of it.

One of the most important challenges of VGI development is uncertainty mentioned by many researchers who have considered it from different aspects (Allingham, 2014; Barron et al, 2013; Elwood et al, 2013; Mohammadi and Malek, 2015). The research conducted on VGI uncertainty have mostly measured the uncertainty of VGI datasets (Vandecasteele, and Devillers, 2015), 2) or concerned with spatial aspect of the data rather than non-spatial aspect (Mülligann et al., 2011; Mooney and Corcoran, 2012). To the best of our knowledge, there are only a very few attempts to introduce mechanisms to lower the uncertainty during the editing process (Grira et al., 2010; Vandecasteele, and Devillers, 2015).

This paper focuses on an approach to reduce uncertainty arisen in assigning tags to features by VGI users during the editing process. The idea is based upon the fact that citizens, who play the role of surveyors in VGI, provide the required non-spatial information based on their perception (Flanagin and Metzger, 2008; Mülligann et al., 2011; Karimipour et al., 2013; Mount, 2013), social and cultural settings (Goodchild, 2007a; Goodchild, 2007b; Coleman et al, 2009; Roche et al., 2012), previous experience (Mülligann et al., 2011), etc. In other words, substituting neo-geographers with experts is equivalent to substituting perception with measurements, or quality with quantity. In the case of experts, the hidden information is about the measuring instruments where there are many standards to

reduce them under an acceptable threshold. As a result, voluntarily generated maps are more suitable for everyday life and recreation, as they are more up-to-date, whereas maps produced by experts are more suitable for engineering purposes.

OpenStreetMap asks contributors to assign (or tag) at least one data type to features, whose meanings may be vague for the user, or distinctions between some of them are not clear, i.e. depend on the users' semantics. As a result, unlike spatial information, data type information are accompanied with hidden semantic. In other words, different people may tag features differently leading to semantic heterogeneity (Vandecasteele, and Devillers, 2015).

Mülligann et al. (2011) is one of the first efforts that concern on the problem of tagging features in VGI. They developed a semantic similarity measure to assist contributors in tagging point features. They use spatial relationships to outline incorrect tags. For example, a pub is usually surrounded with places that can afford drinking alcohol, while waste baskets are distributed uniformly in the city. As another example, tagging two very near features as "fire stations" may be an indicator of a mistake by the contributors.

Ballatore et al. (2013) also developed a semantic network by crawling OSM wiki page to measure the semantic similarity between feature types. The outcome of the paper can be used for recommender systems to tag features in OpenStreetMap, geographic information retrieval, and data mining.

Vandecasteele and Devillers (2015) used the semantic network produced by Ballatore et al. (2013) and combined it with TagInfo to form their database. They developed an open source plugin called OSMantic to recommend tags to users and also warn them when inappropriate tags are used together.

This paper proposes the initial result of a solution for the problem of assigning tags to features in VGI. We believe such uncertainties are the results of perceptual issues arising in serial communication between the system and the user. We categorize the problem, and then utilize semantic modelling to reduce such uncertainties. A hierarchy of feature types is produced. At each step, users are asked a simple question with distinct enough answers, which gradually directs the user to the right type. In addition, users are not forced to tag the features that they really do not have information about. They can proceed in the hierarchy as much as they have information, i.e. as much as they can answer the questions.

The rest of the paper is structured as follows: In Section 2, we scrutinize the perceptual causes of uncertainty that bothers VGI, especially OpenStreetMap. Section 3 describes the proposed approach. Sections 4 presents the initial results of deploying the proposed approach to produce the hierarchy for major linear features of OpenStreetMap, as a successful VGI portal (Haklay & Weber, 2008; Ballatore et al., 2013). Finally Section 5 concludes the paper.

2. PERCEPTUAL ISSUES OF SERIAL COMMUNICATION

Communications may be considered either as parallel or serial. The former makes use of pictorial elements that are perceived all together. On the other hand, means of communication in serial communication are words that are perceived one by one and in a predefined order.

Whenever serial communication is used, measures should be devised to counter the uncertainty. If a neogeographer is asked to digitize a street for a VGI portal, he most probably picks the right drawing tool and draws a simple line. However, the problem arises when he is to tag it with the appropriate term. The perceptual issues of tagging features in VGI are as follows:

(1) What do the VGI portal administrators mean by a certain tag such as "highway" tag? Such terms are among the professional contexts. There are many institutes in the world defining these terms. The worst aspect is that they may have different meanings in different countries or they may have special local meaning in some places. For example, most natural water areas in Newfoundland, Canada are named "ponds". However, in OpenStreetMap, "pond" refers to areas of water created by human activity[1] (Vandecasteele & Devillers, 2015). Even, some countries may have other feature types rather than those available in OpenStreetMap.

(2) A major aspect of well-known VGI portals such as OpenStreetMap is that they support multi-linguality, i.e. the tags are provided in many different languages. In addition to emphasizing our concerns on translating tags in OpenStreetMap, the tags in countries speaking the same language may even differ. For example, the tag "highway=motorway", which is used in UK, is equivalent to "motorway, freeway, and freeway-like road" in Australia, "limited access highway" in Canada, "freeway" in India, and "limited access freeway" in USA[2]!

(3) There are too much tags in OpenStreetMap. The number of tags in OpenStreetMap is much more than a human can perceive in mind for voluntary actions. In addition, users can create their own tags (Mooney and Corcoran, 2012). This is why there are many miss-spelled tags in OpenStreetMap data. Vandecasteele and Devillers (2015) report that there are currently more than 40,000 distinct tags in OpenStreetMap data which is tens of times more than what is really needed.

(4) The terms used to assign tags to features are not clear enough for users (Ballatore et al, 2013). For example, at the first glance, the tags such as "highway=primary", "highway=secondary", "highway=tertiary", "highway=residential", and "highway=living_street" may seem indistinguishable.

Especially, in ancient cities where urban planning has had less opportunity to develop the city in an organized manner, the differentiation of the aforementioned types is a professional task, if not impossible. This problem is regarded as ambiguity (Shi, 2010). On the other hand, contributors of VGI portals are ordinary people without special training in GIS or similar fields. Thus, they tag map features based on their perception or semantic. Naturally, users' semantics are not necessarily the same, which can make VGI imbalanced.

(5) Although OpenStreetMap has done its best to provide clear definitions of different feature types, the extent of some of which may not be clear enough. For example, "highway=tertiary" tag is defined in OpenStreetMap as a tag "used for roads connecting smaller settlements, and within large settlements for roads connecting local centres. In terms of the transportation network, OpenStreetMap tertiary roads commonly also connect minor streets to more major roads.[3]" And, residential road is defined as: "Roads accessing or around residential areas but are not a classified or unclassified highway." It is then mentioned that if you doubt whether the road is residential or unclassified, residential is more specifically defined as: "Street or road generally used only by people that live on that road or roads that branch off it."[4] Although the two definitions are in clear English, they do not share a clear boundary. For example, there are paragraphs describing "highway=tertiary" and "highway=residential_road" tags; but they do not share a clear boundary; i.e. there are many instances that apply to both. As a result, contributors may face problems instantiating real world objects and they use the tags interchangeably depending on their semantics (Vandecasteele & Devillers, 2015). This phenomenon is regarded as vagueness by Shi (2010).

(6) The contributors may lack information that causes imprecision (Shi, 2010). If so, they are unable to provide the portal with the right tag. An illustrative example of imprecision is the location of Eiffel tower. Although answers such as Europe, France, and Paris are all correct with no uncertainty, Europe is the less and Paris is the most precise notion.

Also, Vandecasteele and Devillers (2015) include temporal changes of tags as a source of semantic heterogeneity. We do not agree with them since tags that are not used become deprecated and they can easily be distinguished and replaced with the new tags.

The sextet uncertainty elements caused by users' semantics defined above occur when contributors wish to tag a feature. The effect of them, however, is not equal and it may vary from a person to another. Semantic can help resolve the perceptual problems.

3. HOW TO ALLIVIATE THE EFFECT OF USERS' SEMANTICS?

This paper proposes a solution to alleviate the effect of users' semantics in assigning tags to features in VGI. The solution provides clear definitions of different object types dragging the understanding of the system designers and users together as much as possible. We extract the information of the user, purify it from any irrelevant semantics and use it in VGI. For this, all the possible tags are arranged in a hierarchical structure. The contributor goes through the hierarchy answering some questions asked at each node. The questions are very simple, free of

---

[1] http://wiki.openstreetmap.org/wiki/Tag:landuse%3Dpond

[2] http://wiki.openstreetmap.org/wiki/Highway:International _equivalence

[3] http://wiki.openstreetmap.org/wiki/Tag:highway%3Dtertiary

[4] http://wiki.openstreetmap.org/wiki/Tag:highway%3residential

complex and technical terms, and qualitative. Also, the choices available for every questions are also simple and completely distinct. The benefit of the hierarchical structure is that at each node, the contributor faces a clear question with very few possible answers; whereas in the current approach of OpenStreetMap, the variety of choices without clear distinctions confuses the contributor. This procedure gradually directs the contributor to the right data type. In addition, the contributor can stop descending the hierarchy when she lacks information. This mechanism provides the opportunity to get information from the contributors to the extent they are sure about, i.e., neither more nor less than what they really know!

4. CASE STUDY: TAGGING LINEAR FEATURES IN OSM

To illustrate the proposed idea of the paper, which is also briefly presented in Pazoky et al. (2014), this section describes how the idea is applied on the linear features available at the official wiki page of OpenStreetMap[5]. We produced the hierarchy for aeroway, highway, railway, and waterway types from the 26 available feature types. 38 tags were chosen among these feature types for the case study.

To design the questions and hierarchy, we started the hierarchy with OpenStreetMap categorization of feature types mentioned before. The rest of the hierarchy was produced using a divisive approach, i.e. a member of a category was chosen and its most significant distinction was considered to form the question. However, if the question in its parent node was less general, the two questions were swapped. Then, all the elements were reconsidered and moved if they belong to other nodes.

The resultant hierarchy for the mentioned data types is illustrated in Figure 1. The light green boxes show the provided choices; and the darks green boxes indicate the OpenStreetMap tag associated with that feature type. Furthermore, the blues boxes in Figure 1 are expanded separately in Figures 2 to 5 for "Cars and pedestrians", "Link between two roads", "Rail network", and "water" respectively. The path taken by a hypothetical contributor is shown in Figure 6.

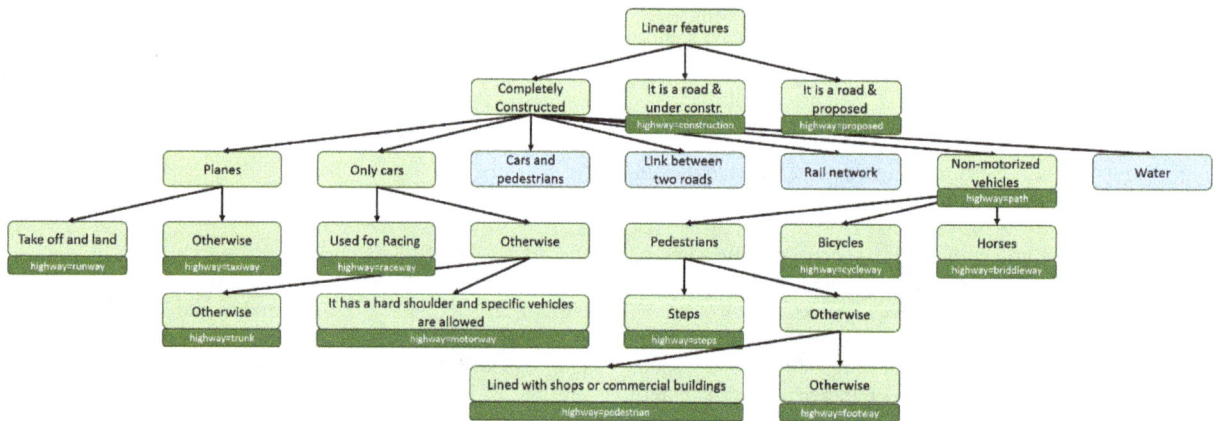

Figure 1. The resultant hierarchy to tag OpenStreetMap features.

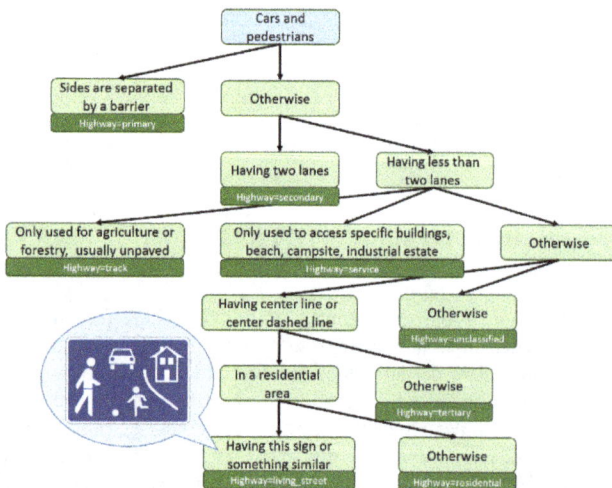

Figure 2. Expanded view of "Cars and pedestrians" box in Figure 1.

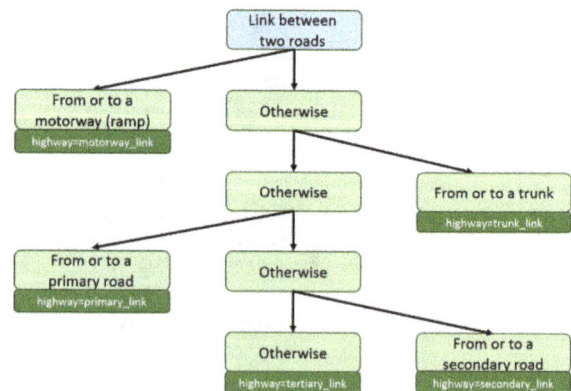

Figure 3. Expanded view of "Link between to roads" box in Figure 1.

[5] http://wiki.openstreetmap.org/wiki/map_features

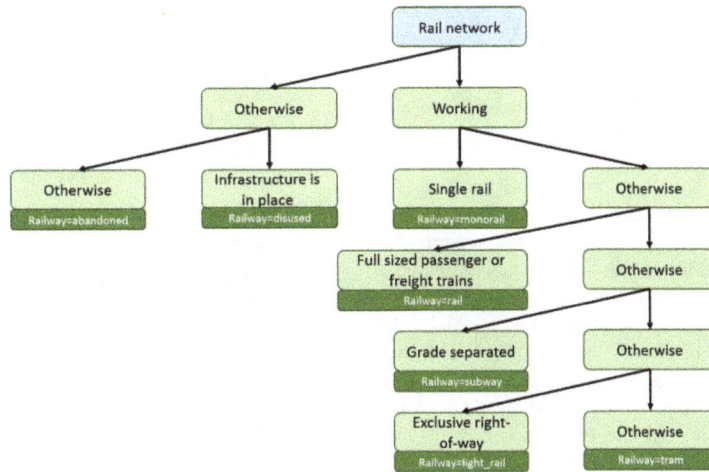

Figure 4. Expanded view of "Rail network" box in Figure 1.

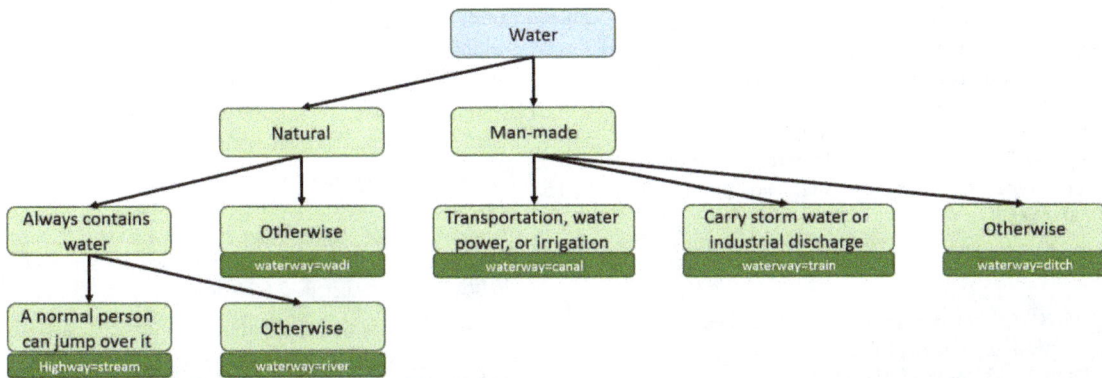

Figure 5. Expanded view of "Water" box in Figure 1.

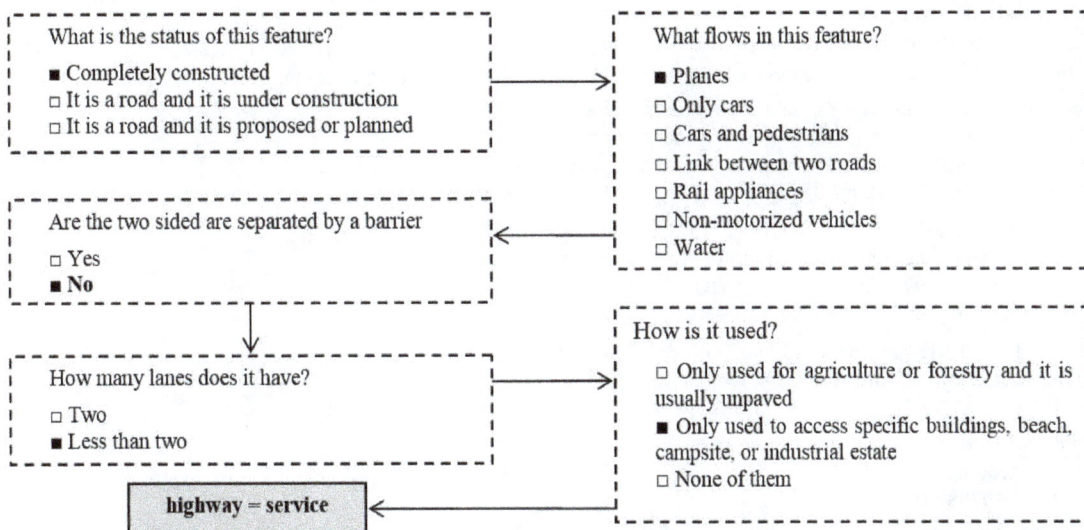

Figure 6. The questions and answers our hypothetical contributor has gone through to reach the highway=service tag.

5. CONCLUSION AND FUTURE WORK

VGI, as a prominent innovation of the past decade in GIScience (Goodchild, 2009a), has been very fruitful in gathering geospatial information from the general public. In this paper, we introduced users' semantics or differences between the perceptions of people on a term as a significant source of uncertainty in VGI. This problem is caused by using complex and professional terms in VGI portals such as OpenStreetMap to tag features. We proposed a solution to alleviate the effect of users' semantics on the issue. Using the hierarchy provided in the solution, contributors are faced with a sequence of questions answering of which leads them to the next questions until they reach the leaf. This way, professional terms are avoided and a thick barrier between the possible answers is established, which result in clear distinct choices. The case study showed the professional terms with vague boundaries are reachable through the hierarchy. This solution lets the VGI portals receive less uncertain information from untrained and inexperienced contributors. It can also be regarded as a step to develop ways to diminish perceptual uncertainties of VGI.

We are going to develop an application implementing our approach to be testable by different people. Then, we intend to ask people from around the world to tag map objects using our approach and see if it really works and gather their feedbacks to improve the usability, user-friendliness, misclassifications, etc. Another goal of us is to expand the hierarchy to accommodate all the feature types of OpenStreetMap, i.e. point, linear, and polygonal objects.

REFERENCES

Allingham, C. (2014). A review of quality of geo-data from user's perspective. Universal Journal of Geoscience 2(2), pp. 70-75. DOI=10.13189/ujg.2014.020205.

Ballatore, A., Bertolotto, M., & Wilson, D. C. (2013). Geographic knowledge extraction and semantic similarity in OpenStreetMap. *Knowledge and Information Systems, 37*(1), 61-81.

Barron, C., Neis, P., and Zipf, A. (2013). A Comprehensive Framework for Intrinsic OpenStreetMap Quality Analysis. *Transactions in GIS.* DOI=10.1111/tgis.12073.

Coleman, D. J., Georgiadou, Y., and Labonte, J. (2009). Volunteered geographic information: the nature and motivation of producers. *International Journal of Spatial Data Infrastructures Research, 4*(1), pp. 332-358.

Elwood, S., Gooodchild, M. F., and Sui, D. (2013). Prospects of VGI research and the emerging fourth paradigm. In: *Crowdsourcing Geographic Knowledge,* Springer Netherlands, pp. 361-375. DOI= 10.1007/978-94-007-4587-2_20.

Flanagin, A. J., and Metzger, M. J. (2008). The credibility of volunteered geographic information. *GeoJournal, 72*(3-4), 137-148.

Goodchild, M. F. (2007a). Citizens as sensors: The world of volunteered geography. *GeoJournal* 69(4), pp. 211-221. DOI=10.1007/s10708-007-9111-y.

Goodchild, M. F. (2007b). Citizens as voluntary sensors: spatial data infrastructure in the world of Web 2.0. *International Journal of Spatial Data Infrastructure Research,* 2, pp. 24-32.

Goodchild, M. F. (2009a). Geographic information systems and science: today and tomorrow. *Annals of GIS,* 15(1), pp. 3-9. DOI=10.1080/19475680903250715.

Goodchild, M. F. (2009b). NeoGeography and the nature of geographic expertise. *Journal of Location Based Services,* 3(2), pp. 82-96. DOI=10.1080/ 17489720902950374.

Grira, J., Bédard, Y., and Roche, S. (2010). Spatial data uncertainty in the VGI world: going from consumer to producer. *Geomatica,* 64(1), pp. 61-71.

Haklay, M., & Weber, P. (2008). Openstreetmap: User-generated street maps.*Pervasive Computing, IEEE, 7*(4), 12-18.

Karimipour, F., Esmaeili, R., and Navratil, G. (2013). Cartographic representation of spatial data quality parameters in volunteered geographic information, *Proceedings of the 26th International Cartographic Conference (ICC 2013),* (Dresden, Germany, August 25-30, 2013).

Mohammadi, N., & Malek, M. (2015). Artificial intelligence-based solution to estimate the spatial accuracy of volunteered geographic data. *Journal of Spatial Science,* 60(1), 119-135.

Mount, J. (2013). The role of context in spatial decision-making in GIScience. *Ph.D. Dissertation, University of Iowa.*

Mülligann, C., Janowicz, K., Ye, M., & Lee, W. C. (2011). Analyzing the spatial-semantic interaction of points of interest in volunteered geographic information. In *Spatial information theory* (pp. 350-370). Springer Berlin Heidelberg.

Mooney, P., & Corcoran, P. (2012). The annotation process in OpenStreetMap. *Transactions in GIS, 16*(4), 561-579.

Pazoky, S., Karimipour, F., and Hakimpour, F. (2014). An Ontological Solution for Perceptual Uncertainties of VGI. In: *Proceedings of the Extended Abstracts of 8th International Conference on Geographic Information Science (GIScience 2014),* Vienna, Austria, 23-26 September 2014, 302-305.

Roche, S., Mericskay, B., Batita, W., Bach, M., and Rondeau, M. (2012). Wikigis basic concepts: Web 2.0 for geospatial collaboration. *Future Internet,* 4(1), pp. 265-284.

Shi, W. (2010). Principles of modeling uncertainties in spatial data and spatial analyses. CRC Press.

Vandecasteele, A., & Devillers, R. (2015). Improving Volunteered Geographic Information Quality Using a Tag Recommender System: The Case of OpenStreetMap. In *OpenStreetMap in GIScience* (pp. 59-80). Springer International Publishing.

35

TRUSTING CROWDSOURCED GEOSPATIAL SEMANTICS

P. Goodhue, H. McNair , F. Reitsma

Dept. of Geography, University of Canterbury, Christchurch, New Zealand
(paul.goodhue, hamish.mcnair)@pg.canterbury.ac.nz, femke.reitsma@canterbury.ac.nz

KEY WORDS: Crowdsourcing, Semantics, Ontology, Data quality, Trust

ABSTRACT

The degree of trust one can place in information is one of the foremost limitations of crowdsourced geospatial information. As with the development of web technologies, the increased prevalence of semantics associated with geospatial information has increased accessibility and functionality. Semantics also provides an opportunity to extend indicators of trust for crowdsourced geospatial information that have largely focused on spatio-temporal and social aspects of that information. Comparing a feature's intrinsic and extrinsic properties to associated ontologies provides a means of semantically assessing the trustworthiness of crowdsourced geospatial information. The application of this approach to unconstrained semantic submissions then allows for a detailed assessment of the trust of these features whilst maintaining the descriptive thoroughness this mode of information submission affords. The resulting trust rating then becomes an attribute of the feature, providing not only an indication as to the trustworthiness of a specific feature but is able to be aggregated across multiple features to illustrate the overall trustworthiness of a dataset.

1. INTRODUCTION

Crowdsourced geospatial information will transform data collection and its legitimacy is contingent on the trustworthiness of the information submitted. We can assess this from a range of perspectives: the spatio-temporal nature of the information, the social aspect of the information (e.g. the provider of the information), and the semantics of the information. In this paper we focus on the semantics of crowdsourced geospatial information, and how we can define a measure of trust using these semantics. Assessing the trust of the semantics associated with crowdsourced geospatial information ensures we know to what degree it is useful. This appraisal can be performed via comparisons of the intrinsic and extrinsic properties of geospatial features to ontologies to produce a feature level trust rating. This trust rating can then be used as an attribute to aid in discovery and analysis of the information, and be aggregated across multiple features to provide an indication of trustworthiness for the data set as whole.

2. RELATED WORK

Research on the semantics of geospatial information find it supports better discovery (Egenhofer 2002), use (Reeve and Han 2005), and access (Yue et al. 2007, Janowicz et al. 2012). Today the increase in use of semantics for search and discovery has led to a proliferation of websites that use ontologies in their backend, e.g. ImageNet (Deng et al. 2009). Tie this to collectively developed tags or annotations (Marcheggiani et al. 2007), and the potential for crowdsourced geospatial datasets being integrated into all stages of the spatial data supply chain is substantial.

Using the crowd involves significant cognitive diversity. To ensure that the semantics expressed in the attributes associated with crowdsourced geospatial features are accurate, we can either leverage this cognitive diversity and apply statistical methods for measuring the agreement of concepts thereby embracing the diversity of responses (Narock and Hitzler 2013), or constrain user input. Semantics can be used to determine the

difference among concepts in geospatial datasets (Kuhn 2005). These differences may be subtle variations in class, e.g. a Munro is a Mountain in some instances, to more distinct differences such as a Coconut Palm and a Date Palm. Measures of semantic similarity have been developed to ascertain how close concepts are to each other (e.g. Janowicz et al. 2011, Sizov 2010), and can be used to compare ontologies and improve spatial datasets (Ballatore et al. 2013).

Measures of trust for crowdsourced geospatial information usually consider the spatial and social aspects of the crowdsourced geospatial information (Goodchild and Li, 2012), but often overlook the semantic aspect. Measures of spatio-temporal trust of crowdsourced geospatial information determine the spatial accuracy of the information (Haklay 2010), such as its shape, orientation and location. Social measures of the trust of crowdsourced geospatial information are based on the reputations of the information producers (Bishr and Janowicz, 2010). For example, the USGS National Mapping Corps project uses the reputation of information producers to determine trust of the information (McCartney et al. 2013).

Some consideration has been given to certain aspects of the semantics of crowdsourced geospatial information, such as the work by Vandecasteele and Devillers (2013), who proposed the use of semantic similarity to constrain the crowdsourced information by notifying the producer of the information when two attributes were too similar or dissimilar. The concept of semantic similarity can be used at the feature level (i.e. on individual pieces of information) to measure the trust of crowdsourced geospatial information by comparing the feature to trusted features of a similar type, both within the crowdsourced dataset or with external datasets (Ramos et al. 2013). Likewise Bordogna et al. (2014) semantically assessed the quality of crowdsourced geospatial information through a linguistic approach using a hierarchical structure of attribute categories (e.g. tags) within the information. Features given a general category are deemed to be less accurate than features given a specific category, and feature assigned multiple categories are also deemed inaccurate as this shows uncertainty in the producer's ability to categorise the feature. We build on these methods with further methods for using semantics to assess the trustworthiness of crowdsourced geospatial information.

3. CROWDSOURCED SEMANTICS

How do we trust the semantics of information provided by the crowd? Improving the knowledge of the semantic quality of the information is achieved through assessing or constraining the semantic aspect of the information. Constraining the semantic aspect of the information can improve the quality of the information but can limit the knowledge captured with the information, therefore semantically unconstrained information coupled with semantic quality assessments can provide us with diverse and trustworthy geospatial information. Semantic quality assessments of crowdsourced geospatial information can be performed on the intrinsic or extrinsic semantics of the information, where the intrinsic semantics involves the feature and its consistency both internally and with linked ontologies, and where extrinsic semantics investigate the feature and its consistency within a wider context of related information and ontologies.

A theoretical example of a crowdsourcing project that would benefit from improved semantic trust is a project that crowdsources the location of fruit producing trees on public land, and some information that describes the trees. The crowdsourced geospatial information would include a point feature depicting the trees location and information about the tree that would describe the type of tree, the quality and quantity of the fruit on the tree, accessibility of the tree (e.g. across a stream or up a bank) and other observable attributes of the tree such as height and diameter. This information would be linked to an ontology describing fruit producing trees and other external data and ontologies to help to determine the semantic trust of the information. Although projects exist that crowdsource fruit tree information (e.g. www.ediblecities.org), little is done with regards to determining the trust of the information, especially the semantic trust of the information. By determining the semantic trust of the crowdsourced information, anyone wishing to use the information can trust that the information accurately describes a tree in the real world.

Intrinsic semantic assessments of crowdsourced geospatial information assess the feature within the context of its containing geospatial dataset. Intrinsic assessments are based on comparing the feature type and attributes with an ontology. Through intrinsic semantic assessments we can determine the internal consistency of the feature and whether or not it conforms to the semantics of its related ontology. A feature that is internally consistent and conforms to an ontology based on its attributes and feature type would be assigned a high rating of trust for its intrinsic semantic component. For example, a crowdsourced fruit tree feature could contain an attribute describing the type of tree the feature represents. Through assessments of this attribute and a tree ontology we could determine if the feature the producer has created describes a fruit producing tree in the real world, as outlined in figure 1. If a producer creates a tree feature and describes its type as "Apple", the tree ontology would tell us that "Apple" is a type of fruit producing tree and therefor the features type is consistent with fruit trees and is somewhat trustworthy. Alternatively, if the producer was to submit a tree feature with a type of "Rose", we would assign this feature a low trust rating because although

"Rose" is a type of plant, it would not fit into a subclass of fruit trees.

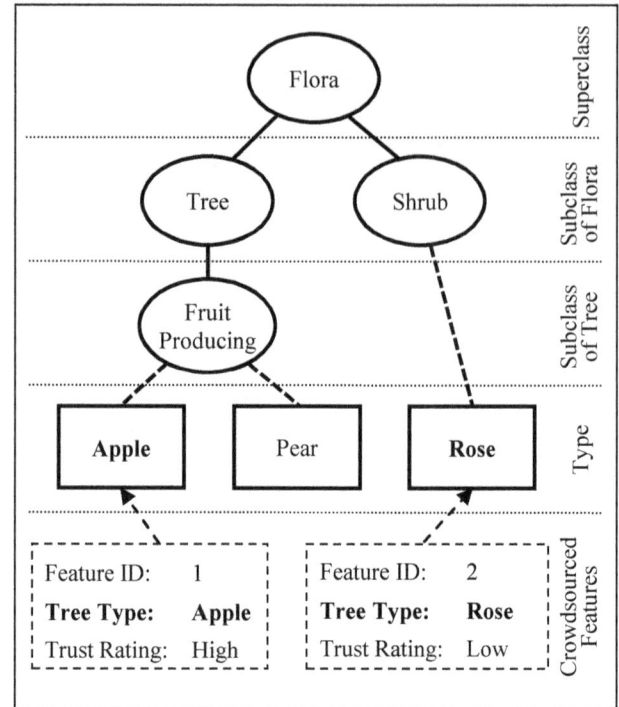

Figure 1. Assessing crowdsourced features with an ontology to determine if their Tree Type attribute is a type of fruit tree.

Each feature is associated with attributes. These attributes may be assessed based on the semantics of their data type or object type. For example, a point feature representing a fruit tree in our example crowdsourcing project may include attribute data with a tree type of "Coconut" and with a height of 100m. Comparing our geospatial dataset against a tree ontology, we would find that Coconut trees are a type of fruit producing tree but do not usually grow taller than 30m. We would assign a low trust rating to this information because the attribute data is inconsistent and therefore untrustworthy. Object properties may also be evaluated for trustworthiness. A feature representing a single fruit tree could not have a geometry type of "Line", and if a member of the crowd created such a feature, this information would also be deemed untrustworthy.

Semantic trust can also be measured through extrinsic assessments, that is, comparing the crowd created information with external datasets. Extrinsic semantic assessments measure the trust of crowdsourced geospatial information by determining how a feature fits into its spatio-temporal surroundings by comparing the semantics of the feature to the semantics of its surroundings. For example, if a point feature was created as part of our example project of a Coconut tree in Antarctica today, intrinsically the information could be trustworthy as the Coconut tree feature may have the appropriate attributes, however comparing our piece of crowdsourced information against other ontologies regarding the climatic conditions Coconut trees grow in, the location of that point feature and the climate at that location, we would not trust this information. Taking into account the spatial and temporal characteristics is important for incorporating contextual knowledge about the crowdsourced information. Adding this additional level of trust to the crowdsourced information makes the information more usable and helps to define the line between the information being completely untrustworthy (e.g. the fruit tree does not exist in the real world) or somewhat untrustworthy but still usable (e.g. the fruit tree exists but is not fruiting at the time of year the information states that it is). Extrinsic semantic assessments can become complex as they do not focus solely on the feature and its meaning, but require descriptions of the ontological relationships between the feature and the wider environment.

Methods for assessing the semantic quality of crowdsourced geospatial information become more complex as the information becomes thematically richer. In cases where the crowdsourced information is constrained, the attribution is common throughout the dataset and the features are likely to be both internally consistent and semantically similar to other features. But by semantically constraining crowdsourced information we risk losing knowledge of the feature that the producer could have otherwise supplied with the feature. Measures of semantic trust are needed for semantically unconstrained crowdsourced information to leverage the diversity and cognition of the crowd while maintaining trust in the crowdsourced information. Alongside measures of the semantic trust of crowdsourced geospatial information are measures of the trust of its spatio-temporal and social components. Intrinsic and extrinsic semantic assessments of crowdsourced geospatial information forms the semantic component of a larger crowdsourcing model that also encapsulates assessments of the intrinsic and extrinsic aspects of the spatio-temporal and social components of the crowdsourced geospatial information. Although measuring the semantic trust of crowdsourced geospatial information provides us with a rating of trust of the information, measures of the spatio-temporal and social trust of the information help to provide an overall trust rating for the information. The wider crowdsourcing model that the semantic assessments form a part of assesses both the crowdsourced information itself and how it fits into its surroundings. The author is also considered in order to determine their influence on the trust of the information. By focussing on the information itself, the semantic and spatio-temporal components of the crowdsourcing model complement each other and the use of both helps to improve the trust of the information further by generating trust ratings that represent all aspects of the information itself.

4. DIRECTION OF RESEARCH

Previous applications of trust models for spatial datasets (Malaverri et al., 2012) and crowdsourced geographic features (Bishr & Mantelas , 2008; Celino, 2013) have produced a scalar value as a proxy for trustworthiness. This can then be employed in much the same way as metadata or provenance information to give an indication of data quality. However, aside from just determining trustworthy and untrustworthy data the use of a single metric provides several notable advantages over traditional, text based metadata. Firstly, for those unfamiliar with the use of traditional datasets (such as members of the crowd) it provides a simple indication as to whether a dataset or feature is appropriate without having to understand the intricacies of metadata documentation. Further to this, it provides scope for the aggregation of feature level trust ratings to give a representation of trustworthiness for a dataset as a whole, as outlined in figure 2. As crowdsourced datasets often have multiple and varied sources this numeric approach is simpler to implement and understand than attempting to aggregate the individual metadata records of each feature into a

coherent article. Finally, quantifying trust not only makes for a convenient feature attribute, but this attribute – being a number as opposed to text – is easily incorporated into computerised systems, giving the data a degree of self-description.

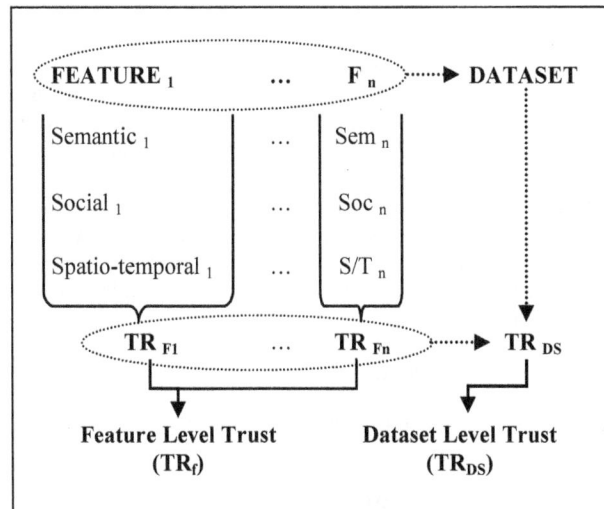

Figure 2. Formation of trust ratings corresponding to feature (TR_F) and dataset (TR_{DS}) levels; in the same manner as a dataset is an accumulation of features a datasets trust rating is an aggregate of feature trust ratings.

Self-describing data increases functionality by providing improved search and implementation capabilities. For example, if a user requires a map of the most trustworthy of fruit tree features (in, say, the interest of maximising time spent collecting fruit versus finding trees) they can perform a simple attribute based search – features with high trust ratings. Alternatively, to enable the utilisation of an entire dataset, a feature's influence can be made proportional to its trust rating. If one were to analyse distribution of fruit trees (to, say, evaluate the impact of birds dispersing seed as opposed to gardeners purchasing plants) the model could be structured to increase the influence of features in the analysis based on their trust ratings. This approach is similar to spatial analysis techniques where the weighting of inputs is based on the perceived quality of the data from which they were derived (MacCormack & Eyles, 2010).

An additional strength of numeric trust ratings is in their ability to aid in the calibration and strengthening of the models which produce them. Comparison of crowdsourced datasets with their authoritative counterparts have been used to prove accuracy and completeness (Haklay, 2010). However, by comparing the analysis of a crowdsourced dataset (via the aforementioned weighted approach) with equivalent analysis of an authoritative dataset it is possible to identify under which scenarios various aspects of the trust model should project the greatest level of influence. Such an approach allows for not only the determining of the importance of the intrinsic and extrinsic assessments of semantic trust, but also the role of semantic trust within the wider trust model as a function of the dataset and what it represents. Such calibration provides for a more robust assessment and subsequently improving overall accuracy and functionality of trust models and the ratings they produce.

5. CONCLUSION

Crowdsourced geospatial information makes use of the diverse knowledge of the crowd, but for the information to be useful it must be trusted. Measures of the trust of crowdsourced geospatial information focus on different aspects of the information, such as the spatio-temporal, social and semantic aspects. The trust of the semantic aspect of crowdsourced geospatial information is often overlooked by crowdsourcing applications, but in unconstrained data models it can strengthen the understanding of the overall trust of the information. Through intrinsic and extrinsic semantic assessments of the quality of crowdsourced geospatial information we can measure the semantic trust of the information, and by coupling these assessments with a wider trust model of the spatio-temporal and social aspects, we can improve the overall trust of the information.

ACKNOWLEDGEMENTS

We gratefully acknowledge the support of the CRCSI, project 3.02.

REFERENCES

Ballatore, A., Bertolotto, M., & Wilson, D. C. (2013). Geographic knowledge extraction and semantic similarity in OpenStreetMap. Knowledge and Information Systems, 37(1), pp.61-81.

Bishr, M., & Janowicz, K. (2010). Can we trust information? - the case of volunteered geographic information. CEUR Workshop Proceedings, 640.

Bishr, M., & Mantelas, L. (2008). A trust and reputation model for filtering and classifying knowledge about urban growth. GeoJournal, 72(3-4), 229-237.

Bordogna, G., Carrara, P., Criscuolo, L., Pepe, M., & Rampini, A. (2014). A linguistic decision making approach to assess the quality of volunteer geographic information for citizen science. Information Sciences, 258, pp.312-327.

Celino, I. (2013). Human computation VGI provenance: semantic web-based representation and publishing. Geoscience and Remote Sensing, IEEE Transactions on, 51(11), 5137-5144.

Deng, J., Dong, W., Socher, R., Li, L., Li, K., & Fei-Fei, L. (2009). ImageNet: A large-scale hierarchical image database, Proc. CVPR 2009, IEEE (2009), pp.248-255.

Egenhofer, M J (2002). Toward the semantic geospatial web. In Proceedings of the 10th ACM international symposium on Advances in geographic information systems (GIS '02). ACM, New York, NY, USA, 1-4.

Goodchild, M. F., & Li, L. (2012). Assuring the quality of volunteered geographic information. Spatial Statistics, 1, pp.110-120.

Haklay, M. (2010). How good is volunteered geographical information? A comparative study of OpenStreetMap and Ordnance Survey datasets. Environment and Planning B: Planning and Design, 37, pp.682-703.

Janowicz K, M. Raubal, and W. Kuhn (2011). The semantics of similarity in geographic information retrieval, Journal of Spatial Information Science, 2, pp.29-57.

Janowicz K, S. Scheider, T. Pehle, and G. Hart (2012). Geospatial Semantics and Linked Spatiotemporal Data – Past, Present, and Future. Semantic Web 0 1–0. 1. IOS Press.

Kuhn, W. (2005). Geospatial Semantics: Why, of What, and How? Journal on Data Semantics 3, pp.1-24.

MacCormack, K. E., & Eyles, C. H. (2010). Enhancing the Reliability of 3D Subsurface Models through Differential Weighting and Mathematical Recombination of Variable Quality Data. Transactions in GIS, 14(4), 401-420.

Malaverri, J. E., Medeiros, C. B., & Lamparelli, R. C. (2012, March). A provenance approach to assess the quality of geospatial data. In Proceedings of the 27th Annual ACM Symposium on Applied Computing (pp. 2043-2044). ACM.

Marcheggiani, E., Nucci, M., Tummarello, G., & Morbidoni, C. (2007). Geo Semantic Web Communities for Rational Use of Landscape Resources. Ontologies for Urban development: Conceptual Models for Practitioners, Turin, Italy, pp.100-113.

McCartney, E., Bearden, M., & Newell, M. (2013). Crowd-Sourcing the Nation: Now a National Effort. Retrieved from http://www.usgs.gov/newsroom/article.asp?ID=3664#.VP5pNfmUc40

Narock, T., & Hitzler, P. (2013). Crowdsourcing Semantics for Big Data in Geoscience Applications. Semantics for Big Data: Papers from the AAAI Symposium. Presented at the AAAI 2013 Fall Symposium Series Semantics for Big Data, Arlington, VA, November 15-17, 2013.

Ramos, J. M., Vandecasteele, A., & Devillers, R. (2013). Semantic Integration of Authoritative and Volunteered Geographic Information (VGI) using Ontologies. Association of Geographic Information Laboratories for Europe (AGILE).

Reeve L, and Han, H. (2005). Survey of semantic annotation platforms. In Proceedings of the 2005 ACM symposium on Applied computing, ACM, pp.1634-1638.

Sizov S (2010). GeoFolk: Latent Spatial Semantics in Web 2.0 Social Media, in: Proceedings Web Search and Data Mining, 2010.

Vandecasteele, A., & Devillers, R. (2013). Improving volunteered geographic data quality using semantic similarity measurements. ISPRS-International Archives of the Photogrammetry, Remote Sensing and Spatial Information Sciences, 1(1), pp.143-148.

Yue, P., Di, L., Yang, W., Yu, G., & Zhao, P. (2007). Semantics-based automatic composition of geospatial Web service chains. Computers & Geosciences, 33(5), pp.649-665.

36

TREATMENT OF GEODETIC SURVEY DATA AS FUZZY VECTORS

G. Navratil [*], E. Heer, J. Hahn

Department for Geodesy and Geoinformation, TU Vienna, Austria – (navratil,hahn)@geoinfo.tuwien.ac.at,
elsa.heer@tuwien.ac.at

Commission II, WG II/4

KEY WORDS: Fuzzy Vector, Geodetic Survey Data, Analysis

ABSTRACT

Geodetic survey data are typically analysed using the assumption that measurement errors can be modelled as noise. The least squares method models noise with the normal distribution and is based on the assumption that it selects measurements with the highest probability value (Ghilani, 2010, p. 179f). There are environment situations where no clear maximum for a measurement can be detected. This can happen, for example, if surveys take place in foggy conditions causing diffusion of light signals. This presents a problem for automated systems because the standard assumption of the least squares method does not hold. A measurement system trying to return a crisp value will produce an arbitrary value that lies within the area of maximum value. However repeating the measurement is unlikely to create a value following a normal distribution, which happens if measurement errors can be modelled as noise. In this article we describe a laboratory experiment that reproduces conditions similar to a foggy situation and present measurement data gathered from this setup. Furthermore we propose methods based on fuzzy set theory to evaluate the data from our measurement.

1. INTRODUCTION

Methods for geodetic observations are typically designed in a way that the resulting errors are small in relation to the observed value and bias is eliminated. Thus the assumption is that the resulting error (the noise) is normally distributed with an expectation value of zero. This requires not only carefully developed equipment but also controlled environmental conditions. With human observers it is possible to perform the observations only under reasonable conditions and postpone observations otherwise. This is not possible with automated systems. Such systems are typically installed if fast response times are essential, for example in case of landslides. Problems occur if the environmental conditions are bad.

If only some of the automatically determined measurements are affected by poor environmental conditions, then the application of robust estimation may be able to identify observations with poor quality and provide good results by eliminating observations with a bad quality. However, if the observed situation changes during the poor environmental condition, these changes will only become visible after the environmental conditions improve and this could be already too late. This paper is based on a different method of analysis: Fuzzy logic. We present an experiment that is designed to reproduce foggy conditions and discuss possible approaches for the analysis of observations. Peter Fisher states that fuzzy set theory is a method to deal with vague, poorly defined objects (Fisher, 1999). In the situation discussed here, the object itself is clearly defined but the environmental conditions, which affect the path that the light takes, is not. Thus we assume that fuzzy set theory is a promising approach to this kind of problem.

The remainder of the paper is organized as follows: In section 2 we give a brief introduction into fuzzy numbers and vectors and their analysis. In section 3 we present the experimental setup and show the resulting data. Section 4 contains some results and a discussion of two ideas for other approaches. Section 5 concludes the paper by providing directions for future work.

2. FUZZY NUMBERS AND FUZZY VECTORS

In statistics, an imprecise (fuzzy) number x^* is defined by its one dimensional characterizing function $\xi(\cdot)$, which is a real function of a real variable x with values between 0 and 1 (Menger, 1951, Zadeh, 1965 as quoted by Viertl, 2006a). The value represents the degree of membership. Fuzzy numbers are represented by δ-cuts for numerical treatment, where δ is a real number from the interval $(0,1]$ (see Fig. 1). Each interval for a δ-cut contains any interval with for a larger value of δ.

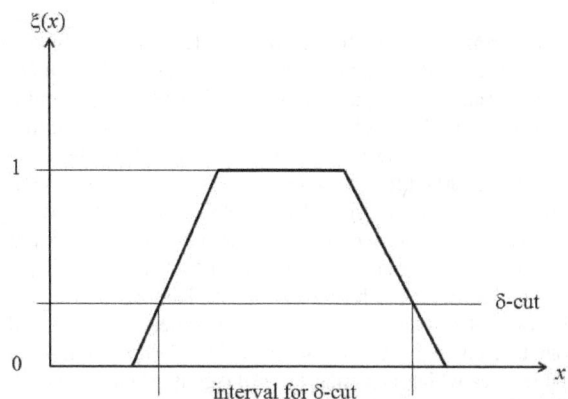

Figure 1. Fuzzy number and δ-cut

---

* Corresponding author

Data analysis then requires that the necessary functions are applied to the fuzzy numbers, i.e., $y^* = f(x_1^*, \ldots x_n^*)$. The result is again a fuzzy number and can be described by δ-cuts (Viertl, 2006b).

When moving from one dimensional fuzzy numbers to n-dimensional fuzzy vectors, different representations are possible (Liang et al., 2009). An n-dimensional fuzzy vector requires an n-dimensional characterizing function ξ: $R^n \rightarrow R$ in the interval (0,1]. A 2-dimensional fuzzy vector ξ_2: $R^2 \rightarrow R$ provides a membership value for each point in a plane. If the vector describes the horizontal extent of a mountain, then each point with $\xi_2=1$ is definitely part of the mountain and each point with $\xi_2=0$ is definitely not part of the mountain. In the area between these extremes, the point may be part of the mountain and the likelihood of the membership increases with the value of ξ_2. It is important to note, hat a 2-dimensional fuzzy vector is not a vector of two fuzzy numbers because this would only be true if the two parameters of the characterizing function can be separated to $\xi_2=\min(\xi(x),\xi(y))$.

Combination of fuzzy numbers or fuzzy vectors can then be done by using the δ-cuts. Fig. 2 shows an example of the combination of two fuzzy numbers with the minimum combination rule. The overlap of the δ-cuts is the result of the combination of the two fuzzy numbers. The same can be done with fuzzy vectors.

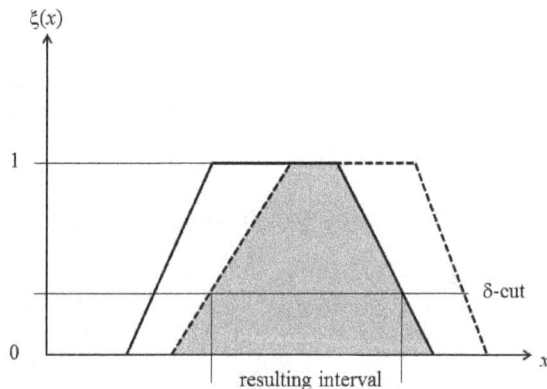

Figure 2. Combination of two fuzzy numbers

3. EXPERIMENTAL SETUP

The experiment was performed by Elsa Heer at TU München. The setup consists of a video theodolite, which was located at two different (known) positions (Fig. 3) and a light emitting target (Fig. 4). The results of a video theodolite observation are crisp and precise measurements and an image. Thus the system had to be disturbed for the experiment. This was done by setting focus and exposure of the video theodolite deliberately wrong. The result of the measurement does not have a clear maximum of intensity. A similar (and more realistic) effect could have been achieved by producing fog in the laboratory. However, this was not possible due to other experimental setups in the room that could have been affected. Fig. 5 shows one of the final images which is similar to an image is taken in a slightly foggy situation. Fig. 6 shows the image of a light source in foggy conditions. The foggy conditions show a similar pattern than the experiment but there are two differences: The area between light and darkness is bigger in Fig. 6 and the radiating pattern from Fig. 5 is missing. The advantage of the results

from the experiment is that the conditions can be easily reproduced.

The goal of this experiment is to compute the coordinates of the target using the directions determined by the theodolite. The theodolite was a Leica Geosystems TPS 1201with an attached CMOS camera.

Geometrically, the setup is a forward intersection. From two known points, the directions to a point with unknown coordinates are determined. The setup results in two lines that need to be intersected. Due to inevitable measurement deviations, the lines are skew lines and the intersection is typically computed by correcting the observations according to the least squares method. The size of the necessary corrections provides an estimate for the quality of the solution. In the setup the angle of intersection is 33 gon. This is not an optimal intersection angle but a realistic approach.

The basic setup thus consisted of four locations:
- 2 positions where the video theodolite was placed for the measurement (instrument positions),
- the position of the target, and
- a reference point to orient the video theodolite (this point is irrelevant for the further discussion because it is only used to align the measurements done with the video theoolite).

The reference point is necessary because theodolites measure horizontal directions with respect to an arbitrary starting direction. In order to use the observations for the determination of coordinates in a given reference frame, the observations need to be oriented, i.e., they need to be rotated such that the reference direction corresponds with the x-axis of the reference frame. This is done by observing the direction to a point where the coordinates are known (the reference point). In our setup the coordinates of all locations were also determined by classical measurements using a theodolite. Therefore, not only the coordinates of the instrument positions and the orientations are known, but also the coordinates of the target position. This guarantees that the quality of the final result can be determined.

Figure 3. Experimental setup: Instrument positions with the target in the background (pictures taken by Elsa Heer)

The intensity of the image can be used to determine the characterizing function. It is obvious, that the typical simplification for the shape of the characterizing function, which is shown in Figures 1 and 2, does not apply for this example. The real shape is more complex and less regular. The circular patterns may even lead to situations where the δ-cuts are not connected areas.

The result of the measurement is now modelled by a fuzzy vector in 3D space. In order to model it, two parameters are necessary to determine the direction of the vector. These parameters can either be modelled as vertical and horizontal angles as observed from the projection centre of the theodolite or as horizontal and vertical displacement in a defined distance from the projection centre (the focal length of the optical system).

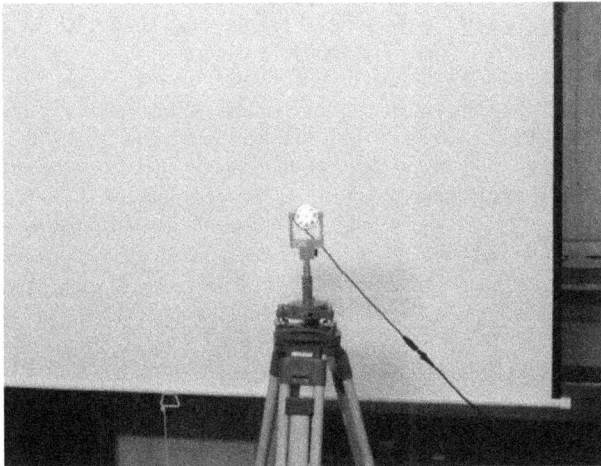

Figure 4. Experimental setup: Target position (picture taken by Elsa Heer)

Figure 5. Image as taken by the video theodolite from point 2

Figure 6. Image of a light source in foggy conditions

4. PRELIMINARY RESULTS

In this section we discuss advantages and disadvantages of three methods that can be used to calculate the coordination of the light source.

4.1 Simple Method

In a first step, a simple method has been tested: In order to make decisions based on fuzzy information, the information is usually defuzzied (the area of the characterizing function is reduced to a point) and then further calculations are applied. The defuzzification can be based on the centroid of the characterizing function. Theoretically, this works if the uncertainty is purely random and thus the centroid is a reasonable approximation of the expectation of the distribution.

The δ-cuts of the characterizing function require a characterizing function in to range of [0,1]. Therefore the intensity of the images (e.g. Fig. 5) has to be normalized. One normalization result is included in Fig. 7. This characterizing function can then be used to compute the δ-cuts.

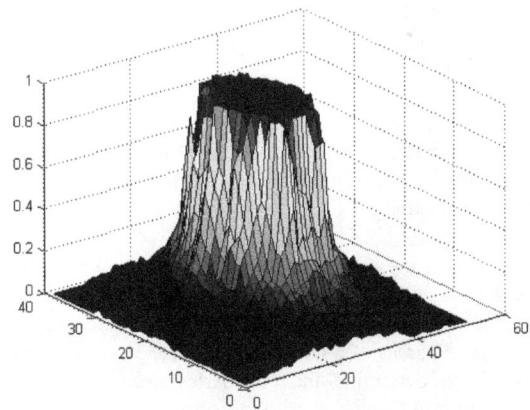

Figure 7. Characterizing function for the direction from point 2 (Heer, 2014)

An open question is, which δ-cuts should be used. In the following, the values 0.25, 0.5, 0.75, and 1 are used for δ. Each of these intersections results in a different area. The centroid of the area is then used to determine the line from the instrument positions to the target position. Once this is done for both instrument positions, the most likely intersections can be computed.

The intersections for the selected δ-cuts are summarized in Table 1. The x- and y-coordinates differ significantly from the results computed with the undisturbed observations (using the full precision of the theodolite) since the distance is short, the atmospheric conditions are controlled, and the theodolite is of high quality, any deviation of more than a few centimetres is a failure. The results made it obvious that this approach is not suitable for the analysis of fuzzy observations, because of differences of max. 1 meter in x direction and of max. 7 meters in y direction.

	x [m]	y [m]	z [m]
δ=0.25	101.286	117.040	49.749
δ=0.5	99.113	110.455	49.869
δ=0.75	99.428	113.649	49.821
δ=1	100.047	123.086	49.668
undisturbed result	101.303	117.754	49.643

Table 1. Estimated point coordinates for different δ-cuts and the accurate result from the undisturbed measurement

4.2 Cone Boundary Intersection Method

Another approach would be intersecting the cones defined by the δ-cuts. The centre of the optical system of the theodolite is known (typically by the coordinates of the point and the vertical offset of the theodolite). The angular readings of the theodolite determine the direction in which the telescope is pointing. Since the video camera is attached to the telescope and the mutual orientation is known, the orientation and position of the image in relation to the centre point is known. A δ-cut consists of pixels with value 1 (the interior of the vector) and 0 (the exterior of the vector). The boundary of the cone is determined by the boundaries between pixels with different values. Each instrument position provides a cone and the intersection of these two cones is the volume of space that agrees with both observations.

A property of the cone intersection is that the direction of highest uncertainty is recognizable. Fig. 8 shows that the intersection area has a small lateral extent and a much larger longitudinal extent.

A problem of the cone intersection is the complexity of the δ-cuts. As shown in Fig. 5, the data are not clearly structured. The rays with high intensity and the concentric circles may lead to disconnected areas of δ-cuts and therefore to complex volumes. Methods to determine the boundaries need to be carefully programmed to avoid numerical problems.

Another problem is the result itself. Surveyors typically work with random variables, which can be modelled as crisp values with an attached standard deviation. All computational algorithms have been optimized for this kind of input and it may require significant changes to include volumetric point information. Defuzzification of the resulting volume is not a good solution either. The volume is connected to a specific δ-cut. There is no information in the model providing probability of specific values. Typically, the volume will get smaller if the value for δ increases with the smallest volume for δ=1. However, only using δ=1 for the computation is dangerous because there may be situations where these cones do not overlap. Then lower numbers for δ have to be used in order to obtain a result. Therefore, in larger evaluations different δ-cuts need to be used for the whole project since it may be difficult to assess the optimal δ-cut at the beginning.

4.3 Grid Ray Intersection Method

Another method represents the cones as a bundle of rays. In order to calculate ray the characterizing function is stored as an image. Thus each pixel belonging to a specific δ-cut could be used to define a ray from the projection centre of the theodolite through the centre of the pixel. When using different δ-cuts, each ray belongs to a specific δ-cut. It may even be argued, that it belongs to any δ-cut with a δ lower than the value used to create the ray. If, for example, the pixel is part of the 0.5-cut but not part of the 0.75-cut, then it definitely is part of the 0.25-cut too, because the area of the 0.25-cut is at least as large of the area of the 0.5-cut and includes the 0.5-cut.

Fig. 9 illustrates the concept of the grid ray intersection method. Each cone is represented by a bundle of rays. These bundles are intersected and the resulting intersection points form the area of intersection, or in 3D the volume of intersection.

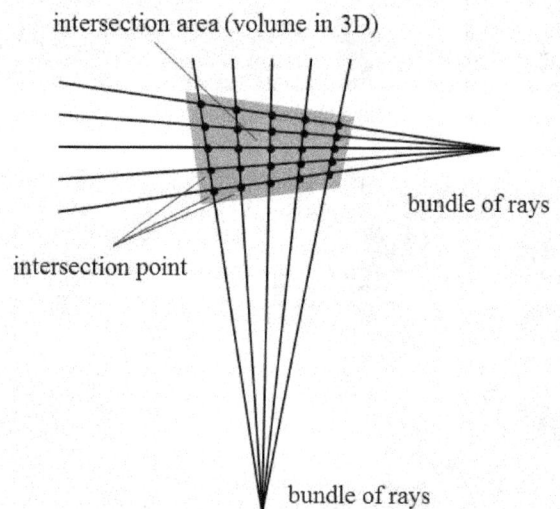

Figure 9. 2D-representation of the grid ray intersection method

In 3D, the rays will not intersect because they will be skew lines. This is true for any two lines from different bundles but some lines will pass each other at a close distance such that they can count as intersecting. It is necessary to identify these lines. It is possible to compute a midpoint between two arbitrary rays (the point in the middle of the closest connection between the

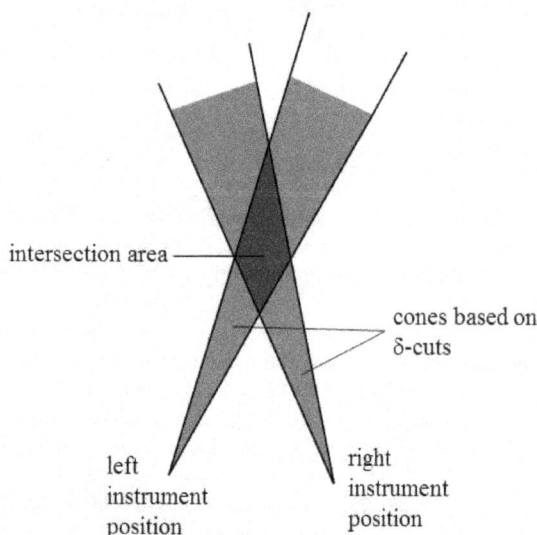

Figure 8. Concept of cone boundary intersection method; the position of the target is somewhere in the intersection area

two rays). The intersection volume can be generated by computing the midpoints for any two rays that 'almost intersect' (pass each other at a close distance). The criterion to determine that two rays are too far apart to be classified as 'almost intersecting' is critical for the evaluation. A too large space will be filled with points if the criterion is too weak and an insufficient number of points or no points at all will be found if the criterion is too strong. The criterion needs to be based on the approximate distance between projection centre and a reasonable starting value could be the lateral distance between two neighbouring rays from one bundle.

Since the basis for the bundle of rays is a regular raster, the following idea could be used: The lateral distance between two rays increases with the distance from the projection centre. The raster has an equally horizontal and vertical resolution and thus neighbouring points have a distance of d in the 4-neighbourhood and a maximum of $d \cdot sqrt(2)$. Half this maximum distance could be used to identify nearby rays. The identification of one pair of nearby rays (a,b) does not preclude the existence of another ray c that is also nearby the ray a. Thus all possible combinations have to be checked.

The result of the process is a set of points in space for each δ-cut. Since a δ-cut includes all rays for δ-cuts with a higher value for δ, the set of intersection points with shrink with an increasing value of δ. The space filled by the set of points is the δ-cut of the intersection.

The image size of the used CMOS camera is 1280x960 pixels. The δ-cut for $\delta=1$ contains approximately 160,000 pixel. Thus 25 billion combinations have to be checked for the smallest cone only. Assuming orthogonal rays and characterizing functions with a square extent of 400x400 pixels, then each ray from bundle 1 is nearby 400 pixels from the other ray. As a result, the point cloud consists of 64 million points. The limiting factor is not the large number of intersections or points, it is the fact that this is only a simple example and the numbers are already quite high.

5. CONCLUSIONS AND OPEN QUESTIONS

During the experiment and analysis it became obvious that the analysis of fuzzy observation data is not simple. The quality of the simple solution is unacceptable. Therefore, other strategies need to be developed and tested. Two strategies have been presented: The first approach uses the boundaries of the δ-cuts for the two observations. The intersection of these boundaries then defines the δ-cut of the final result. The second approach represents the δ-cuts by the contained points instead of their boundary and models the volume of the cones as a bundle of rays thorough these points.

Both solutions have advantages and disadvantages. The cone boundary intersection produces the boundary of the cone this provides a clear topological structure with an interior and an exterior for each δ-cut. However, computations may be demanding due to glancing intersections. The grid ray intersection, on the other hand, has a quite simple computational model. Apart from parallel rays there are not much computational pitfalls. The major problem in this approach is the huge number of computations and the size of the resulting data set. A possible solution could be the reduction of grid resolution. If each 10x10 square is represented by a ray then there are only 1,600 rays left in each bundle and only 2.56

million intersections need to be checked. However, applying this kind of strategy needs further investigation. Especially the dependency between reduction of the resolution and the deterioration of the quality needs to be analysed.

Another open question is the relation to the standard method to compute the result for the forward intersection. It is assumed that random error (noise) affects the observational data. The least squares approach provides best estimates for the noise and thus determines self-consistent measurement data, in this case two lines in 3D which intersect in a point. Is it possible to combine this approach with one of the strategies developed in this paper? Can we achieve better results if—in analogy to what happens to observation values in a least squares approach—we slightly move the cones?

The intersection of two cones has the largest volume if the axes of symmetry of the two cones meet. Thus larger overlapping areas do mean higher consensus between the symmetry axes but this may not necessarily be a sign of better quality in general.

ACKNOWLEDGMENT

TU München supported the experiment by providing the technical equipment, the location, and the measurement expertise. We thank Thomas Wunderlich for providing the video theodolite and Andreas Wagner for the practical support.

REFERENCES

Fisher, P., 1999. Models of Uncertainty in Spatial Data. In: Geographical Information Systems: Principles, Techniques, Management and Applications, Longley, P, Goodchild, M., Maguire, D., and Rhind, D. (eds.), Vol. 1, John Wiley & Sons, Inc. Hoboken, New Jersey, pp. 191-205.

Ghilani, C.D., 2010. Adjustment Computations. 5th ed., John Wiley & Sons, Inc. Hoboken, New Jersey, 647 p.

Heer, E., 2014. Positionsbestimmung mittels unscharfer Information. Bachelor Thesis at TU Vienna, Department for Geodesy and Geoinformation, Vienna, Austria.

Liang, J., Navara, M. & Vetterlein, T., 2006. Different Representations of Fuzzy Vectors. In: *Symbolic and Quantitative Approaches to Reasoning with Uncertainty*, Springer, Berlin Heidelberg, LNCS Vol. 5590, pp. 700-711.

Menger, K., 1951. Ensembles flous et fonctions aleatoires. Comptes Rendus de l'Academie de Sciences Paris, France.

Viertl, R., 2006a. Univariate statistical analysis with fuzzy data. *Computational Statistics & Data Analysis*, 51(1), pp. 133-147.

Viertl, R., 2006b. Description of Data Uncertainty and its Consequences for Data Analysis. Report SM-2006-4, TU Vienna, Vienna, Austria.

Zadeh, L., 1965. Fuzzy Sets. Information and Control, **8**, pp. 338-353.

37

INCLUDING USERS' SEMANTICS IN EVALUATING THE CREDIBILITY OF CROWDSOURCED LANDSCAPE DESCRIPTIONS

A. Forati [a,*], S. Soleimani[b], F. Karimipour [a] and M.R. Malek [b]

[a] Faculty of Surveying and Geospatial Engineering, College of Engineering, University of Tehran, Tehran, Iran
(forati, fkarimipr)@ut.ac.ir
[b] Faculty of Geodesy and Geomatics Engineering, K.N. Toosi University of Technology, Tehran, Iran
ssoleimani@mail.kntu.ac.ir, m.malek@kntu.ac.ir

Commission II, WG II/4

KEYWORDS: Landscape Description, Crowdsourcing, Credibility, Users' Semantics

ABSTRACT

Landscape refers to the visible features of an area of land often considered in terms of their aesthetic appeal. The spatial configurations, composition conditions and perception situations of a landscape may be described by people, which reveal how they see and percept the landscape; thus they are different from person to person depending on the way they think or experience their surroundings. Especially, when the landscape descriptions are acquired through crowdsourcing processes, it is affected by users' semantics, which results in descriptions with varying credibilities. This paper proposes an approach to consider the user's semantics in evaluating the creditability of crowdsourced landscape descriptions.

1. PROBLEM DEFINITION

Landscape refers to the visible features of an area of land often considered in terms of their aesthetic appeal. The notion of landscape is also illuminated by the concepts of land, nature, space, and temporal elements as well as people's attitude towards them (Litton 1968, Derungs and Purves, 2013). There are many different perspectives in order to define landscape, such as recreational, silvicultural, hydrological, ecological, acoustical, and wildlife, to name a few (McGarical, 1995). However landscape definition mostly refers to the visible features of an area of land to be considered as recreational areas.

Landscape description provides insight into the worlds of perception. Knowing how people see and perceive a landscape imparts additional information about different characteristics of landscape that could be used in aesthetic analysis, environmental protection, visual pollution, way finding, and landscaping. The landscape descriptions may be categorized into (1) spatial configurations; (2) composition conditions; and (3) perception situations. *Spatial configurations* of a landscape refer to how the available elements are arranged. Distance between elements, direction of elements respect to an origin, visible sections of each element from specific points of view, and topological relations between elements are among metrics that define landscape configuration (Litton 1968, Yaouanc et al., 2009). Figure 1 illustrates some of the spatial configurations of a landscape. *Composition conditions* mostly quantify the distribution, variety, and amount of features such as plants and animals in an area without considering spatial qualities (McGarigal et al., 1995; Ritchie et al., 2009). As composition metrics includes the environmental structure of landscape, the environmental pollution and even animal migration could be discussed in this category. Finally, *perception situations* and knowledge of landscape determine the way people perceive

various landscapes (Bell, 2012). Observers, depending on their outlooks, have specific feelings when they are in a certain type of landscape, which among others, highly corresponds to personal factors.

The main issue in acquiring the landscape descriptions is that perception of land varies from person to person depending on the way they think or experience their surroundings. There are evidences that parameters like gender, age, education, profession, and even health conditions influence landscape descriptions provided by people (Schirpke et al., 2013).

This is more challenging when the landscape descriptions are acquired through crowdsourcing processes (Howe, 2006), by which a broad spectrum of heterogeneous data resources (which is affected by users' semantics) could be generated at a rapid rate. In addition, as crowdsourced data is mostly based on human experience of geography, deploying perception-based parameters to express their spatial quality is more efficient than measurement-based parameters used in case of official spatial data.

Generally, crowdsourcing and participating is a relatively new method in the field of landscape science. However, Brown and Brabyn (2012) conducted a research on different compositional values of the landscape using public participation GIS. This study attempts to focus on the compositional values such as wilderness, aesthetic, spiritual, recreation, and economic that are important to all people, without considering the semantic of the participants. However, according to different categories of landscape descriptions, the semantic of the observer who describes a view plays a role in her/his descriptions.

Among several perception-based parameters suggested in the literature, Flanagin and Metzger (2008) introduce the concept of *credibility*, as a perceptual variable, for evaluating collaborative productions. "Although there is no clear definition of credibility, it is generally thought to be the believability of a source or message, which is composed of two primary dimensions: trustworthiness and expertise" (Flanagin and Metzger, 2008). This paper proposes an approach to consider the user's semantics in evaluating the credibility of crowdsourced landscape descriptions. We draw inspiration from the Kessler et al. (2013)'s definition of trustworthiness and extend it by relating data trustworthiness, user reputation and accounting social group reputation to.

Distance

Direction

Topology

Figure 1. Some of the configuration conditions in landscape description

2. SEMANTIC ISSUES IN LANDSCAPE DESCRIPTION

The landscape is a combination of natural and conceptual phenomena. Spatial configurations and composition conditions are essential to describe the landscape. However, cross-cultural issues are still questionable in understanding how people perceive the landscapes (Purves, 2008). The landscape description almost acts as a mental tool to help people share their experiments and thoughts (Cohen, 2013).

People may possess different perspectives when facing a specific landscape. Furthermore, their mental possession affects the way they define and describe a landscape. Some assessment tools associated with physical and mental features of landscape, called *metrics* (Yaouanc et al., 2009), could be combined to provide a formal environment for differentiating the structures of landscapes. These metrics could be easily used by all expert and non-expert people. However, the credibility of acquired data would vary due to users' semantics.

As discussed, landscape has different aspects when it comes to be described as a mental and physical phenomenon. For visual aspect, there are some metrics namely, topology relations, direction relations, distance, solid angle, and salience (Soleimani, 2015) that almost depend on user's mindset, experiences, skills, and even culture. To illustrate, thinking about a feature in a landscape as a salient, one is dependent to the observer preference. Moreover, according upon various human's skills, some observer could estimate solid angle and distance more precisely than others. Obviously, such differences affects the credibility of the collected data them.

The crowdsourced data is more based on human cognition and experience than on measurements (Karimipour and Azari, 2015), thus describing a landscape through a crowdsourcing process may undermine the credibility of the results due to varieties in humans' landscape cognition as well as their semantics.

Therefore, in order to evaluate the credibility of users' descriptions of landscapes, one should consider the users' subjective cast of minds. For instance, in case of spatial configuration of landscape, the descriptions provided by the engineer participants may be considered more credible. More precisely, it seems that participant with measuring skills provide more credible "distance" and "area", participant with a geography background may produce more credible "direction", and those who are expert at mapping activities produce more credible topological data.

3. PROPOSED APPROACH

As discussed, the semantics of users, as sensors in crowdsourcing process, have an effective role in data credibility. This paper defines the credibility of landscape description based on the users' semantics. We rank the users for different metrics of landscape description, and correspondingly, the credibility of the data they provide.

A web-based platform (Figure 2) is designed to collect the required information about the landscape description. In order to produce the visual data of a landscape, participants have to fill out the predefined metrics embedded in the designed web-page, such as available salient elements and their estimated visual area. Some other information including coordinates, time, and even weather could be gathered automatically considering the available information in the geo-tag image uploaded by the participant.

Besides, a set of user information is acquired for the desired credibility assessment. As a matter of fact, knowing user's age, gender, education, profession, interests, and disease, to name but a few, could give us an insight on how credible is the data he/she could produce.

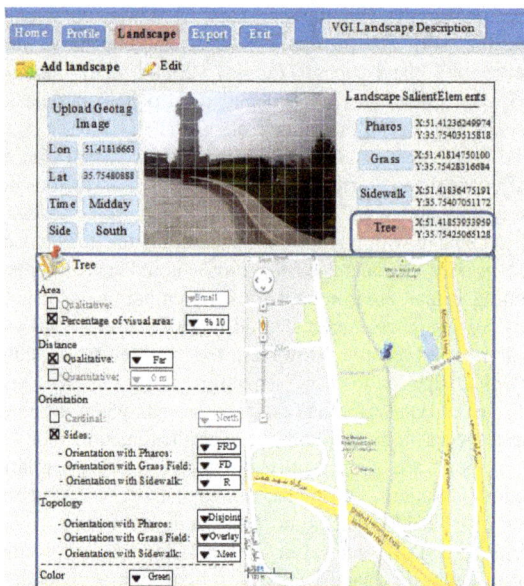

Figure 2. The web-based platform designed to collect the crowdsourced landscape descriptions

The steps of the proposed approach are as follows (Figure 3):

Classification: In order to evaluate the credibility of the data provided, the users are classified into different groups such as civil engineers, surveyors, doctors, artists, to name a few, and initial credibility values are assigned to each group for each aspect of landscape description (i.e., configuration, composition, and perception issues). This specifies the users' ability to describe various aspects of landscape. The credibility of these aspects directly relate to the users semantics and conceptions. For example, we expect that the "distance" provided by an engineer participant who is always interacting with dimensions, incontrovertibly, has more credibility compared to the data acquired by an artist; while the introduced "color" by an artist participant could have a more credibility rather than the one suggested by an engineer.

Credibility evaluation: The crowdsourced landscape description acquired by a user will be assigned the relevant credibility values: For an already registered user, his/her own credibility values will be applied. For a new user, the initial credibility would be the one assigned to the corresponding group for the desired aspect of landscape description; and if the user is not a member of any classes, a fuzzy inference process will be applied to assign relevant initial credibility values, which as we see in the third step, will be updated during the data acquisition process. By integrating the data provided by different users, a final description of that aspect will be specified with a certain credibility.

Update: The above credibility resulted from integrating the crowdsourced data will be used to update the initial credibility values assigned to the individual registered users, as well as different user groups through established methods suggested in the literature (for example, Chatterjee et al., 2008; Adler et al., 2011; Adler and Alfaro, 2007).

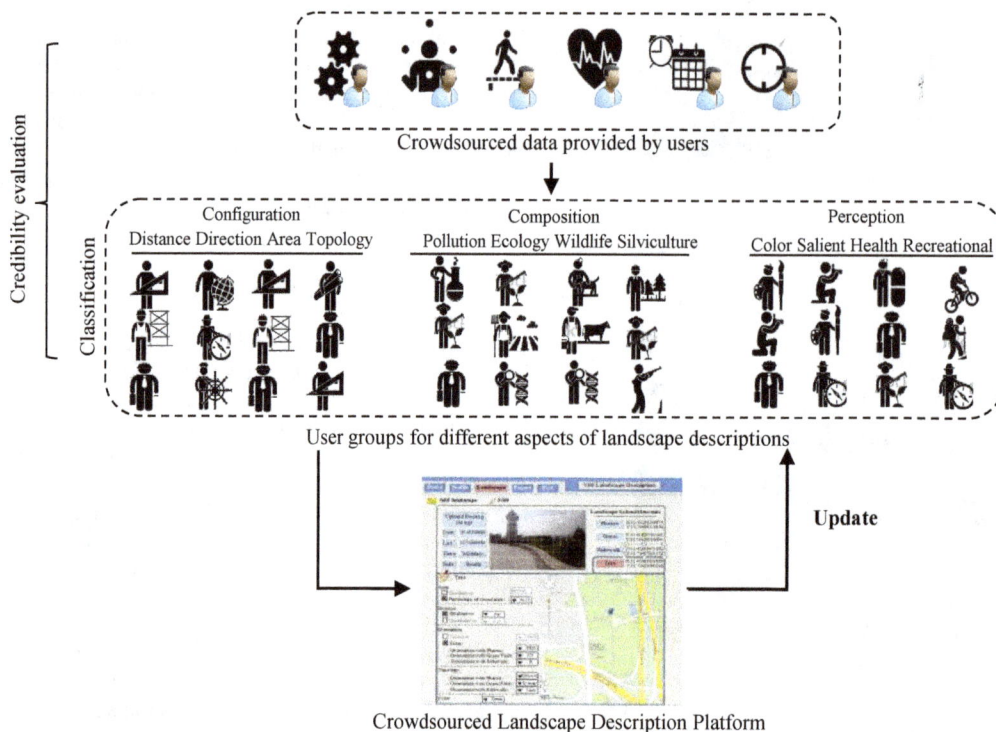

Figure 3. The proposed approach to include the users' semantics in landscape description

4. CONCLUSION

This paper proposes an approach to include the users' semantics in evaluating the credibility of landscape descriptions acquired through crowdsourced processes. The initial outcomes of implementing the proposed approach show an improvement in the final results. However, grouping and assigning proper initial credibility values are the main concerns to be considered in future. We are also working on generalizing the approach to be deployed for other types of crowdsourced data.

REFERENCES

Adler, B.T. and Alfaro., L.D., 2007. A Content-Driven Reputation System for the Wikipedia, Proceedings of the 16th International Conference on World Wide Web, Banff, Alberta, Canada, pp. 261-270.

Adler, B.T., Alfaro, L.D., Mola-Velasco, S.M., Rosso, P. and West, A.G., 2011. Wikipedia Vandalism Detection: Combining Natural Language, Metadata and Reputation Features, Computational Linguistics and Intelligent Text Processing, LNCS Vol. 6609, pp. 277-288.

Bell, S., 2012. Landscape: Pattern, Perception and Process. Routledge.

Brown, G. and Brabyn, L., 2012. An Analysis of the Relationships between Multiple Values and Physical Landscapes at a Regional Scale using Public Participation GIS and Landscape Character Classification. Landscape and Urban Planning, 107(3), 317-331.

Chatterjee, K., Alfaro, L.D. and Pye, I., 2008. Robust Content-Driven Reputation, Proceedings of the 1st ACM Workshop on AISec, Alexandria, Virginia, USA, pp. 33-42.

Cohen, R. (Ed.), 2013. The Development of Spatial Cognition, Psychology Press.

Derungs, C., and Purves, R.S., 2013. From Text to Landscape: Locating, Identifying and Mapping the Use of Landscape Features in a Swiss Alpine Corpus, International Journal of Geographical Information Science, 28(6), pp. 1-22.

Flanagin A and Metzger M., 2008. The Credibility of Volunteered Geographic Information, GeoJournal, 72(3), pp. 137-148.

Howe, J., 2006. The rise of crowdsourcing. Wired magazine, 14(6), 1-4.

Karimipour, F. and Azari, O., 2014. Citizens as Expert Sensors: One Step Up on the VGI Ladder, In: Progress in Location-Based Services, Lecture Notes in Cartography and Geoinformation, Springer, pp. 213-222.

Keßler, C., Theodore, R. and de Groot, A., 2013. Trust as a Proxy Measure for the Quality of Volunteered Geographic Information in the Case of OpenStreetMap, In: Vandenbroucke, D., et al. (eds.), Geographic Information Science at the Heart of Europe, Lecture Notes in Geoinformation and Cartography, Springer, pp. 21-37.

Litton, R.B., 1968. Forest Landscape Description and Inventories - a Basis for Landplanning and Design.

McGarigal, K. and Marks, B.J., 1995. Spatial Pattern Analysis Program for Quantifying Landscape Structure. Technical Report PNW-GTR-351. US Department of Agriculture, Forest Service, Pacific Northwest Research Station.

Purves, R.S. and Edwardes, A.J., 2008. Exploiting Volunteered Geographic Information to Describe Place. Proceedings of the GIS Research UK 16th Annual Conference, pp. 252-255.

Ritchie, L.E., Betts M.G., Forbes, G., and Vernes, K., 2009. Effects of Landscape Composition and Configuration on Northern Flying Squirrels in a Forest Mosaic, Forest Ecology and Management, 257(9), pp. 1920-1929.

Schirpke, U., Tasser, E., and Tappeiner, U., 2013. Predicting Scenic Beauty of Mountain Regions. Landscape and Urban Planning, Vol. 111, pp. 1-12.

Soleimani, S., 2015. Landscape Description in Volunteered Geographic Information (VGI) Using Spatial and Temporal Relationships, MSc Thesis, KNT University, Tehran, Iran (In Persian).

Yaouanc, L., Saux, J.M. and Claramunt, C., 2010. A Semantic and Language-based Representation of an Environmental Scene. Geoinformatica, 14(3), pp. 333-352.

38

SEMI-AUTOMATED DETECTION OF SURFACE DEGRADATION ON BRIDGES BASED ON A LEVEL SET METHOD

A. Masiero[a, *], A. Guarnieri[a], F. Pirotti[a], A. Vettore[a]

[a] Interdepartmental Research Center of Geomatics (CIRGEO), University of Padova,
Viale dell'Università 16, Legnaro (PD) 35020, Italy -
masiero@dei.unipd.it
(alberto.guarnieri, francesco.pirotti, antonio.vettore)@unipd.it

Commission V, WG V/3

KEY WORDS: Deterioration detection, Monitoring, Large infrastructures, Level sets, Image segmentation

ABSTRACT

Due to the effect of climate factors, natural phenomena and human usage, buildings and infrastructures are subject of progressive degradation. The deterioration of these structures has to be monitored in order to avoid hazards for human beings and for the natural environment in their neighborhood. Hence, on the one hand, monitoring such infrastructures is of primarily importance. On the other hand, unfortunately, nowadays this monitoring effort is mostly done by expert and skilled personnel, which follow the overall data acquisition, analysis and result reporting process, making the whole monitoring procedure quite expensive for the public (and private, as well) agencies.

This paper proposes the use of a partially user–assisted procedure in order to reduce the monitoring cost and to make the obtained result less subjective as well. The developed method relies on the use of images acquired with standard cameras by even inexperienced personnel. The deterioration on the infrastructure surface is detected by image segmentation based on a level sets method. The results of the semi-automated analysis procedure are remapped on a 3D model of the infrastructure obtained by means of a terrestrial laser scanning acquisition.

The proposed method has been successfully tested on a portion of a road bridge in Perarolo di Cadore (BL), Italy.

1. INTRODUCTION

Surfaces of human infrastructures are subject to degradation, mostly due to the effect of (natural) climate elements (sun, rain, wind), to human dependent factors (e.g. increase in the volume of traffic) and to aging.

Nowadays, the monitoring activity of the state of these infrastructures is mostly done by visual inspection performed by expert staff: these highly skilled personnel is involved in periodical recognitions, recording and reporting of the updated infrastructure conditions. It is worth to notice that this procedure typically provides subjective results (conditions are typically subjectively reported on questionnaires) at a quite high cost, related to the relatively frequent use of expert staff.

This paper proposes a different approach in order to reduce the cost of infrastructure monitoring.

The goal of surface infrastructure monitoring is that of periodically updating the detected conditions on a digital representation of the infrastructure, in order to ease the control of the degradation temporal evolution by specific human operators (Costantino and Angelini, 2013a, Costantino and Angelini, 2015, Camarda et al., 2010, Guarnieri et al., 2013).

In this work, the digital 3D representation of the infrastructure is obtained by means of terrestrial laser scanning (TLS). Notice that the use of TLS devices is usually reserved to skilled personnel, hence this is usually a quite expensive operation. For this reason, the use of TLS in this work is limited to una tantum acquisition typically done at the beginning of the monitoring. Nevertheless,

other TLS acquisitions can be considered in the following, when required by the occurrence of specific conditions.

Instead, the degradation detections and condition updates are obtained by the analysis of photos acquired by standard digital cameras (not necessarily professional, e.g. cameras embedded in smartphones). Then, the degradation results are mapped onto their correspondent positions on the TLS-based 3D model of the infrastructure.

The main advantages of this integrated strategies approach are as follows:

- The use of digital cameras is commonly considered a much simpler operation than TLS acquisition, hence it can be performed by not highly skilled personnel. This allows a significant monitoring cost reduction, and/or the possibility of more frequent degradation observations.

- Despite some work has been done in order to exploit the luminance of laser scanner radiation in order to classify materials (Costantino and Angelini, 2013b), actually the analysis of the surface degradation status is usually simpler on digital images than on laser scanning data.

- The reprojection of the detected deterioration results on the 3D model of the infrastructure help the operator to realize the overall conditions of the infrastructure (e.g. the areas where deterioration is more remarkable) and to follow the temporal evolution of the degradation.

A quite common choice is to exploit level set methods as an efficient image segmentation tool (Sethian, 1999, Sethian, 2001,

---
*Corresponding author.

Osher and Fedkiw, 2003): accordingly to other previous works (Cerimele and Cossu, 2007, Cerimele and Cossu, 2009), this paper considers the use of level set methods applied to the L^*, a^*, b^* color space. As shown in Section 2., the use of L^*, a^*, b^* instead of the RGB color space allows to successfully reduce the correlation between different color channels.

Details on the whole segmentation procedure are provided in the next section, while the experimental results on a portion of a road bridge in Perarolo di Cadore (BL), Italy, are shown in Section 3. Finally, some conclusions are drawn in Section 4.

2. ASSESSMENT OF DEGRADED SURFACE AREAS

The data analysis procedure works as follows: once the images are acquired they are separately processed by an ad hoc algorithm. Notice that if the number of images is sufficiently large and with a proper overlapping level, then a 3D photogrammetric reconstruction would be possible as well. Despite this case would be more informative, actually it typically requires an higher level of accuracy during the image acquirement process. Therefore, in order to keep the image capture task as simple as possible, in the proposed method the data analysis is performed on each single image.

The image analysis procedure is user-assisted in its first step (step (a) of the following procedure), whereas the following ones are performed automatically:

(a) The user pinpoints the matches between a set of points in the image and in the 3D infrastructure representation: these correspondences are used for determining the infrastructure surfaces to be controlled and to assess the local map between the image points and their 3D coordinates.

(b) Automatic modal decomposition of the image values (according to a convenient L^*, a^*, b^* representation) on the region of interest.

(c) Region segmentation (i.e. determination of deteriorated areas) based on a level sets method.

(d) Remapping of the segmentation curves on the 3D representation.

The above steps will be detailed in the following.

2.1 Identification of the regions of interest and of the camera pose

The user manually select a set of matching points on the image and on the 3D model. Let $\{M_i\}_{i=1,...,n}$ be the set of (non-coplanar) 3D points and $\{m_i\}_{i=1,...,n}$ the corresponding 2D coordinates on the image plane. The distortion caused by the lens of the acquisition camera is assumed to be negligible or already estimated and properly corrected. Hence, without loss of generalization, hereafter the coordinates $\{m_i\}_{i=1,...,n}$ on the image plane are assumed to be distortion free.

A subset of the $\{m_i\}$ points is used by the user to outline polygonal curves for delimiting the areas to be analyzed, e.g. areas to be analyzed are delimited by the user by using closed polygonal curves. Each of these areas is assumed to be well approximated by a planar surface. The j-th area is delimited by the points $\{m_{ij}\}_{i=1,...,n_j}$, with $n_j \geq 3$, for each j.

Camera position and orientation with respect to the reference system used to express the $\{M_i\}$ points can be computed in closed-form as described in the following (Ma et al., 2003).

According to the pinhole camera model, undistorted camera measurements can be modeled as follows:

$$m_i = \frac{P_{12}[M_i^\top \ 1]^\top}{P_3[M_i^\top \ 1]^\top} ,\qquad(1)$$

where P_{12} and P_3 are the first two rows and the third row, respectively, of the camera projection matrix P. The projection matrix P can be expressed in terms of the matrix of inner parameters K and of the camera position t and orientation matrix R with respect to the global reference system: $P = K[R \ - Rt]$.

By simple matrix manipulations of (1), it immediately follows that the value of the matrix P can be estimated (in closed–form) by solving a simple linear system:

$$\begin{bmatrix} -M_i^\top & 1 & 0 & 0 & u_i M_i^\top & u_i \\ 0 & 0 & -M_i^\top & 1 & v_i M_i^\top & v_i \end{bmatrix} p = 0 ,\qquad(2)$$

for $i = 1,\ldots,n$, where p is a unit vector containing the normalized values of P, and $m_i = [u_i \ v_i]^\top$. A scaled version of P can be obtained by simply rearranging the terms in p.

Once P has been computed (up to a scale factor), the matrices K and $[R \ - Rt]$ can be computed as the results of the QR factorization, (Golub and Loan, 1989), of the first three columns of P. The scale factor of P can be obtained by imposing that the term on the last column and last row of K is equal to one. Finally, t can be computed by pre-multiplying the forth column of P with $-R^{-1}K^{-1}$.

The above estimation can be improved by using bundle adjustment–like procedures in order to obtain more accurate estimations (Ma et al., 2003), if needed.

Points $\{m_i\}_{i=1,...,n}$ and $\{M_i\}_{i=1,...,n}$ will be used also in subsection 2.5 to map the 2D segmentation results onto the 3D bridge points.

2.2 L^*, a^*, b^* representation

Images acquired by standard cameras are usually represented in the RGB (red, green, blue) color space. Despite being convenient for the visualization on standard displays, this space is not the most suitable when dealing with segmentation purpose. Indeed in the RGB representation each color channel is typically highly correlated with the others, making the analysis of the three channels mostly redundant.

In order to tackle this issue, the L^*, a^*, b^* representation for the image colors is considered as it allows to obtain much less correlated color channels.

The conversion between RGB and L^*, a^*, b^* color space is usually described by passing through the XYZ representation:

$$\begin{bmatrix} X \\ Y \\ Z \end{bmatrix} = \begin{bmatrix} 0.4125 & 0.3576 & 0.1804 \\ 0.2127 & 0.7152 & 0.0722 \\ 0.0193 & 0.1192 & 0.9502 \end{bmatrix} \begin{bmatrix} R \\ G \\ B \end{bmatrix}\qquad(3)$$

and

$$
\begin{aligned}
L^* &= 116 f(Y/Y_n) - 16 & (4) \\
a^* &= 500[f(X/X_n) - f(Y/Y_n)] & (5) \\
b^* &= 200[f(Y/Y_n) - f(Z/Z_n)] & (6)
\end{aligned}
$$

where X_n, Y_n, Z_n represent the (normalized) values representing the white color in the XYZ representation, and the function $f(\cdot)$ is defined as follows:

$$
f(t) = \begin{cases} t^{1/3} & \text{if } t > (6/29)^3 \\ \left(\frac{29}{6}\right)^2 \frac{t}{3} + \frac{4}{29} & \text{otherwise.} \end{cases} \quad (7)
$$

Figure 1: Example of region to be segmented.

In order to validate the above considerations on correlations among RGB and L^*, a^*, b^* channels, the Pearson correlation coefficient has been computed as well. In the case of the image shown in Fig. 1, this coefficient results to be 0.97 for the red and green color channels (Fig. 2), whereas it is 0.03 for the L^* and a^* channels (Fig. 3). Correlation between the red and green channel is also apparent in Fig. 2, while L^* and a^* are weakly correlated as shown in Fig. 3. Similar considerations can be repeated for the the other possible couples of channels.

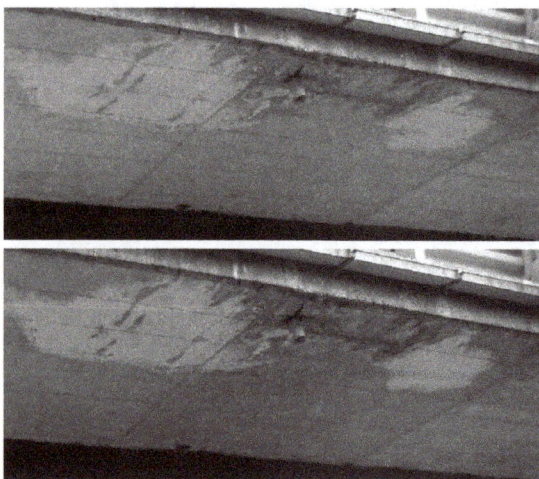

Figure 2: Red (top) and green (bottom) color channels for the image in Fig. 1. Brighter pixels are associated to higher values of the considered variables.

To simplify the notation hereafter a one dimensional signal will be considered (i.e. the lightness L^*). Nevertheless the approach can be extended to the three dimensional signal case.

2.3   Lightness multimodal density

The goal of this step is to estimate the probability density of the lightness L^* values in the region of interest: such density is usu-

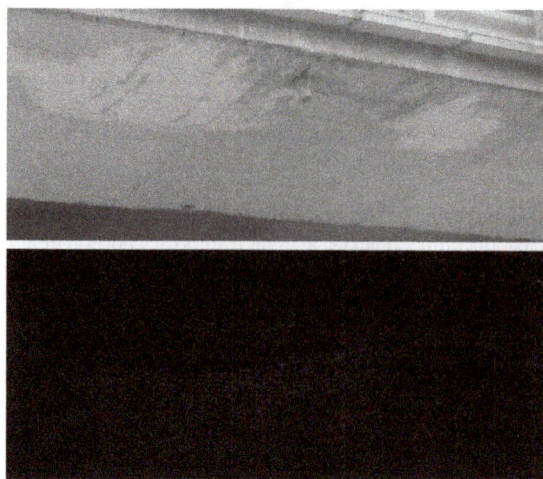

Figure 3: L^* (top) and a^* (bottom) channels for the image in Fig. 1. Brighter pixels are associated to higher values of the considered variables.

ally a multimodal density, where each mode is typically associated to a region with different characteristics. Hence, the rational is that of obtaining a rough segmentation of the region in different areas according to the distribution of the lightness values in different modes.

The histogram of the lightness values will be considered as a sample approximation of the probability density. Fig. 4 shows the histogram corresponding to the multimodal lightness density of image in Fig. 1.

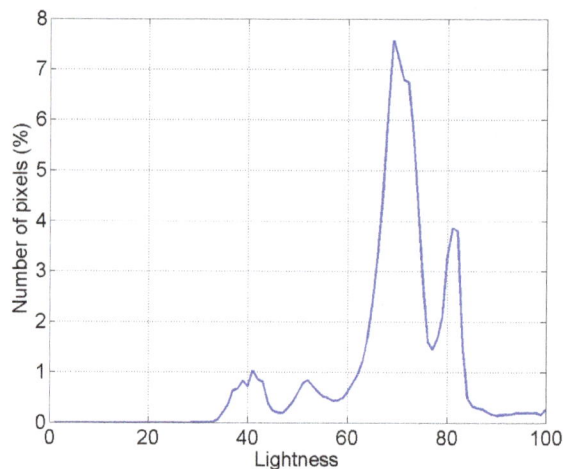

Figure 4: Example of histogram of lightness values, corresponding to the image in Fig. 1.

It is assumed that the mode with the largest number of associated pixels corresponds to the non-deteriorated region, this way, it is possible to easily distinguish between non-deteriorated and deteriorated regions (those corresponding to the larger mode and to the others, respectively). It is worth to notice that under this assumption it is not necessary to determine the total number of significant modes in the image (everything out of the largest mode is considered as a deteriorated region). Similar considerations allow to draw analogous conclusions if the algorithm can associate the most significant mode to deteriorated regions.

Otherwise, when neither of the above cases can be considered, associating a different region to each mode (i.e. to each local maximum in the probability density), it is possible to segment regions characterized by different statistics, but without automatically distinguishing which correspond to the deteriorated ones

Notice that the above assumption and/or restriction will be removed in the future development of this project: assuming to monitor the same infrastructure for a certain amount of time, several temporal samples will be available. The analysis of the current temporal sample will take advantage of those of the previous ones: this way, limitations on the analysis procedure will be removed thanks to the use of previously acquired information.

Once the number of modes m has been determined (for instance by using local maxima detection), the separation between different modes is done on the histogram based on the Otsu's m-thresholding method (Otsu, 1975, Otsu, 1979). Otsu's method provides the threshold τ_i to separate the histogram bins associated to the i-th detected mode from those associated to the $(i+1)$-th, for $i = 1, \ldots, m-1$.

In order to simplify the presentation, without loss of generalization hereafter only two regions are considered, i.e. $m = 2$, separable by means of the threshold τ. The generalization to a generic value of m is immediate.

2.4 Region segmentation

Region segmentation considered in this subsection is the final step to be done in order to separate deteriorated areas from the non-deteriorated ones. The approach considered here is level set segmentation-like (Sethian, 1999, Sethian, 2001, Osher and Fedkiw, 2003).

First, a rough segmentation is obtained by means of the model decomposition presented in the previous subsection: the number of segmented areas to be considered is set equal to the number of detected modes m. Then, a pixel is assigned to the i-th segmented area if its lightness value has been assigned in the previous subsection to the i-th mode, for $i = 1, \ldots, m$.

Unfortunately, the above rough segmentation is usually very noisy, hence the following level set method is applied:

- The closed curve Γ separating the two regions of interest on the image is implicitly described as

$$\Gamma = \{(x, y) \mid \phi(x, y) = 0\}, \quad (8)$$

where (x, y) are the coordinates on the image domain and the level set function $\phi(x, y)$ will be described in the following.

- The level set function is initialized as $L^*(x, y) - \tau$, then it is evolved by considering both the positions where edges are more probable in the images (i.e. where it is more likely to have the border between the two regions) and the regularity (i.e. smoothness) of the curve:

$$\frac{\partial \phi}{\partial t} = -S \cdot \nabla \phi + k|\nabla \phi|, \quad (9)$$

where the first term on the right side of the above equation, $S \cdot \nabla \phi$, corresponds to the introduction of an external vector field related to the image edges, whereas the second term, $k|\nabla \phi|$, aims at increasing the curve smoothness.

- In this work the external vector field S corresponds to an edge detection function. Let I_G be the original image filtered by a Gaussian and LoG be the Laplacian of Gaussian filter:

$$I_G = G * I, \quad (10)$$

and

$$LoG(x, y) = -\frac{1}{\pi \sigma^4} \left(1 - \frac{x^2 + y^2}{2\sigma^2}\right) e^{-\frac{x^2+y^2}{2\sigma^2}}, \quad (11)$$

where $*$ stands for the 2-dimensional convolution and the Gaussian filter G is defined as follows

$$G(x, y) = \frac{1}{2\pi\sigma^2} e^{-\frac{x^2+y^2}{2\sigma^2}}. \quad (12)$$

Let $I_L = LoG * I$, then, S on the point (x, y) is defined as follows:

$$S(x, y) = I_L(x, y) \frac{\nabla I_G(x, y)}{|\nabla I_G(x, y)|}, \quad (13)$$

that is, the magnitude of the external field in (x, y) is determined by $I_L(x, y)$ (i.e. the value of the original image I filtered with the LoG operator), whereas the direction of the the external field is determined by the filtered gradient of the image, $\nabla I_G(x, y)$, normalized by its length. Where the gradient is equal to 0, $S(x, y)$ is set to be a zero vector as well.

2.5 Remapping of the segmented regions

Once the regions have been segmented the resulting boundaries have to be remapped on the 3D space in order to spatially positioning them on the infrastructure.

By assumption each of the areas to be segmented can be well approximated as a planar surface, hence a local two-dimensional coordinate system can be defined on such surface.

Consider the j-th analyzed area, delimited by the points $\{m_{ij}\}_i$, $i = 1, \ldots, n_j$, with $n_j \geq 3$, for each j. Since the considered area can be approximated as a planar surface, each of its point can be expressed as a linear combination of 3 non–collinear points in $\{m_{ij}\}_i$, $i = 1, \ldots, n_j$. Without loss of generalization, let m_{i1}, m_{i2}, m_{i3} be non-collinear.

Furthermore, let $u = m_{i1}$ and let U be the result of the Gram-Schmidt orthogonalization of the matrix $[m_{i2} - m_{i1} \quad m_{i3} - m_{i1}]$.

Then, each point M^* of the planar surface of interest can be expressed as follows:

$$M^* = u + Ux, \quad (14)$$

$x \in \mathbb{R}^2$, where, with a slight abuse of notation, hereafter x corresponds to the point coordinates according to the new reference system.

Let M^* be the 3D point associated to a 2D point m^* on the segmentation boundary. Accordingly to (1), the 3D position M^* can be obtained from the following:

$$(m^* P_3 - P_{12}) \begin{bmatrix} M^* \\ 1 \end{bmatrix} = 0. \quad (15)$$

Substituting (14) in the above equation,

$$(m^* P_3 - P_{12}) \left(\begin{bmatrix} u \\ 1 \end{bmatrix} + \begin{bmatrix} U \\ 0 \end{bmatrix} x \right) = 0, \quad (16)$$

and, after simple matrix manipulations of the above equation:

$$x = -\left((m^*P_3 - P_{12})\begin{bmatrix} U \\ 0 \end{bmatrix}\right)^\dagger (m^*P_3 - P_{12})\begin{bmatrix} u \\ 1 \end{bmatrix}.$$
(17)

Finally, M^* can be obtained from (14).

3. RESULTS

The proposed method has been tested on a road bridge located in Perarolo di Cadore, a small village in the Italian alps. The considered bridge is shown in Fig. 5, whereas the cloud points of a portion of the bridge (acquired by means of TLS) is shown in Fig. 6.

Figure 5: Road bridge located in Perarolo di Cadore considered in as test case in this work.

Figure 6: Cloud points of a portion of the road bridge of Fig. 5.

The image segmentation results obtained (by means of the level set method described in the previous section) for the image in Fig. 1 are shown in Fig. 7. The region considered in the analysis of this image is a portion of the planar surface under the bridge, delimited by green lines in the figure. Instead, segmented regions are delimited by red lines: the regions corresponds to degraded areas, where degradation has been probably caused by percolation issues. It is worth to notice that segmentation shown in the figure corresponds to the separation between the two largest modes in Fig. 4: this is sufficient to distinguish the areas where degradation is more clear, whereas, if needed, segmentation based on the other modes with lower intensity in Fig. 4 can be used to distinguish other areas in the figure with other (i.e. lower) levels of degradation.

Similarly, the results for other portions of the bridge are shown in Fig. 8 and 9 (Fig. 8 and 9 are composed by two sub-figures: the original image is shown on the top, whereas the obtained results are shown in the bottom).

Figure 7: Example of segmentation results, corresponding to the image in Fig. 1. The region considered for the analysis is delimited by the green lines, whereas boundaries of the segmented areas are indicated by red lines.

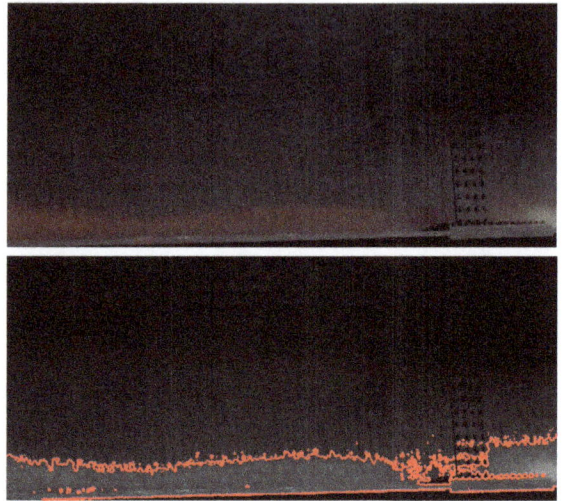

Figure 8: Top: original image. Bottom: example of segmentation results.

Once the 2D regions have been segmented, their boundaries can be remapped on the 3D space accordingly to the procedure described in subsection 2.5. For instance, the segmentation boundaries in Fig. 7 can be remapped as shown in Fig. 10 (in order to improve the readability of the figure, x points (computed as in (17)) are shown instead of M^*).

Actually, the segmentation boundaries can be remapped on the 3D-model as well: Fig. 11 shows the curve formed by the segmentation boundary points $\{M^*\}$ (obtained from (14)) superimposed on the cloud points data. In order to ease the visualization of the results the figure shows a zoom on a specific (and significant) portion of the original image, viewed from an observation direction approximately orthogonal to the analyzed surface.

Similarly, Fig. 12 shows the segmentation boundaries for Fig. 8 remapped on the 3D-model.

4. DISCUSSION AND CONCLUSIONS

This paper proposed a semi-automated method that exploits level set image segmentation in order to estimate the boundaries of degraded regions on complex infrastructures, e.g. the bridge considered in this work.

As shown in the obtained results, the method allows to obtain appropriate 2D region boundaries (Fig. 7 and 7), which can be remapped on the 3D-model as well (Fig. 10, 11 and 12).

Despite our results are quite encouraging, some work has still to be done in order to make the whole procedure as simple as

Figure 9: Top: original image. Bottom: example of segmentation results.

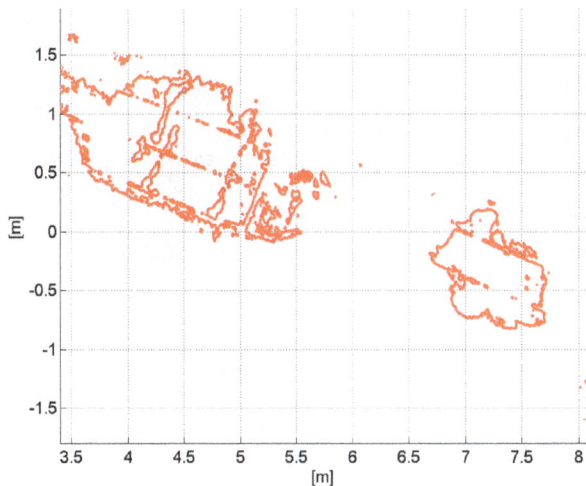

Figure 10: Example of segmentation results: boundaries of detected degradation areas of Fig. 7 remapped in the 3D space (in order to improve the readability of the figure, x points are shown instead of M^*).

possible to the user. In particular, the case of the analysis of periodicals recordings of the same infrastructure will be considered in the future. The rational in such case is that of properly exploiting information already available from previous analyzed data to reduce the interaction with the user, i.e. to make the procedure more autonomous.

Furthermore, this work relies on the use of differences in the lightness values in order to distinguish different areas. Despite this is quite intuitive, it might be subject to errors in certain cases. Future investigation will be dedicated to the use of different image analysis techniques (e.g. (Geladi and Grahn, 1996, Facco et

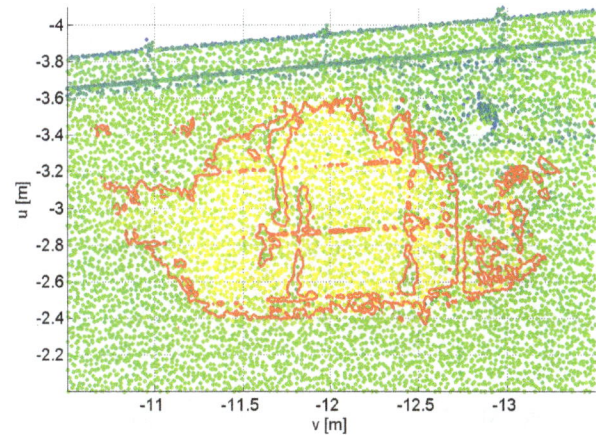

Figure 11: Example of segmentation results: a portion of the segmentation boundaries (red lines) for Fig. 1 remapped in the 3D space. Boundary lines are superimposed to the cloud points acquired by means of TLS. Observation direction is approximately orthogonal to the analyzed surface.

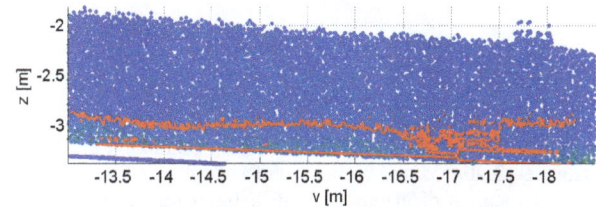

Figure 12: Example of segmentation results: a portion of the segmentation boundaries (red lines) for Fig. 8 remapped in the 3D space. Boundary lines are superimposed to the cloud points acquired by means of TLS. Observation direction is approximately orthogonal to the analyzed surface.

al., 2011, Facco et al., 2013)) in order to provide more robust results from this point of view. Also different level set segmentation methods can be considered in order to improve this work from this point of view, for instance the method proposed in (Li et al., 2011) allows to partially compensate intensity inhomogeneities in the image (e.g. due to lightness changes in the image).

It is also worth to notice that one might consider the possibility of computing the segmentation of degraded areas directly on the intensities of TLS data. Actually, this is possible, however such intensities are influenced by several factors, which make them typically not so reliable. For instance, while in Fig. 11 intensities of TLS data are quite reliable (they allow to approximately distinguish the degraded areas), TLS data shown in Fig. 12 are not useful for our aim. According to this consideration, the use of TLS data to determine the accuracy of the positions of the computed boundaries might be not so reliable. Instead, TLS data can be used to compare the distance between the remapped planar surface where M^* points lie on and the corresponding 3D TLS point positions: this distance, that in our simulations is typically of few centimeters, clearly depends on the accuracy of the method, on the measurement error of TLS data points and on how close is the real surface to a plane.

Our future investigations will also consider the evaluation of possible issues related to the use of reconstructed 3D model (Tucci et al., 2012). *This method has been tested also in relation to a didactic activity which will take place at the Joint Summer School of Alpine Research 2015, applied to terrestrial laser scanning data sets, and further results in this sense will enable to assess its suitability in various fields of application.*

REFERENCES

Camarda, M., Guarnieri, A., Milan, N., Vettore, A. et al., 2010. Health monitoring of complex structure using tls and photogrammetry. Proceedings of the International Archives of the Photogrammetry, Remote Sensing and Spatial Information Sciences, Newcastle upon Tyne, UK pp. 21–24.

Cerimele, M. M. and Cossu, R., 2007. Decay regions segmentation from color images of ancient monuments using fast marching method. Journal of Cultural Heritage 8(2), pp. 170–175.

Cerimele, M. M. and Cossu, R., 2009. A numerical modelling for the extraction of decay regions from color images of monuments. Mathematics and Computers in Simulation 79(8), pp. 2334–2344.

Costantino, D. and Angelini, M., 2013a. Topographic survey for structural monitoring, case: Quadrifoglio condominium (lecce). ISPRS - International Archives of the Photogrammetry, Remote Sensing and Spatial Information Sciences XL-5/W3, pp. 179–187.

Costantino, D. and Angelini, M., 2015. Three dimensional integrated survey for building investigations. Journal of Forensic Sciences.

Costantino, D. and Angelini, M. G., 2013b. Qualitative and quantitative evaluation of the luminance of laser scanner radiation for the classification of materials. ISPRS-International Archives of the Photogrammetry, Remote Sensing and Spatial Information Sciences 1(2), pp. 207–212.

Facco, P., Masiero, A. and Beghi, A., 2013. Advances on multivariate image analysis for product quality monitoring. Journal of Process Control 23(1), pp. 89–98.

Facco, P., Masiero, A., Bezzo, F., Barolo, M. and Beghi, A., 2011. Improved multivariate image analysis for product quality monitoring. Chemometrics and Intelligent Laboratory Systems 109, pp. 42–50.

Geladi, P. and Grahn, H., 1996. Multivariate image analysis. John Wiley & Sons Inc., New York (U.S.A.).

Golub, G. and Loan, C. V., 1989. Matrix Computations. Johns Hopkins University Press: Baltimore, MD.

Guarnieri, A., Milan, N. and Vettore, A., 2013. Monitoring of complex structure for structural control using terrestrial laser scanning (tls) and photogrammetry. International Journal of Architectural Heritage 7(1), pp. 54–67.

Li, C., Huang, R., Ding, Z., Gatenby, J. C., Metaxas, D. N. and Gore, J. C., 2011. A level set method for image segmentation in the presence of intensity inhomogeneities with application to mri. Image Processing, IEEE Transactions on 20(7), pp. 2007–2016.

Ma, Y., Soatto, S., Košecká, J. and Sastry, S., 2003. An Invitation to 3D Vision. Springer.

Osher, S. and Fedkiw, R., 2003. Level set methods and dynamic implicit surfaces. Vol. 153, Springer Science & Business Media.

Otsu, N., 1975. A threshold selection method from gray-level histograms. Automatica 11(285-296), pp. 23–27.

Otsu, N., 1979. A threshold selection method from gray-level histograms. Systems, Man and Cybernetics, IEEE Transactions on 9(1), pp. 62–66.

Sethian, J., 1999. Level Set Methods and Fast Marching Methods: Evolving Interfaces in Computational Geometry, Fluid Mechanics, Computer Vision, and Materials Science. Cambridge Monographs on Applied and Computational Mathematics, Cambridge University Press.

Sethian, J., 2001. Evolution, implementation, and application of level set and fast marching methods for advancing fronts. Journal of Computational Physics 169(2), pp. 503 – 555.

Tucci, G., Cini, D. and Nobile, A., 2012. A defined process to digitally reproduce in 3D a wide set of archaeological artifacts for virtual investigation and display. Journal of Earth Science and Engineering 2(2), pp. 118–131.

39

ASSESSMENT OF THE VOLUNTEERED GEOGRAPHIC INFORMATION FEEDBACK SYSTEM FOR THE DUTCH TOPOGRAPHICAL KEY REGISTER

Magdalena Grus [a], Daniël te Winkel [a]

[a] Kadaster, the Netherlands, magdalena.grus@kadaster.nl and daniel.tewinkel@kadaster.nl

Commission II, WG II/4

KEY WORDS: quality, crowdsourcing, feedback system, Topographical Key Register, LEAN, automatic generalization.

ABSTRACT

Since Topographical Key Register has become an open data the amount of users increased enormously. The highest grow was in the private users group. The increasing number of users and their growing demand for high actuality of the topographic data sets motivates the Dutch Kadaster to innovate and improve the Topographical Key Register (BRT). One of the initiatives was to provide a voluntary geographical information project aiming at providing a user-friendly feedback system adjusted to all kinds of user groups. The feedback system is a compulsory element of the Topographical Key Register in the Netherlands. The Dutch Kadaster is obliged to deliver a feedback system and the key-users are obliged to use it. The aim of the feedback system is to improve the quality and stimulate the usage of the data. The results of the pilot shows that the user-friendly and open to everyone feedback system contributes enormously to improve the quality of the topographic dataset.

1. INTRODUCTION

1.1 Kadaster and the Topographical Key Register (BRT)

The Dutch Kadaster is a non-departmental public body, operating under the political responsibility of the Minister of Infrastructure and the Environment. One of its statutory tasks is to maintain a number of registrations. Other activities are customised work and advice, information provision and international activities. One of the maintained key registers is the Topographical Key Register (BRT), which consists of digital topographic data sets at different map scales (1:10k, base data set and derived / generalized data sets at scale 1:50k, 1:100k, 1:250k, 1:500k and 1:1.000k). The law requires an actuality of less than two years for the whole range of the BRT product family (te Winkel, 2015). Governmental organizations are obliged to use the available BRT data sets for the exchange of geographical information. This is known as the "collect once, use many" principle.

1.2 Quality principles

The data quality of the Topographical Key Register is internally and externally controlled according to ISO19113 standards. The internal control takes place by means of the "control protocol" document which has to be published at least once a year to the public. The controlled elements are: logical consistency, positional accuracy, thematic accuracy, actuality, completeness and the feedback system. These elements are checked by the topographer during the update of the data set, the automatic validation procedures and the 5% quality metric at the end of the updating process. Furthermore, once every three years the quality of BRT has to be tested and evaluated by an independent expert against the same elements.

1.3 Quality and Feedback System

The feedback system is an important and compulsory tool to ensure the quality of the Topographical Key Register. The feedback process facilitates the correcting of any error detected in the key register by users of the data. The aim of the feedback system is to improve the quality of the data in the BRT. The Dutch Kadaster as an owner of Topographical Key Register has legal obligation to provide a feedback system. In the same time the key-users (municipalities, governmental organisations) are obliged to give a feedback when they have reasonable doubt that authentic data in the BRT is incorrect.

The private users, in contrast to the governmental key-users, did not have any legal obligation to provide feedback about errors in the topographic register.

1.4 Open data and its influence

In 2012 Topographical Key Register became open data under the CC-BY licence. This means that all the products of the BRT family became freely available to everyone to use and republish without restrictions from copyright and patents. The CC-BY licence bounds the user to provide the name of the creator and attribution parties with the created product (te Winkel, 2015).

It is since this decision that the amount of private users has grown enormously. Before open data the BRT was mainly used by governmental organisations. In the first year after open data (2012) commercial companies started to explore the possibilities of the BRT and in 2013 we saw a strong increase of the use by private persons (Figure 1) (Bregt et al., 2013; 2014). In this figure you see the distribution of users in user groups. In 2011 the use of the BRT for education / research was combined with the use by the government.

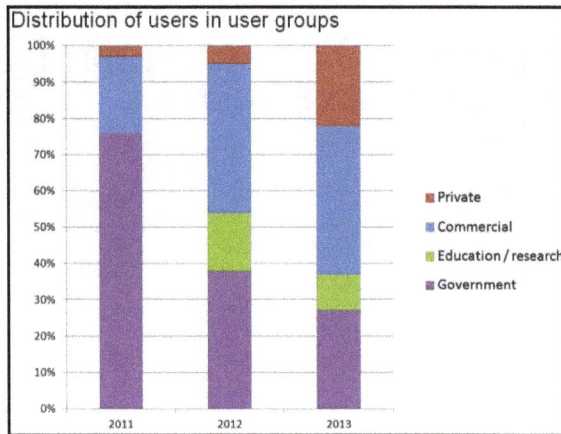

Figure 1. The users of the BRT separated in user groups

The introduction of the BRT as open data has led to a rapid growth in the use of the topographic products. The wider use of the data has led to an increased demand for better actuality, which exceeds the demand for data quality (te Winkel, 2015). The new users expect that a change in the real world is directly translated to the available maps and services. In several user meetings the increase of actuality of the topographic data sets was one of the main issues for improvement. In 2011 the actuality of the base data set (1:10k) just met the requirement of an actuality less than two years. For the small scale datasets the actuality was between two and ten years.

1.5   LEAN methodology

To increase the actuality of the BRT, Kadaster used the LEAN methodology to change the production process of the base data set. One of the elements of LEAN methodology is the focus on the direct creation of value for the customer. All steps in the process that do not add value for the customer are referred to as waste and are a target for elimination. This methodology changed the production process and the mindset of the employees.

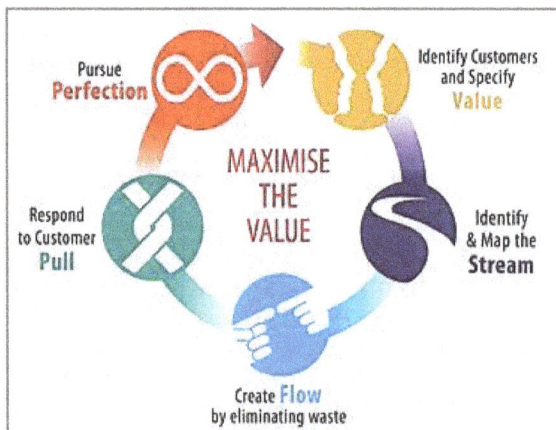

Figure 2. The principles of the LEAN methodology

It is difficult to see the waste in a production process when you look at all the production steps separately. Often the waste is not a step in the process, but, for example, a transfer of the product from one person to another person or a step that adds no value for the customer. By visualizing the value stream of the process, you can identify the waste and find a solution to eliminate it.

Also the customer was asked to participate actively in improving the production process. Every change in the specifications of the BRT product family or in the production process has to be evaluated by the customer from his users perspective. This careful approach may take some time, but in the end the customer is more satisfied, involved and more willing to accept the change.

1.6   Automatic generalization

The introduction of automatic generalization in September 2013 gave the actuality of the 1:50k scale data set a boost (te Winkel, 2015). The new process is developed between 2010 and 2013 in several iterations. In the development of this automated process Kadaster asked the customer to give feedback for the next development iteration. The iterative approach resulted in a product that was accepted and appreciated by the user. The result is a fully automated generalization production workflow that generalizes the 1:50k map series from the 1:10k base data set in less than three weeks.

The workflow starts with a validation end enrichment of the source data. This step is necessary to resolve remaining errors in the base data set and to add information used by the automatic process that is not (yet) present in the base data set. Detected errors are repaired in the data set and are reported to the production process of the 1:10k map for a substantial solution in the long term. The generalization process itself consists of three subsequent steps: model generalization, geometric generalization (displacement) and graphic generalization (cartographic conflict resolution). These three steps are developed as separate tools. The advantage of this modular approach is the potential to replicate the models, adapt parameter values or even substitute parts of the process and produce other map series (i.e. a 100k map series).

Because of the reduction in processing time, it is now possible to generalize the small scale data sets directly in flow after the finishing of the base data set. Therefore the actuality of the automatically derived data set is the same as the actuality of the base data set and meets the requirement of an actuality less than two years. Nowadays the base data set and derived map series are released simultaneously with the same actuality five times a year. For the customer this means more actual maps more frequent.

2. MAIN BODY

2.1 Pilot Feedback system for the BRT

In October 2013 the Dutch Kadaster started a research to develop and implement a user-friendly feedback system. One of the aims was to provide an easy to use and user friendly system for all users willing to give a feedback about the Topographical Key Register. The already existing feedback system was outdated and was not adjusted to the new user groups and this new purpose. Another goal was to check the potential of the crowdsourcing concept and to evaluate the quality of the data collected by volunteers. This research consists of four steps which are visible in Figure 3.

Step 1: Collection of a group of volunteers.
The request for the volunteers was placed on the Kadaster website and on social media (LinkedIn group for the BRT, Twitter).

Step 2: Building a user-friendly feedback system.
Kadaster developed a feedback system that worked on all device types (PC, tablets and smartphones) and on the most popular operating systems (iOS, Android, Windows).

Step 3: Collection and data validation.
To stimulate and effectively motivate the group of users without feedback obligation, we decided to validate immediately all delivered errors.

Step 4: Correct errors as soon as possible.
By directly updating our map when a reported error was accepted, Kadaster wanted to show its appreciation to the user group for cooperating in this research.

Figure 4. The status map application

By means of this status application the user's community is being informed about how Kadaster handles their findings. This makes the whole feedback process transparent to all. The users can view their own and others signals plotted on the map with additional information from Kadaster when or if they will be corrected.

2.3 Results of the pilot

Within two months (November and December 2013) Kadaster has managed to collect 130 volunteers who expressed their willingness to participate in the pilot. About 70% of them were not representing a governmental organisation (Figure 5).

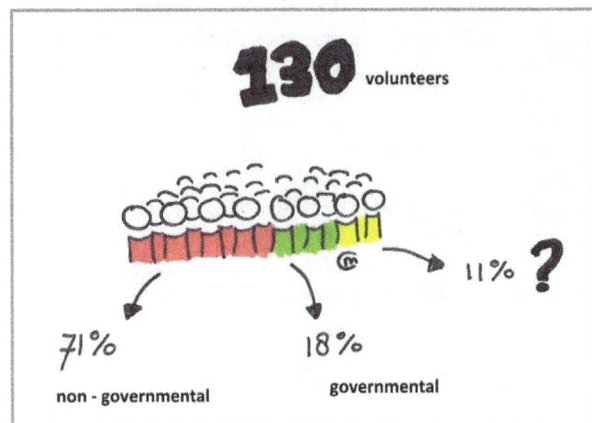

Figure 3. Feedback process flow during the pilot

2.2 Status and feedback over feedback

One of the crucial elements of the pilot was an interface showing the status of the received feedback. The status map application is a communication platform to inform detectors about the status of their feedback and to give comments regarding those feedbacks (Figure 4).

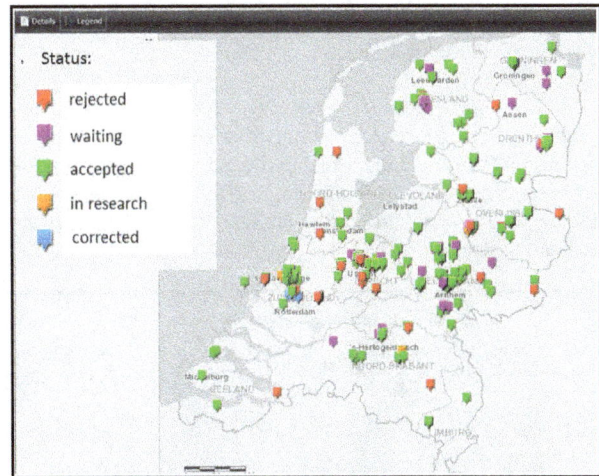

Figure 5. Different groups of volunteers

The request for the participants was placed on Kadaster website en on social media (LinkedIn group for the BRT, Twitter). The volunteers got an e-mail with the instructions how to use the "feedback application" and give a feedback. They were asked to use the provided hyperlink, choose the place they where familiar with and verify it with the data available/presented on the map in the application (Grus, 2014). By placing a point on the map it was possible to indicate the place with a potential error. Also the user was asked to provide some extra information about the user self (for example the type of user (private, government, education) and contact information).

From group of the volunteers around 80% actively participated in the pilot and gave a feedback about errors in the Topographical Key Register.

In two months time Kadaster have received around 369 feedbacks from de selected group of users. To give a better picture, in 2013 by means of the old feedback system Kadaster was receiving around 28 feedbacks, in 2012 – 10 and in 2011- 8 (Figure 6). This pilot attracted more attention than the existing feedback system.

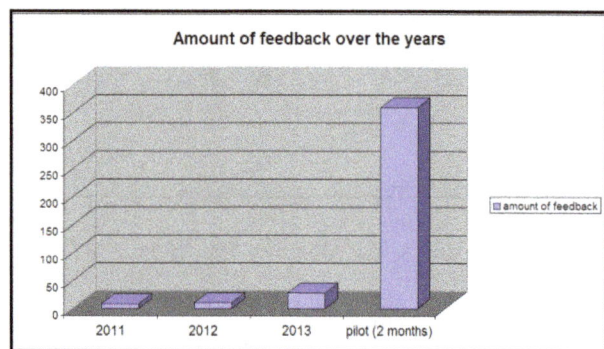

Figure 6. Differences in amount of feedback per year

In the last phase of the pilot we have send a questionnaire to all volunteers to evaluate the user-friendliness of the feedback system. The users were very enthusiastic about the pilot and very satisfied with the new feedback system.

2.4   Quality of the data

All the reported errors were checked by a group of qualified topographers. About 70% of the feedbacks received during the pilot (around 200 reported errors) were approved and as correct and they were used to adjust the digital topographic data set. 20% of the reported errors could not be verified, because recent aerial photographs were not available. When the new aerial photographs become available, also these reported errors are checked.
Only 10% of the provided feedback were rejected. One of the reasons to reject a feedback is the internal topographic rules. For instance when a reported missing object is not part of the specifications of the Topographical Key Register.

3.   CONCLUSIONS

Actuality is one of the most important quality elements for the digital topographic data sets. Growing demand for the higher actuality of BRT products is a good motivator to look for new potential improvements. The improvement of the production process with the LEAN methodology and automatic generalisation resulted in an increased actuality. The position of the customer and his active contribution has become important element in the improvement of the production process.
The group of 71% non-governmental participants during the pilot, proves that the crowdsourcing as a source of product improvement has a huge potential. The results from the pilot prove also that an open to everyone, user-friendly and transparent feedback system can significantly contribute to improve quality of the digital topographic datasets.

REFERENCES

Bregt, A.K., Castelein, W., Grus, Ł., 2013. *De effecten van een open basisregistratie topografie (BRT)*. Wageningen University, Wageningen, The Netherlands.

Bregt, A.K., Grus, Ł., Eertink, D., 2014. *Wat zijn de effecten van een open basisregistratie topografie na twee jaar*. Wageningen University, Wageningen, The Netherlands.

Grus, M., 2014. *Pilot terugmelding BRT- Evaluatie Rapport*. Apeldoorn: Kadaster.

Te Winkel, D., 2015. International Workshop on Spatial Data and Map Quality "*The Dutch Key register Topography as open data*", Malta, January 2015.

PERMISSIONS

All chapters in this book were first published in ISPRS, by Copernicus Publications; hereby published with permission under the Creative Commons Attribution License or equivalent. Every chapter published in this book has been scrutinized by our experts. Their significance has been extensively debated. The topics covered herein carry significant findings which will fuel the growth of the discipline. They may even be implemented as practical applications or may be referred to as a beginning point for another development.

The contributors of this book come from diverse backgrounds, making this book a truly international effort. This book will bring forth new frontiers with its revolutionizing research information and detailed analysis of the nascent developments around the world.

We would like to thank all the contributing authors for lending their expertise to make the book truly unique. They have played a crucial role in the development of this book. Without their invaluable contributions this book wouldn't have been possible. They have made vital efforts to compile up to date information on the varied aspects of this subject to make this book a valuable addition to the collection of many professionals and students.

This book was conceptualized with the vision of imparting up-to-date information and advanced data in this field. To ensure the same, a matchless editorial board was set up. Every individual on the board went through rigorous rounds of assessment to prove their worth. After which they invested a large part of their time researching and compiling the most relevant data for our readers.

The editorial board has been involved in producing this book since its inception. They have spent rigorous hours researching and exploring the diverse topics which have resulted in the successful publishing of this book. They have passed on their knowledge of decades through this book. To expedite this challenging task, the publisher supported the team at every step. A small team of assistant editors was also appointed to further simplify the editing procedure and attain best results for the readers.

Apart from the editorial board, the designing team has also invested a significant amount of their time in understanding the subject and creating the most relevant covers. They scrutinized every image to scout for the most suitable representation of the subject and create an appropriate cover for the book.

The publishing team has been an ardent support to the editorial, designing and production team. Their endless efforts to recruit the best for this project, has resulted in the accomplishment of this book. They are a veteran in the field of academics and their pool of knowledge is as vast as their experience in printing. Their expertise and guidance has proved useful at every step. Their uncompromising quality standards have made this book an exceptional effort. Their encouragement from time to time has been an inspiration for everyone.

The publisher and the editorial board hope that this book will prove to be a valuable piece of knowledge for researchers, students, practitioners and scholars across the globe.

LIST OF CONTRIBUTORS

E. Nocerino, F. Menna and F. Remondino
3D Optical Metrology unit, Bruno Kessler Foundation (FBK), via Sommarive 18, Trento 38123, Italy

F. Fassi
Politecnico di Milano, ABC Dep. 3DSurvey Group, via Ponzio 31, Milano 20133, Italy

P. Pahlavani, R.A. Abbaspour and A. Zare Zadiny
Dept. of Surveying and Geomatics Eng., College of Engineering, University of Tehran, Tehran, Iran

M. Modiri
Professor at Malek Ashtar University of Technology, Esfahan, Iran

M. Masumi
M.s degree of Information Technology Management

A. Eftekharic
M.s of degree of Remote Sensing, Dept. of surveying and Geomatics engineering, University of Tehran, Tehran, Iran

V. Mousavi, M. Khosravi, M. ahmadi, N. Noori, A. Hosseini naveh and M. Varshosaz
K.N.Toosi University of Tecknology, Faculty of Surveying Engeeniring, Tehran,Iran

S. Niazi, M. Mokhtarzade and F. Saeedzadeh
Faculty of Geodesy and Geomatics Engineering, K. N. Toosi University of Technology, Valiasr street, Tehran, Iran

Farhad Samadzadegan, Farzaneh Dadras Javan and Hadiseh Hasani
School of Surveying and Geospatial Engineering, College of Engineering, University of Tehran, Tehran, Iran

Behshid Khodaei
Miaad Andishe Saz, Research and Development Company, Tehran, Iran

M. Poor Arab Moghadam
MSc. Student in GIS Division, School of Surveying and Geospatial Eng., College of Eng., University of Tehran, Tehran, Iran

P. Pahlavani
Assistant Professor, Center of Excellence in Geomatics Eng. in Disaster Management, School of Surveying and Geospatial Eng. College of Eng., University of Tehran, Tehran, Iran

M. Nekouei Shahraki and N. Haala
Institute for Photogrammetry (ifp), University of Stuttgart, Geschwister-Scholl-Str. 24D, 70174 Stuttgart, Germany

Sara Rahimi and Hossein Arefi
School of Surveying and Geospatial Engineering, University of Tehran, Tehran, Iran

Reza Bahmanyar
Institute of Remote Sensing Technology (IMF), German Aerospace Center (DLR), Wessling, Germany

S.M.R. Moosavi
Dept. of Geomatic Engineering, Islamic Azad University, Larestan, Iran

A. Sadeghi-Niaraki
GIS Dept., Geoinformation Technology Center of Excellence, Faculty of Geodesy & Geomatics Eng, K.N.Toosi Univ. of Tech., Tehran, Iran

Ph. Engela and B. Schweimlera
Faculty of Landscape Sciences and Geomatics, Hochschule Neubrandenburg, University of Applied Sciences

H. Rastiveis
School of Surveying and Geospatial Engineering, Faculty of Engineering, University of Tehran, Tehran, Iran

M. Moradi
School of Surveying and Geospatial Engineering, College of Engineering, University of Tehran, Iran

M. R. Delavar
Center of Excellence in Geomatic Engineering in Disaster Management, School of Surveying and Geospatial Engineering, College of Engineering, University of Tehran, Iran

A. Moradi
Department of Social Sciences, Farhangian University, Resalat Branch, Zahedan, Iran

Mirahmad Mirghasempour
Dept. of Civil Engineering, Shahid Rajaee Teacher Training University, Tehran, Iran

Ali Yaser Jafari
Dept. of Architecture and Urbanism, Shahid Rajaee Teacher Training University, Tehran, Iran

Mohammad Omidalizarandi and Ingo Neumann
Geodetic Institute, Leibniz Universität Hannover, Germany

M.Modiri
Professor at Malek Ashtar University of Technology, Esfahan, Iran

H.Enayati
M.s degree of Photogrammtry, Khaje Nasir university Tehran, Iran

M. Ebrahimikia
M.s degree of Photogrammtry, The university of Tehran, Iran

H. Rastiveis, E. Hosseini-zirdoo and F.Eslamizadea
Dept. of Geomatics Engineering, School of Eng., University of Tehran

N. Zarrinpanjeh
Islamic Azad University of Qazvin

F. Dadrassjavan
University of Tehran

H. Fattahi
Tehran Traffic Control Company

H. Tamiminia and A. Safari
School of Surveying and Geospatial Engineering, Dept. of Remote Sensing, College of Engineering, University of Tehran, Iran

S. Homayouni
Dept. of Geography, Environmental Studies and Geomatics, University of Ottawa, Ottawa, Canada

S. Vaezi, M.S. Mesgari and F.Kaviary
Dept. of Geomatic Engineering, K. N. Toosi University of Technology, Valy-Asr Street, Mirdamad Cross, Tehran, Iran

J. Wasowski
CNR-IRPI, National Research Council, 70126 Bari, Italy

F. Bovenga
CNR-ISSIA, National Research Council, 70126 Bari, Italy

R. Nutricato and D. O. Nitti
GAP srl, c/o Politecnico di Bari, 70126 Bari, Italy

M. T. Chiaradia
Dipartimento Interateneo di Fisica, Politecnico di Bari, 70126 Bari, Italy

Mohammadreza Sahelgozin and Abbas Alimohammadi
GIS Dept., Geoinformation Technology Center of Excellence, Faculty of Geodesy & Geomatics Engineering K.N.Toosi University of Technology, Tehran, Iran

M. Zeinolabedini and A. Esmaeily
Dept. of Water Resources Engineering, Graduate University of Advanced Technology, Kerman, Mahan, 7631133131

D. Wagner
University of Tehran, Tehran, Iran

V. Coors and N. Alam
Hochschule für Technik Stuttgart – University of Applied Sciences, Stuttgart, Germany

M. Wewetzer and M. Pries
Beuth Hochschule für Technik – University of Applied Sciences, Berlin, Germany

Sh. Shataee
Associate Professor, Forestry Department, Forest Sciences Faculty, Gorgan University of Agricultural Sciences and Natural Resources, 386, Gorgan, Iran

J. Mohammadi
Assistance Professor, Forestry Department, Forest Sciences Faculty, Gorgan University of Agricultural Sciences and Natural Resources, 386, Gorgan, Iran

Nurollah Tatar, Mohammad Saadatseresht, Hossein Arefi and Ahmad Hadavanda
School of Surveying and Geospatial Information Engineering, College of Engineering, University of Tehran

M. Saadatseresht and M. Hasanlou
School of Surveying and Geospatial Engineering, University of Tehran, Tehran, Iran

A.H. Hashempour
Robotic Photogrammetry Research Group, Close Range Photogrammetry Lab., University of Tehran, Tehran, Iran

A. Kheiri
Faculty Technical and Engineering, Eslamic Azad University of Larestan, Iran

F. Karimipour and M. Forghani
Faculty of Surveying and Geospatial Engineering, College of Engineering, University of Tehran, Tehran, Iran

Nurollah Tatar, Mohammad Saadatseresht, Hossein Arefi and Ahmad Hadavand
School of Surveying and Geospatial Information Engineering, College of Engineering, University of Tehran

F. Samadzadegan and H. Hasani
School of Surveying and Geospatial Engineering, College of Engineering, University of Tehran, Tehran, Iran

S. Youneszadeh and P. Pilesjo
Dept. of Physical Geography and Ecosystem Science, Lund University, Lund, Sweden

N. Amiri
Dept. of Geoinformatics, Munich University of Applied Sciences, Munich, Germany
Faculty of Geoinformation science and Earth observation, ITC, University of Twente, Enschede, The Netherlands

A. Şekertekin, Ş. H. Kutoglu and A. M. Marangoz
BEU, Engineering Faculty, Geomatics Engineering Department 67100 Zonguldak, Turkey

S. Kaya
ITU, Civil Engineering Faculty, 80626 Maslak Istanbul, Turkey

H. González-Jorge, L. Díaz-Vilariño, J. Martínez-Sánchez and P. Arias
Applied Geotechnologies Group, Dept. Natural Resources and Environmental Engineering, University of Vigo, Campus Lagoas-Marcosende, CP 36310 Vigo, Spain

S. Lagüela
Department of Cartographic and Terrain Engineering, University of Salamanca, 05003 Ávila, Spain

S. Hassany Pazoky, F. Karimipour and F. Hakimpour
Faculty of Surveying and Geospatial Engineering, College of Engineering, University of Tehran, North Kargar St., Tehran, Iran

P. Goodhue, H. McNair and F. Reitsma
Dept. of Geography, University of Canterbury, Christchurch, New Zealand

G. Navratil, E. Heer and J. Hahn
Department for Geodesy and Geoinformation, TU Vienna, Austria

A. Forati and F. Karimipour
Faculty of Surveying and Geospatial Engineering, College of Engineering, University of Tehran, Tehran, Iran

S. Soleimani and M.R. Malek
Faculty of Geodesy and Geomatics Engineering, K.N. Toosi University of Technology, Tehran, Iran

A. Masiero, A. Guarnieri, F. Pirotti and A. Vettorea
Interdepartmental Research Center of Geomatics (CIRGEO), University of Padova

Magdalena Grus and Daniël te Winkel
Kadaster, the Netherlands, Viale dell'Università 16, Legnaro (PD) 35020, Italy

Index